Stability of Time Dependent and Spatially Varying Flows

Stability of Time Dependent and Spatially Varying Flows

Proceedings of the Symposium on the Stability of
Time Dependent and Spatially Varying Flows
Held August 19–23, 1985, at NASA Langley Research Center,
Hampton, Virginia

Edited by
D.L. Dwoyer and M.Y. Hussaini

With 145 Figures

Springer-Verlag
New York Berlin Heidelberg
London Paris Tokyo

D.L. Dwoyer
NASA Langley Research Center
Hampton, VA 23665, U.S.A.

M.Y. Hussaini
Institute for Computer Applications
  in Science and Engineering (ICASE)
ICASE/NASA
NASA Langley Research Center
Hampton, VA 23665, USA

Library of Congress Cataloging in Publication Data
Stability of time dependent and spatially varying
  flows.
  Papers presented at a workshop sponsored by the
Institute for Computer Applications in Science and
Engineering and NASA Langley Research Center,
Aug. 19–23, 1985.
  1. Fluid dynamics—Congresses. 2. Aerodynamics—
Congresses. I. Dwoyer, Douglas L. II. Hussaini,
M. Yousuff. III. Institute for Computer Applications
in Science and Engineering. IV. Langley Research
Center.
TA357.S696   1987   620.1′064   86-27893

© 1987 Springer-Verlag New York Inc.
All rights reserved. This work may not be translated or copied in whole or in part. Any questions concerning permissions or copyright to reuse or reproduce the materials in this volume should be directed to the individual authors.

Printed and bound by R.R. Donnelley & Sons, Harrisonburg, Virginia.
Printed in the United States of America.

9 8 7 6 5 4 3 2 1

ISBN 0-387-96472-X Springer-Verlag New York Berlin Heidelberg
ISBN 3-540-96472-X Springer-Verlag Berlin Heidelberg New York

Introduction

This volume is the collection of papers presented at the workshop on "The Stability of Spatially Varying and Time Dependent Flows" sponsored by the Institute for Computer Applications in Science and Engineering (ICASE) and NASA Langley Research Center (LaRC) during August 19–23, 1985. The purpose of this workshop was to bring together some of the experts in the field for an exchange of ideas to update the current status of knowledge and to help identify trends for future research. Among the invited speakers were D.M. Bushnell, M. Goldstein, P. Hall, Th. Herbert, R.E. Kelly, L. Mack, A.H. Nayfeh, F.T. Smith, and C. von Kerczek. The contributed papers were by A. Bayliss, R. Bodonyi, S. Cowley, C. Grosch, S. Lekoudis, P. Monkewitz, A. Patera, and C. Streett.

In the first article, Bushnell provides a historical background on laminar flow control (LFC) research and summarizes the crucial role played by stability theory in LFC system design. He also identifies problem areas in stability theory requiring further research from the view-point of applications to LFC design. It is an excellent article for theoreticians looking for some down-to-earth applications of stability theory.

The next seven articles deal with the stability of spatially varying incompressible flows, followed by two articles on compressible flows. Nayfeh, who is among the first to study the stability of nonparallel flows, reviews rather comprehensively the theoretical studies of the secondary instability appearing in the early stages of transition in a boundary layer flow. He divides these investigations into two types—the mutual interaction type including resonant and nonresonant interactions, and the parametric type. He compares and contrasts these approaches in the context of the spring pendulum problem. Herbert's treatment of the secondary instability by Floquet theory falls into the latter category. He presents a general formulation of the theory incorporating the spatial growth of a Blasius boundary layer, and he considers detuned modes (in addition to subharmonic and fundamental ones) for combination resonance.

The spatial growth of a boundary layer is one of the key factors in the receptivity problem. The asymptotic analysis of Goldstein clearly establishes this factor as responsible for the interaction between a large-scale external disturbance and a comparatively small-scale Tollmien-Schlichting wave in the boundary layer. A similar receptivity problem is treated numerically by Gatski and Grosch. They solve the two-dimensional, time-dependent incompressible Navier-Stokes equations by a second-order accurate finite-difference method for the unsteady flow past a semi-infinite flat plate with a sharp leading edge. Their preliminary results demonstrate how an external disturbance can engender instability waves in the leading edge region. Lekoudis studies the quasi-parallel linear spatial stability of the three-dimensional boundary layer on an ellipsoid at incidence, and shows that his results agree qualitatively with experiment.

The nonlinear stability of separating flows such as the leading edge separation bubble, mid-chord or trailing-edge separation, and flows past surface-mounted obstacles, is of practical importance. Smith presents a progress report on his continuing work in this area. This work is unique, addressing questions which are truly nonlinear and nonparallel and holds promise of a unified theory of instability and transition. The article of Bodonyi and Smith is in a similar vein and provides results of practical significance for short-scale Rayleigh instabilities in the flow past surface-mounted obstacles.

Mack's masterly review of linear compressible stability theory is both succinct and complete. He emphasizes those particular aspects which distinguish the compressible from the incompressible theory. He also addresses the possibility of a multiplicity of inviscid and viscous solutions in the compressible case and their possible physical significance. He points out that even the incipient stages of transition at supersonic and hypersonic speeds are not well understood. The work of Bayliss et al. is an attempt to investigate subsonic stability phenomena by means of a full numerical solution of the two-dimensional compressible Navier-Stokes equations. They study the nonlinear growth of unstable waves in a flat plate boundary layer and their control by active heating and cooling of the plate.

The Taylor-Gortler instability mechanism associated with the curved Stokes layers is the focus of Hall's article. He relates this problem to the instability of a fluid layer heated time periodically from below. A similar instability can occur in general flows which are spatially varying and periodic in time. The next three articles deal with classical plane Stokes layers: The paper by von Kerczek discusses the modifications to the stability characteristics resulting from a sinusoidal time modulation of the well-known Hagen-Poiseuille flow, Eckman layer, and the flow of a liquid film down a vertical plate. Monkewitz and Bunster present both experimental and theoretical results on Stokes layers for Reynolds numbers (based on boundary layer thickness and free stream amplitude) up to 600. Their theoretical model is a quasi-steady approach involving nonlinear interactions between modes with a rather heuristic assumption that the difference in phase speeds and growth rates is small. Cowley's approach is based on a multiple scale expansion for asymptotically large Reynolds numbers; this yields to leading order a Rayleigh eigenvalue problem. He concludes that high-frequency perturbations to a Stokes layer can grow and that the most significant growth rates tend to occur as the outer flow is decelerating. This agrees qualitatively with experimental observation.

The Landau-Stuart-Watson equation has been crucial to many nonlinear stability theories. It has explained many instability phenomena of physical interest. Recently, it has become a subject of study in its own right, being a prototypical equation giving rise to bifurcation and low-order spatial chaos. Sirovich and Newton have devised a clever combination of analysis and numerics which yields an efficient method for studying its stability and bifurcation properties over a wide parameter range. Their article gives a flavor of their recent investigations.

Full numerical solutions of the Navier-Stokes equations are playing an increasingly important role in the study of stability and transition to turbulence. The next two papers are typical of such studies: Ghaddar and Patera's study of stability and resonance in grooved-channel flows is both original and systematic. Streett and Hussaini's article is a preliminary demonstration of work in progress pertaining to the Taylor experiment on Couette flow between concentric rotating cylinders with fixed end walls.

Last, but not the least, is the experimental investigation by Williams of the vortical structures in the breakdown stage of transition in a flat plate boundary layer. Such experiments supplemented by simulation are crucial in unravelling the delicate mechanisms involved in the transition process.

Finally, the editors would like to take this opportunity to thank the participants in the workshop for their contributions, and the staff of Springer-Verlag for their assistance in putting together this volume.

DLD, MYH

Contents

Introduction v

Contributors xi

Application of Stability Theory to Laminar Flow Control
—Progress and Requirements
D.M. Bushnell and M.R. Malik 1

On Secondary Instabilities in Boundary Layers
Ali H. Nayfeh 18

Floquet Analysis of Secondary Instability in Shear Flows
Thorwald Herbert, Fabio P. Bertolotti and German R. Santos 43

The Generation of Tollmien–Schlichting Waves by Long
Wavelength Free Stream Disturbances
M.E. Goldstein 58

Numerical Experiments on Boundary-Layer Receptivity
Thomas B. Gatski and Chester E. Grosch 82

The Linear Stability of the Incompressible Boundary Layer on an
Ellipsoid at Six Degrees Incidence
S.G. Lekoudis 97

Non-Linear Effects and Non-Parallel Flows: The Collapse of
Separating Motion
F.T. Smith 104

On Short-Scale Inviscid Instabilities in the Flow Past Surface-
Mounted Obstacles
R.J. Bodonyi and F.T. Smith 148

Review of Linear Compressible Stability Theory
Leslie M. Mack 164

Wave Phenomena in a High Reynolds Number Compressible
Boundary Layer
A. Bayliss, L. Maestrello, P. Parikh and E. Turkel 188

Instability of Time-Periodic Flows
Philip Hall 206

Stability Characteristics of Some Oscillatory Flows–Poiseuille, Ekman and Films
Christian von Kerczek 225

The Stability of the Stokes Layer: Visual Observations and Some Theoretical Considerations
P.A. Monkewitz and A. Bunster 244

High Frequency Rayleigh Instability of Stokes Layers
Stephen J. Cowley 261

Ginzburg–Landau Equation: Stability and Bifurcations
L. Sirovich and P.K. Newton 276

Stability and Resonance in Grooved-Channel Flows
Nasreen K. Ghadder and Anthony T. Patera 294

Finite Length Taylor Couette Flow
C.S. Streett and M.Y. Hussaini 312

Vortical Structures in the Breakdown Stage of Transition
D.R. Williams 335

Contributors

Bayliss, A.
Department of Electrical
 Engineering and Computer
 Science
Northwestern University
Evanston, Illinois 60201, USA

Bertolotti, Fabio P.
Department of Engineering
 Science and Mechanics
Virgina Polytechnic Institute and
 State University
Blacksburg, Virginia 24061, USA

Bodonyi, R.J.
Department of Mathematical
 Sciences
Indiana-Purdue University at
 Indianapolis
Indianapolis, Indiana 46223, USA

Bunster, A.
Department of Mechanical,
 Aerospace and Nuclear
 Engineering
University of California
Los Angeles, California 90024
 USA

Bushnell, D.M.
NASA Langley Research Center
Hampton, Virginia 23665-5225
 USA

Cowley, Stephen J.
Department of Mathematics
University College London
London WC1E 6BT, UK

Gatski, Thomas B.
NASA Langley Research Center
Hampton, Virginia 23665-5225
 USA

Ghaddar, Nasreen K.
Department of Mechanical
 Engineering
Massachusetts Institute of
 Technology
Cambridge, Massachusetts 02139
 USA

Goldstein, M.E.
NASA Lewis Research Center
Cleveland, Ohio 44135, USA

Grosch, Chester E.
Old Dominion University
Institute of Oceanography
Norfolk, Virginia 23508, USA
and
ICASE, NASA Langley Research
 Center
Hampton, Virginia 23665-5225
 USA

Hall, Philip
Mathematics Department,
University of Exeter
Exeter, UK

Herbert, Thorwald
Department of Engineering
  Science and Mechanics
Virginia Polytechnic Institute and
  State University
Blacksburg, Virginia 24061, USA

Hussaini, M.Y.
ICASE
NASA Langley Research Center
Hampton, Virginia 23665-5225
USA

von Kerczek, Christian
Mechanical Engineering
  Department
The Catholic University of
  America
Washington, D.C. 20064, USA

Lekoudis, S.G.
Office of Naval Reseach
Arlington, Virginia 22217-5000,
  USA

Mack, Leslie M.
Jet Propulsion Laboratory
California Institute of Technology
Pasadena, California 91109, USA

Maestrello, L.
NASA Langley Research Center
Hampton, Virginia 23665-5225
USA

Malik, M.R.
High Technology Corporation
Hampton, Virginia 23666-7262
USA

Monkewitz, P.A.
Department of Mechanical,
  Aerospace and Nuclear
  Engineering
University of California
Los Angeles, Calfornia 90024
USA

Nayfeh, Ali H.
Department of Engineering
  Science and Mechanics
Virginia Polytechnic Institute and
  State University
Blacksburg, Virginia 24061, USA

Newton, P.K.
Division of Applied Mathematics
Brown University
Providence, Rhode Island 02912
USA

Parikh, P.
Vigyan Research Associates, Inc.
Hampton, Virginia 23666-7262
USA

Patera, Anthony T.
Department of Mechanical
  Engineering
Massachusetts Institute of
  Technology
Cambridge, Massachusetts 02139
USA

Santos, German R.
Department of Engineering
  Science and Mechanics
Virginia Polytechnic Institute and
  State University
Blacksburg, Virginia 24061, USA

Sirovich, L.
Division of Applied Mathematics
Brown University
Providence, Rhode Island 02912
  USA

Smith, F.T.
Mathematics Department
University College
London WC1E 6BT, UK

Streett, C.S.
NASA Langley Research Center
Hampton, Virginia 23665-5225
  USA

Turkel, E.
ICASE, NASA Langley Research
  Center
Hampton, Virginia, 23665-5225,
  USA
and
Department of Mathematics
Ramat-Aviv, Tel-Aviv, Israel
Tel-Aviv University

Williams, D.R.
Fluid Dynamics Research Center
Mechanical and Aerospace
  Engineering Department
Illinois Institute of Technology
Chicago, Illinois 60616, USA

Application of Stability Theory to Laminar Flow Control—Progress and Requirements

D.M. Bushnell and M.R. Malik

Abstract

Paper briefly summarizes the current status of linear stability theory as applied to laminar flow control for aerodynamics. Results indicate that the conventional "N factor" method of correlating stability theory and transition has a broad application range, including low- and high-speeds, two- and three- dimensional mean flow and TS, Gortler and crossflow disturbance modes. Linear theory is particularly applicable to the laminar flow control problem as, for system efficiency, control must be exercised and disturbances maintained in the linear regime. Current areas of concern for LFC, which require further stability theory research, include TS-crossflow interaction, combined disturbance fields (roughness, waviness, noise) and suction-induced disturbances. Some results on wave-interactions are presented.

Introduction

Laminar flow control i.e., delay of boundary-layer transition, could (and in some cases does) provide major increases in performance for both aerodynamic and hydrodynamic systems. Since viscous effects account for 50 percent of CTOL drag, the application of LFC to wings, body, nacelles, and empannage could result in large net performance improvements.

Applicable LFC techniques in air include mean field control such as (a) extensive regions of favorable pressure gradient and/or wall cooling (for two-dimensional mean flow), (b) suction, and (c) streamline curvature including convex longitudinal, transverse and in-plane curvature. A more recent approach to the LFC problem is still in the very early research stage and involves direct control of the fluctuation/disturbance field as opposed to the conventional indirect control through modification of the mean flow profiles. Typical direct control techniques include simultaneous spatial and temporal gradients imposed by (a) wall temperature, (b) wall motion, or (c) wall mass transfer.

Applications of immediate interest for LFC include (1) general aviation, (2) executive class jets, (3) CTOL and SST transports, (4) missiles, and (5) torpedoes. LFC provides design options for larger range, reduced volume, increased speed, increased sensor/weapon effectiveness as well as fuel conservation and cost savings. Of particular interest is the reduction of radiation equilibrium temperature through LFC at SST conditions, allowing the use of aluminum at higher Mach numbers. References 1 and 2 (and the 900 Refs. therein) comprise a reasonably current summary and bibliography of LFC.

LFC research dates from the late 1930's and initially considered the use of favorable pressure gradient control ("natural laminar flow") for the unswept wings of the period. This research was largely experimental and only made possible by the development of low-disturbance wind tunnels (through the efforts of Dryden and others). These facilities increased the experimentally observed flat plate transition Reynolds number an order of magnitude and allowed the experimental confirmation (at NBS in the U.S.) of the TS stability theory developed earlier in Germany. Flight experiments confirmed the existence of extensive regions of laminar flow but this initial research stage was essentially terminated in the early 50's by (a) the advent of jet aircraft with their higher speeds and attentent wing sweep, and (b) problems with wing surface deformation under load. In addition, since LFC is a sensitive stability problem it is affected by a myraid of possible "spoilers" (Fig. 1) and therefore the maintenance and reliability issues keep appearing (without an adequate treatment) throughout.

Since, with sweep, favorable pressure gradient is destablizing due to the crossflow instability mode, LFC in the swept wing case reappeared in the form of suction. Initial efforts, while successful even in flight, were terminated due essentially to the low price of fuel. However, suction LFC reemerged following the oil embargo and fuel/energy crisis of the 70's. After initial applications-orientated research for suction LFC in the late 70's and early 80's, the major (aeronautical) LFC interest focused upon the so-called "hybrid" system, which combines suction in the wing leading edge (high pressure gradient/high crossflow region of the wing), with pressure gradient control in the vicinity of the wing box.

Typical performance levels obtained thus far in LFC flight tests include (a) transition at midchord on unswept wings with transition Reynolds numbers the order of 15×10^6, (b) transition Reynolds number decreasing with increasing sweep (no suction case), typical values being 10×10^6 at 17°, 6×10^6 at 21° and 3×10^6 at 25°. With suction; full chord laminar flow was obtained up to Reynolds numbers of 36×10^6 (unswept) and 47×10^6 (30° sweep). Recent studies of transition locus on general aviation aircraft (Refs. 3 and 4) indicate that the thicker and smoother aircraft skins in current use have obviated many of the earlier problems associated with "natural" laminar flow.

The results from over 40 years of laminar flow control research indicate that, for subsonic/unswept wings, pressure gradient (natural) laminar flow control is a fact and is currently in use. For the transonic speed/swept wing case suction control "works", but residual maintenance and reliability/cost questions remain. The full suction case is such a major step that the current approach-of-choice for the swept wing is an attempt to combine, for low-to-moderate sweep, suction in the leading edge with pressure gradient control over the wing box, the so-called hybrid approach.

The purpose of the present paper is to briefly summarize the relationship of stability theory to transition prediction and laminar flow control system design. A major result of the present work is the identification of LFC problem areas requiring further research in the area of stability theory.

Transition Prediction

The transition process which LFC seeks to "control" is composed of several distinct physical processes. These include the "subcritical region" where disturbances are forced upon the viscous flow. Except for high amplitude forcing (e.g., non-linear bypass) disturbances are damped in this region of the flow. The resulting disturbance level in the critical region is all important in determining the subsequent behavior of the transition process. Beyond the critical region both existing and newly forced disturbances are amplified. For reasonable (low) amplitudes this amplification is exponential in space and time (as given by linear theory). In the usual case this exponential growth (linear region) comprises the bulk of the transition region. The particular normal modes responsible for the amplification are a function of the particular mean flow under consideration. For boundary-layer flows these instabilities include TS (viscous) waves, Rayleigh (inflectional) waves (e.g., instability due to crossflow or at high Mach numbers) and, for curved streamlines, Gortler waves.

As the "transition point" (appearance of Emmons spots) is approached the linear processes (exponential growth) which are responsible for most of the overall amplification begin to become (finally) non-linear and secondary instabilities and spectral broadening (and usually linear growth!) occurs. Although this non-linear region generally subtends a relatively small portion (spatially) of the entire transition process, a truely <u>predictive</u> method for transition should include the physical richness of the various secondary and tertiary (and beyond) instability processes. The unraveling of this non-linear region is an area of considerable current research (e.g., Ref. 5), but the work has not yet progressed far enough to allow incorporation into design-type transition prediction techniques. Furthermore, this work has only addressed the situation where two-dimensional TS waves constitute the primary instability mechanism.

To summarize, the bulk of the transition process (in the absense of the Morkovin, "High Intensity Bypasses," Ref. 6) is described by linear theory, with the actual disturbance level (and hence the actual locus of non-linear onset/eventual transition location) a function of the initial and ambient disturbance levels. Although both the internalization of ambient fluctuation fields and the non-linear "end game" immediately proceeding transition are the subject of intense current research, the state-of-the-art in transition prediction is essentially the well known e^N method, originated by Smith (Ref. 7) and Van Ingen (Ref. 8). Although this technique is fraught with inaccuracies and deficiencies it does, to first order, at least include the influence of changes in the mean flow upon the instability process.

The e^N method is akin to evaluating the output of an amplifier knowing only the gain and not the input. Therefore, the technique is only expected to yield reasonable results for flows that (a) have similar initial disturbance level and spectra, and (b) are dominated by a particular disturbance mode rather than a combination thereof (e.g., wave-wave interactions are not included). References 9 and 10 provide particularly cogent insights into the e^N method and the general problem of transition prediction. The e^N approach involves three steps: (1) computation of the mean velocity profiles with sufficient accuracy for determination of second derivatives, (2) computation of the normal instability modes appropriate to the physics of the problem, and (3) a search/integration routine to determine the mode with the largest integrated growth. Three "black box" codes exist to pursue such calculations for two-dimensional and axisymmetric boundary layers (Ref. 11) and low-speed (Ref. 12) and high-speed (Ref. 13) swept-wing flows. As opposed to the early works on the e^N method these techniques all utilize non-similar boundary-layer computations and, as closely as possible within the linear theory framework, the correct physics for the problem.

The original e^N research determined that transition occurred when N was the <u>order of 9</u> (with nominal "scatter" from 7 to 11) for low-speed two-dimensional wing flows and also for the low-speed Gortler problem, i.e., the same nominal numbers applied for both the TS and Gortler instability modes. For many flows, particularly adverse pressure gradients, the spatial shift in predicted transition location between an N of 7 and an N of 11 is minimal, due to the locally high growth rates. Research over the past several years by M. R. Malik and others at NASA Langley has attempted to determine the limits of applicability of the e^N approach for a broader range of flows including three-dimensional and high Mach number cases.

The extension of the e^N method to the compressible two-dimensional (axisymmetric) case is documented in Reference 14. The local Mach numbers were supersonic but not hypersonic and therefore the dominant disturbance was still the first oblique TS mode. The geometry was a sharp small angle cone at zero angle of attack. Data were available for 2 low-stream disturbance cases, F-15 flight experiment, and results from the Langley pilot M ~ 3.5 quiet tunnel. The resulting N factors are shown on Figure 2. Of interest from these calculations is (a) the peak N factors at transition (peak of the curves shown) are still the order of 9 to 11 and (b) the flight and quiet tunnel data are essentially in agreement in terms of the e^N formalism. If transition data from conventional high-speed facilities with their large amplitude stream acoustic disturbances had been used to determine N the resulting values would have been the order of 2. This again highlights the limitation of the e^N approach. However, low disturbance ground facilities are evidently capable of simulating transition processes which are indicative of flight conditions (e.g., Fig. 2), which is the major purpose of ground facilities in the first place.

A second extension of the e^N method investigated at NASA Langley involved the compressible version of the Gortler problem (Ref. 15). Experimental data suitable for such a correlation were available from the Langley supersonic low-disturbance facility development program. Since the major stream disturbance in supersonic and hypersonic tunnels is noise radiated from the turbulence in the nozzle wall boundary layers, the only real fix to the problem of a low disturbance M > 1 wind tunnel is to keep the nozzle wall boundary layer laminar. This is accomplished by utilizing massive slot suction upstream of the throat to bleed off the pre-existing turbulent wall flow entering from the stagnation chamber. The favorable pressure gradient associated with the nozzle expansion then serves to control the TS growth in the subsequent nozzle wall boundary layer and the performance of the facility in terms of the ability to maintain a low stream disturbance/laminar wall boundary-layer condition is dictated by the Gortler instability processes in the vicinity of the nozzle reflex contour (concave curvature) required to create inviscid wave cancellation/uniform flow. Transition data from several nozzles were analyzed using the compressible linear theory for Gortler waves. The results for the Mach 3.5 two-dimensional nozzle case are shown on Figure 3. Again, the peak N factors are the order of 9 to 11 with the most unstable (spanwise) wavelength the order of δ.

Two three-dimensional boundary layer transition processes have been examined thus far in the on-going Langley program, the rotating disk and the subsonic swept wing. Both flight and low disturbance wind tunnel data are in hand for the supersonic swept wing case, but the stability analysis/comparisons are not yet complete. The rotating disk (Ref. 16) is an interesting case as the transition process is of the crossflow variety, and is quite similar to the swept wing. A significant difference between the swept wing and disk, as noted in Reference 16, is the presence of the Coriolis force (due to rotation).

The rotating disk case is the first instance, at least in the Langley program, where the extreme importance of in-plane curvature effects surfaced. Conventionally, curvature effects of usual concern are either longitudinal or transverse. The effect of the former is quite large and can lead to either major disturbance damping or augmentation, depending upon the sign of the curvature. Also for the concave longitudinal case curvature is responsible for the Gortler wave mode. Transverse curvature, for the external flow case, also can lead to damping of disturbances with large enough δ/r values leading to relaminarization. The importance and even the existence of in-plane curvature effects are much less widely recognized or appreciated. For the rotating disk case the in-plane curvature effects, present due to flow divergence, were found to be first order. As shown on Figure 4 the inclusion of curvature and Coriolis terms reduces the N factor at transition by a factor of 2 (from 23 down to 11) and also changes the critical Reynolds number.

The other three-dimensional case computed thus far in the Langley program is the swept wing. Two different regimes occur on the wing, a near leading edge region, where in-plane curvature (of both the mean streamlines and the disturbance fronts) is again a first order effect, and farther back where the in-plane curvature effect is usually negligible. The major instability mode for the swept wing case is the crossflow induced disturbance vortex. Of particular interest is the observation in the Langley swept wing work (e.g., Refs. 17 and 18) that the most unstable crossflow waves are not always the stationary ones which are seen in the surface flow visualization. For many flow situations (depending upon sweep and pressure gradient history) the most unstable waves (yielding the highest N factors) are traveling waves. Sample computations for the near leading edge case are shown on Figure 5 (Ref. 17). When in-plane curvature effects are included the N factor decreases from the order of 17 to values of the order of 11. As noted previously, the most unstable waves are not at zero frequency. The computed N values for transition regions away from the immediate leading-edge region are shown on Figure 6 (Ref. 18), where the N values are obtained in this particular case from low-speed flight experiments. Again the dominant disturbances are (non-zero frequency) crossflow-induced vortices, with N factors varying from 7 to 11.

To summarize this section on transition prediction through the use of stability theory, the state of the art is still the e^N approach. However, several recent studies, as briefly reviewed herein, indicate that this approach is applicable over a surprisingly broad range of speed and disturbance modes. When the background disturbance level is of the order of .05 percent, then the N factor is the order of 9 to 11 for both Gortler and TS modes at both low speed and supersonic conditions. The same statement also holds for the crossflow mode at low speeds. The calculations and comparisons for the supersonic crossflow case are still in progress. As noted herein, it is necessary to include the applicable first order physics such as in-plane curvature, and Coriolis and non-stationary waves in the N factor calculation.

As a general observation, there are 3 broad classes of transition information. In class 1, large stream or surface disturbance levels are present (e.g., $U'_\infty \sim O(1\%)$, roughness, turbine blades in gas turbines, propeller wash regions) and the N factors are quite low (less than 5). Class 2 is composed of the data examined herein such as low-disturbance tunnel data and open ocean/conventional flight data. The N value for this class is the order of 9 to 11. In class 3 are grouped flow situations with inordinately low ambient/surface disturbance levels such as super careful ground or flight experiments and data on sailplane surfaces where the unit Reynolds numbers are low, surfaces are smooth, and propulsion units (and their vibration/noise) are absent. For these data the N value can approach 15 to 20.

Obviously something considerably better than the e^N method should, can, and will be developed in due course, incorporating the on-going research on both receptivity and secondary/tertiary/non-linear (end-game) instabilities. The application of such advanced methods will be exceedingly difficult due to the concomitant requirement for extensive and detailed information on the ambient and surface disturbance environment. This information simply does not yet exist in any but the most rudimentary form and is unlikely to be available in the foreseeable future. The day-to-day and even minute-to-minute changes which occur in any physical situation will set a lower bound on the precision of such ambient disturbance information, even if available, and therefore transition prediction will always be an "inexact" science. "The highest transition Reynolds numbers were usually observed during cloudy overcast days at lower temperatures early in the morning" (Ref. 19).

Wave-Interactions in Boundary Layers

A three-dimensional boundary layer is generally rich in instability modes. What modes are actually excited depends a great deal upon the "particular" forcing present. This is the classical "receptivity" problem (Ref. 6). The presence of a finite-amplitude disturbance in a boundary layer could also lead to the excitation of disturbance modes which may otherwise be damped or weakly unstable according to linear stability theory. Examples of such interactions are the possible excitation of TS instability in the presence of crossflow vortices on a swept wing or the excitation of TS instability in the presence of Gortler vortices on a concave wall like the one used in the Langley Mach 3.5 quiet tunnel. The obvious question to be asked is what amplitude the primary instability mode will have to attain for such interactions to take place.

We intend to address the wave-interaction problem within the framework of Navier-Stokes equations. Time-dependent, three-dimensional, Navier-Stokes equations are solved by the Fourier-

Chebyshev spectral method of Reference 20. The particular three-dimensional boundary layer considered is that formed on a rotating disk. As a start, we consider interaction between two crossflow waves. This type of interaction was present on a swept plate with imposed pressure gradient (Refs. 21 and 22). At time t = 0, we use as initial condition, the linear eigen function of two disturbances imposed upon the von Karman three-dimensional mean flow for the rotating disk. The Reynolds number of the flow is R = 500. The first disturbance is the most amplified stationary disturbance of nondimensional wavelength λ = 17.2. This disturbance amplifies according to linear stability theory with a temporal growth rate ω_i = .00719. The initial amplitude of this disturbance is taken to be ϵ_1 which is the ratio of the (absolute) maximum of the eigen function to the (absolute) maximum of the mean flow. The second disturbance has a wavelength of $\lambda/2$ and decays with a temporal rate ω_i = -.000865. The amplitude of this second disturbance is ϵ_2 which is assigned a value of .0001. We ask the question that what value of amplitude ϵ_1 will excite an instability of wavelength $\lambda/2$.

The results for ϵ_1 = .05 are presented in Figure 7(a) where the ratio of the disturbance energy to the (von Karman) mean flow energy is plotted against time. The results were obtained by using an 8 x 4 x 33 grid. A strong instability instantly develops for disturbance of wavelength $\lambda/2$ while from linear theory this disturbance should decay. Up to a time of about t = 100, the growth rate of the primary disturbance follows the linear theory prediction. After t = 100, the secondary disturbance begins to affect the growth rate of the primary disturbance. The results for ϵ_1 = .01 and .003 are presented in Figures 7(b) and 7(c), respectively. For ϵ_1 = .01, the $\lambda/2$ disturbance initially decays for some time and then begins to grow. For ϵ_1 = .003 the $\lambda/2$ disturbance follows approximately the linear theory decay rate up to t $\simeq$ 120 and then slowly begins to grow. In both cases the primary disturbance follows the linear theory growth rate. It can be concluded from these calculations that for a wave excitation of crossflow/crossflow type to take place the amplitude of the primary wave should be the order of 1 percent. These results are somewhat qualitative in that the Navier-Stokes code did not account for Coriolis force and in-plane curvature effects which were found to be very significant in linear stability analysis (Ref. 16). Further results with these effects included and TS/crossflow type interactions will be presented elsewhere by the second author.

Application of Stability Theory to Laminar Flow Control

Laminar flow control is, in essence, a stability problem. In fact, LFC constitutes one of the most technologically significant applications of aerodynamic stability theory. From a practical standpoint it is not feasible to avoid all disturbance growth i.e., keep the flow subcritical. Such an approach simply requires/consumes too much control power. In addition, if this is accomplished through thinning the boundary layer the flow becomes hypersensitive to the residual roughness of even "smooth" surfaces and a "high intensity bypass" to early transition can be set up. Therefore, the LFC philosophy is to allow some disturbance growth. However, bitter experience and careful calculation indicates that disturbances must be maintained in the linear range and that control at an early state of growth is much more efficient systemwise. As the disturbance levels rise, the multifarious nature of the transition process requires inordinately large control levels.

Therefore, LFC design requires, in the "clean" case of a single disturbance mode with low background disturbance, the application of linearized methods in the linear disturbance range, an ideal application of stability theory. (However, as will be clear in the ensuing discussion, LFC is not always a "clean" problem.) LFC design is therefore not a problem of transition prediction, but one of ensuring that transition is never even approached. A simplex solution to this design problem is simply to "derate" the e^N approach, i.e., select a lower N value and utilize sufficient control power to keep N within bounds. However, if the ambient disturbance levels are large, or intermodel/wave-wave interaction occurs, then transition may still occur, even if the local N value is quite low.

Stability research, essentially using the derated e^N method, has included/investigated the following effects/flows for LFC design: (a) two-dimensional and infinite swept tapered wings; (b) incompressible and compressible flows; (c) pressure gradient, wall cooling and wall suction control; (d) finite amplitude effects in two dimensions; (e) in-plane and body curvature; (f) suction strips/discrete suction vs. area suction, (g) optimum placement of suction; (h) concave curvature/Gortler problem; (i) interrelationship of temporal and spatial stability in three-dimensional mean flows, (j) and a beginning on the crossflow-TS and Gortler-TS interaction problems. Due to the interest in application of a "hybrid" approach for LFC on swept wings a major problem of interest at the present time is the physics of crossflow-TS interaction. In fact, for "natural" transition, even with "control," important wave-wave interactions of even the same mode can be important. These interactions are often suppressed in the "unnatural" laboratory studies where waves of specific frequency and orientation are excited, for clarity, by vibrating ribbons, etc. Not only are the crossflow-TS wave interactions themselves of interest but also the sensitivity of such interactions to the effects of (1) Mach number, (2) acoustics (particularly fuselage noise radiation), (3) waviness, (4) gaps and steps, and (5) suction surface induced disturbances, i.e., sensitivities which have been studied, both theoretically and experimentally, for the single mode case must be revisited for multiple/interacting modes.

There is also interest, particularly for smaller aircraft such as general aviation and executive jets, in fuselage "natural laminar flow." Here the areas of concern which should be addressed by stability theory include (a) crossflow effects induced by the non-axisymetric nature of conventional fuselage forebodies and (b) the mitigation of the effects of unavoidable gaps and bumps etc., through careful geometric tailoring.

Laminar flow control for supersonic vehicles such as the SST introduces a new set of requirements for further stability theory research. Of particular concern is the occurrence of suction-induced shock waves which can amplify incident (embedded) disturbance fields and, for subsonic leading edges, focus toward the rear of the wing. Also, there is a need for examination of the single, and combined, effects of waviness, steps/gaps, roughness and other transition "spoilers" at supersonic speeds to determine "design allowables". The available high-speed flight transition data suggests a "unit Reynolds number" trend which may be due to atmospheric particle-bow shock interactions and subsequent convection of disturbances produced by such an interaction into the boundary layer. This possible stream-disturbance generation mechanism, unique to high-speed flows, should also be studied as a possible limit or modifier to the performance of LFC at high speeds.

Concluding Remarks

Laminar flow control constitutes a tremendously important and major application of stability theory. Efficient LFC system design requires that disturbance levels be kept in the linear range even in the presence of such transition spoilers as roughness, waviness, and acoustic disturbances. For background disturbance levels characteristic of low-disturbance tunnels and many flight conditions the e^N method with $N \sim 0$ (9-11) provides a useful engineering prediction of transition for two- and three-dimensional flows, low and high speeds and for TS, Gortler and crossflow disturbance modes. Further stability research required for the LFC design problem includes (a) crossflow-TS interactions (for "hybrid" systems), (b) acoustic receptivity at $M \sim 0$ (1), (c) the localized influence of single and multiple suction holes, (d) disturbance amplification through shock waves, (e) effects of wall waviness and steps/gaps and their mitigation through geometrical tailoring, and (f) the effects of combined spoilers (acoustic disturbance incident upon a wavy surface etc.).

References

1. Bushnell, Dennis M.; and Tuttle, Marie H.: Survey and Bibliography on Attainment of Laminar Flow Control in Air Using Pressure Gradient and Suction. Volume I. NASA RP-1035, September 1979.
2. Tuttle, Marie H.; and Maddalon, Dal V.: Laminar Flow Control (1976-1982). A Selected, Annotated Bibliography. NASA TM-84496, August 1982.
3. Holmes, Bruce J.; Obara, Clifford J.; and Yip, Long P.: Natural Laminar Flow Experiments on Modern Airplane Surfaces. NASA TP-2256, June 1984.
4. Obara, Clifford J.; and Holmes, Bruce J.: Flight-Measured Laminar Boundary-Layer Transition Phenomena Including Stability Theory Analysis. NASA TP-2417, April 1985.
5. Herbert, T.: Secondary Instability of Shear Flows. Special Course on Stability and Transition of Laminar Flow, von Kaman Institute, Rhode-St-Genese, Belgium, March 1984, pp. 7-1 - 7-13.
6. Morkovin, Mark V.: Critical Evaluation of Transition From Laminar to Turbulent Shear Layers With Emphasis on Hypersonically Traveling Bodies. AFFDL-TR-68-149, U.S. Air Force, March 1969.
7. Smith, A. M. O.; and Gamberoni, N.: Transition, Pressure Gradient and Stability Theory. Rep. No. ES 26388, Douglas Aircraft Co., Inc., August 31, 1956. (Also available in IX Congress International de Mecanique Appliquee, Tome IV, Universite de Bruxelles, 1957, pp. 234-244.)
8. Van Ingen, J. L.: A Suggested Semi-Empirical Method for the Calculation of the Boundary Layer Transition Region. Univ. of Tech., Dept. Aero. Eng., Report VTH-74, Delft, Holland, 1956.
9. Mack, L. M.: Boundary-Layer Linear Stability Theory. Special Course on Stability and Transition of Laminar Flow, von Karman Institute, Rhode-St- Genese, Belgium, March 1984, pp. 3-1 - 3-81.
10. Mack, L. M.: Transition Prediction and Linear Stability Theory. In AGARD Conference Proceedings No. 224, pp. 1-1 - 1-22, Nato, Paris, 1977.
11. Gentry, A. E.; and Wazzan, A. R.: The Transition Analysis Program System: Volume II - Program Formulation and Listings. McDonnel Douglas Corporation Report No. J7255/02, 1976.
12. Srokowski, A. J.; and Orszag, S. A.: Mass Flow Requirements for LFC Wing Design. AIAA Paper No. 77-1222.
13. Malik, M. R.: COSAL - A Black-Box Compressible Stability Analysis Code for Transition Prediction in Three-Dimensional Boundary Layers. NASA CR-165925, May 1982.

14. Malik, M. R.: Instability and Transition in Supersonic Boundary Layers. <u>Laminar Turbulent Boundary Layers</u>. (Ed. E. M. Uram and H. E. Weber), Energy Resources Technology Conference, New Orleans, LA, February 12-16, 1984.

15. Beckwith, I. E.; Malik, M. R.; and Chen, F.-J.: Nozzle Optimization Study for Quiet Supersonic Wind Tunnels. AIAA Paper 84-1628. Presented at the AIAA 17th Fluid Dynamics, Plasmadynamics, and Lasers Conference, Snowmass, Co, June 25-27, 1984.

16. Malik, M. R.; Wilkinson, S. P.; and Orszag, S. A.: Instability and Transition in Rotating Disk Flow. AIAA Journal, Vol. 19, 1981, pp. 1131-1138.

17. Malik, M. R.; and Poll, D. I. A.: Effect of Curvature on Three-Dimensional Boundary Layer Stability. AIAA Journal, Vol. 23, 1985, pp. 1362-1369.

18. Hefner, J. N.; and Bushnell, D. M.: Status of Linear Boundary Layer Stability Theory and the e^N Method With Emphasis on Swept Wing Applications. NASA TP-1645, 1980.

19. Pfenninger, W.; and Bacon, J. W., Jr.: Amplified Laminar Boundary Layer Oscillations and Transition at the Front Attachment Line of a 45° Swept Flat-Nosed Wing With and Without Boundary Layer Suction. <u>Viscous Drag Reduction</u>. (Ed. C. S. Wells), Plenum Press, 1969.

20. Malik, M. R.; Zang, T. A.; and Hussaini, M. Y.: A Spectral Collocation Method for the Navier-Stokes Equations. J. Computational Physics, in press.

21. Reed, H.: Disturbance-Wave Interactions in Flows With Crossflow. AIAA Paper No. 85-0494.

22. Saric, W. S.; and Yeates, L. G.: Generation of Crossflow Vortices in a Three-Dimensional Flat-Plate Flow. IUTAM Symposium on Laminar-Turbulent Transition, Novosibirsk, U.S.S.R., July 9-13, 1984, Springer-Verlag.

- ROUGHNESS
  - DISCRETE
  - DISCONTINUOUS
  - TWO-DIMENSIONAL
  - THREE-DIMENSIONAL
  - STEPS
  - GAPS
  - PARTICLE IMPACT/EROSION
  - CORROSION
  - LEAKAGE

- WALL WAVINESS
  - TWO-DIMENSIONAL
  - THREE-DIMENSIONAL
  - SINGLE WAVE
  - MULTIPLE WAVE
  - DISTORTION UNDER LOAD

- SURFACE AND DUCT VIBRATION

- ACOUSTIC ENVIRONMENT
  - ATTACHED FLOW
  - SEPARATED FLOW
  - PROPULSION SYSTEM
  - VORTEX SHEDDING

- STREAM FLUCTUATIONS AND VORTICITY
  - PROPELLER WAKES
  - OCEAN SURFACE
  - BODY WAKES (FISH/AIRCRAFT)
  - HIGH SHEAR AREAS (WEATHER FRONTS/ JET STREAM EDGES/OCEAN CURRENTS)

- PARTICLES
  - ICE CLOUDS
  - RAIN
  - ALGAE
  - SUSPENSIONS
  - FAUNA (INSECTS, FISH, ETC.)

- LFC SYSTEM-GENERATED DISTURBANCES
  - VORTEX SHEDDING (BLOCKED SLOTS, HOLES, PORES)
  - ACOUSTIC OR CHUGGING
  - PORE DISTURBANCES
  - NON-UNIFORMITIES

Fig. 1. Possible stream/wall disturbances critical to boundary layer transition.

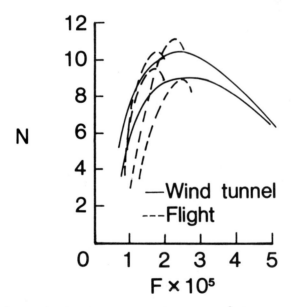

Fig. 2. Computed N factors at transition for a 10^0 sharp cone in a "quiet" wind tunnel and in flight at supersonic Mach numbers. Note that the peak of each curve lies between $N \simeq 9\text{-}11$.

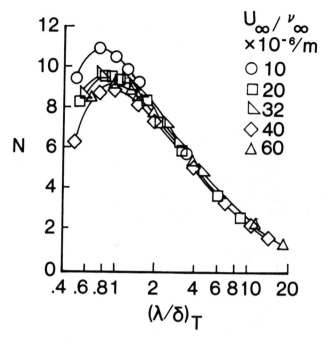

Fig. 3. Computed N factors at transition for the concave wall of the Langley Mach 3.5 "quiet" tunnel. Görtler instability plays a dominant role in the transition process.

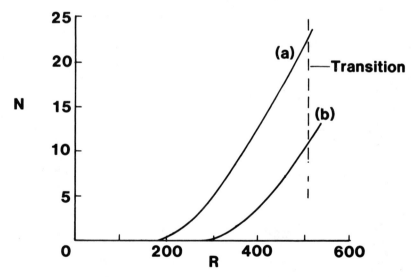

Fig. 4. N factor distribution for rotating disk flow:
(a) Orr-Sommerfeld equation
(b) Sixth-order equation to account for Coriolis force and streamline curvature.

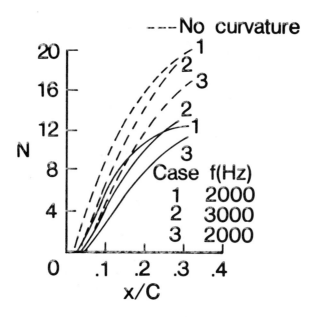

Fig. 5. Computed N factors with and without curvature effects for a swept cylinder for three different sweep angles: case 1, θ = 30°; case 2, θ = 55°; case 3, θ = 60°.

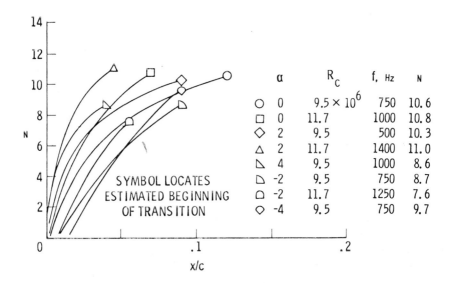

Fig. 6. N factor distributions for cranfield 45° swept wing.

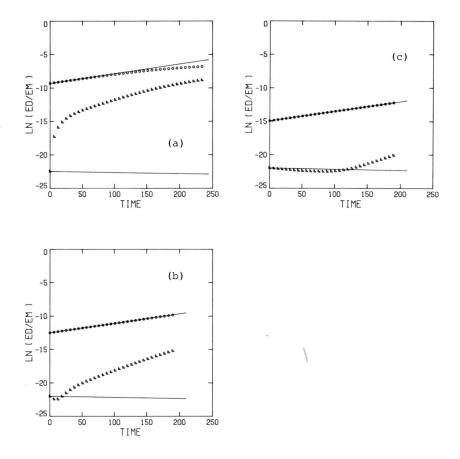

Fig. 7. Crossflow/crossflow interaction in rotating disk flow. The primary wave has a wavelength λ_1 = 17.2 and initial amplitude ε_1: (a) ε_1 = .05, (b) ε_1 = .01, (c) ε_1 = .003. The secondary wave has a wavelength $\lambda_2 = \lambda_1/2$ and initial amplitude ε_2 = .0001. Solid lines represent linear theory results.

On Secondary Instabilities in Boundary Layers

Ali H. Nayfeh

One of the routes to transition consists of a cascade of successive instabilities. The first instability is usually called primary instability and it may consist of a two- or three-dimensional Tollmien Schlichting wave, a Gortler vortex, or a cross-flow instability. The next two instabilities in the cascade are usually called secondary and tertiary instabilities. Two approaches of treating secondary instability are reviewed and compared. The first approach is a mutual interaction approach which accounts for the influence of the secondary instability on the primary instability. The second approach is usually called parametric instability. It does not account for this influence and leads to linear equations with periodic or quasi-periodic coefficients. Whereas the mutual interaction approach is not limited by the amplitude of the secondary instability, the parametric approach is limited to infinitesimal amplitudes of the secondary instability. The role of the Squire mode in the secondary instability on a flat plate is discussed.

1. Introduction

There is experimental evidence that one of the major routes to transition consists of a cascade of successive instabilities. The first instability is usually called primary instability and it may take the form of a two- or three-dimensional Tollmien-Schlichting (T-S) wave, or a Gortler vortex, or a cross-flow instability. The next two instabilities are usually called secondary and tertiary instabilities. When the primary instability is a T-S wave, the secondary instability may take the form of a T-S wave or a streamwise vortex. When the primary wave is a Gortler vortex, the secondary instability may take the form of a T-S wave or a streamwise vortex. When the primary instability is a cross-flow instability, the secondary instability may take the form of a T-S wave or a cross-flow instability.

Recent carefully controlled experiments [1-6] on the transition from laminar to turbulent flows were conducted in low-turbulence windtunnels. Two-dimensional disturbances of known frequencies were generated by vibrating ribbons in the Blasius boundary layer on a flat plate. The evolutions of these disturbances downstream were monitored by hot-wire measurements and flow visualization. The spectrum, growth, and amplitude distributions at different locations were determined. These experiments have identified a major route from laminar to turbulent conditions. Downstream of the vibrating ribbons, two-dimensional Tollmien-Schlichting (T-S) waves were observed to travel with the flows. For low vibration amplitudes of the ribbons, the T-S waves grow and decay in accordance with linear stability theory. On the other hand, for large vibration amplitudes, the waves grow downstream to appreciable amplitudes and secondary instabilities set in. These instabilities consist of three-dimensional waves that have nearly periodic spanwise deformations, producing spanwise alternating peaks and valleys (Λ-shaped vortex loops) with enhanced and reduced oscillation amplitudes. Figure 1 shows a reproduction of a

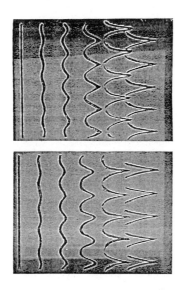

Figure 1. Streaklines in Blasius boundary layer. The drawings are a reproduction of photos taken by Saric and co-workers at the VPI Stability Wind Tunnel. Flow is from left to right. Top, subharmonic disturbance. Bottom, fundamental disturbance.

recent photograph of a smoke-wire flow visualization taken by Saric and his co-workers in the VPI & SU wind-tunnel. At low T-S amplitudes, one observes the arrangement in Figure 1a. The Λ-shaped vortex loops are staggered, repeating every $2\lambda_x$, where λ_x is the wavelength of the T-S waves, indicating a subharmonic of the T-S waves. The hot-wire measurements record a frequency that is one-half the T-S wave (i.e., subharmonic). This subharmonic instability was first identified by Kachanov, Kozlov and Levchenko [4] through the spectra of the hot-wire signals. Later, Kachanov and Levchenko [5] measured the global and local growth characteristics of this instability, including its growth and amplitude distribution. In some observations [6], $\lambda_z > \lambda_x$ and in others [1,6] $\lambda_z < \lambda_x$, where λ_z is the spanwise wavelength of the staggered vortices. At higher T-S amplitudes, one observes the arrangement in Figure 1b. The Λ-shaped vortex loops are such that the peaks and valleys are aligned [1-3], repeating every λ_x. The fixed hot-wire measurements record signals having a frequency that is equal to that of the T-S wave (i.e., fundamental). Typically, $\lambda_z < \lambda_x$. The three-dimensional secondary waves grow rapidly, forming concentrated shear layers at the peaks. The resulting highly inflectional instantaneous velocity profiles develop a tertiary instability in the form of small-scale high frequency disturbances (so-called spikes), leading to the ultimate breakdown of the laminar flow and the development of a turbulent flow. The breakdown sequence of a primary T-S wave, the appearance of an aligned peak-valley splitting, the appearance of spikes, and breakdown to turbulence was first observed by Klebanoff and co-workers [1,2] and it is usually called the "K-type breakdown"; it has also been observed in plane Poiseuille flow [7,8].

A number of experimental studies investigated the influence of steady streamwise vortices on the transition from laminar to turbulent flow. Aihara [9], Tani and Komoda [10], Tani and Sakagami [11], and Tani and Aihara [12] studied the influence of steady streamwise vortices on two-dimensional Tollmien-Schlichting waves generated by a vibrating ribbon. They considered the case in which the vortices were generated naturally on a concave surface (Görtler vortices) as well as the case in which the vortices were generated artificially by a row of wings on a concave surface. They measured the distributions of the mean velocity and wave intensity across the boundary layer for three spanwise positions at a number of streamwise stations. They concluded that the Görtler vortices indirectly affect transition by inducing a spanwise variation in boundary-layer thickness, at least when the radii of curvature are not extremely small. However, they did not present any measurements of the growth rates of the Tollmien-Schlichting waves. Wortman [13] investigated the development of natural transition downstream of Görtler vortices. Using the tellurium method, he determined the direction and relative magnitude of the unsteady velocities from the streaklines by confining his observations to the vicinity of the starting point of the streaklines. He observed a steady second-order instability that destroys the symmetry of the Görtler vortices. He suggested that this instability is caused by secondary vortices having spanwise wavelengths that are twice those of the Görtler vortices. Then he observed a third-order instability consisting of regular three-dimensional oscillations.

The experimental observations stimulated many theoretical investigations into secondary instabilities in boundary layers. Excluding purely numerical solutions of the full Navier-Stokes equations [14,15], these investigations can be divided into two broad groups. The first group treats the mutual interaction of the primary and secondary instabilities. This group can be subdivided further into non-resonant and resonant interaction of waves. The second group treats the influence of the primary instability on the secondary instability but neglects the influence of the secondary instability on the primary instability. It leads to linear equations with periodic or quasi-periodic coefficients and hence it is usually called parametric instability.

Benney and Lin [16] and Benney [17,18] proposed a non-resonant interaction model to explain the aligned peak-valley splitting observed by Klebanoff and his co-workers [1,2]. They considered the interaction of a two-dimensional wave of the form

$$u = a_1 \zeta_{11}(y) e^{i\alpha(x-c_1 t)} + cc$$

with a three-dimensional wave of the form

$$u = a_2 \zeta_{12}(y) \cos\beta z \, e^{i\alpha(x-c_2 t)} + cc$$

where a_1 and a_2 are constants. For simplicity, they assumed that $c_1 = c_2$, while recognizing that they differ by 10% to 15%. Allowing these waves to interact at second order, they found that the secondary flow contains terms that are proportional to $a_1 a_2 \cos\beta z$ and $a_1 a_2 \sin\beta z$ and terms that are proportional to $a_2^2 \cos 2\beta z$ and $a_2^2 \sin 2\beta z$. These secondary flows correspond to streamwise vortices. Thus they succeeded in showing qualitatively that the observed systems of vortices [1,2] are a result of the interaction of two- and three-dimensional waves having the above forms. However, this model cannot furnish an estimate of preferred spanwise periodicity because by choosing the periodicity of the three-dimensional wave, any desired spanwise wavelength can be generated.

Stuart [19] removed the assumption of equal phase speeds c_1 and c_2 and carried out the expansion to third order for the case of temporal stability, thereby determining coupled Landau equations describing the variations of a_1 and a_2 with t. Using the Benney-Lin model and relaxing the assumption of equal phase speeds, Antar and Collins [20] presented numerical results for the secondary flows for the Blasius flow as well as the Falkner-Skan flows. Herbert [21] reformulated the method of Landau constants for two interacting waves. Then he specialized the results to the Benney-Lin model and presented numerical calculations for the neutral surface of Tollmien-Schlichting waves of finite amplitude. Volodin and Zel'man [22] used the method of averaging to determine the Landau equations governing the spatial modulations of two interacting two-dimensional waves and used the phase plane to investigate the stationary states.

In the resonant sub-group, Raetz [23,24] and Stuart [25] established the occurrence of triad resonances for certain waves which are neutrally stable according to the linear theory, and Lekoudis [26] established triad resonances over swept wings. Craik [27] established the occurrence of triad resonances for certain unstable waves on a flat plate. Specifically, he found that a two-dimensional wave having the form

$$u = a\zeta_{10}\zeta_n(y)e^{i(2\alpha x - 2\omega t)}$$

forms a triad resonance with the two three-dimensional waves

$$u = a_n\zeta_{1n}(y)e^{i(\alpha x \pm \beta z - \omega t)}$$

Then, using temporal-stability theory, he derived equations governing the nonlinear evolution of the amplitudes a and a_n. He found that the amplitudes become infinite in a finite time, an explosive instability. Craik [28] extended the temporal instability of triad resonances to third order and examined the instability of shear flows. Lekoudis [29] derived the nonlinear equations describing the spatial and temporal evolution of the amplitudes of the triad waves by relaxing Craik's assumption of perfect resonance. Craik and Adam [30] and Craik [31] obtained solutions for the nonlinear evolution equations for resonant wave triads. Benney and Gustavsson [32] analyzed the interaction of a two-dimensional T-S wave with the Squire mode (i.e., vortical mode).

In contrast with the Benney-Lin model, which is incapable of estimating a preferred spanwise variation of the streamwise vortices, the triad-resonance model appears to favor the selective growth of a three-dimensional wave, which in turn generates a specific spanwise variation of the streamwise vortices. To check this hypothesis, Craik calculated the spanwise wavelength favored by the traid resonance for the conditions of the experiments of Klebanoff, Tidstrom, and Sargent [2]. Their experiments were carried out in a boundary layer on a flat plate. An artificial disturbance of known frequency was introduced in the boundary layer by a vibrating ribbon. They considered two cases: one with the frequency 145 c/s and one with the frequency 65 c/s. Although the disturbance was predominantly two-dimensional, it possessed a small three-dimensional component with a fixed spanwise wavelength, owing to their placement of small pieces of tape equally spaced (1" apart) under the vibrating ribbon. Downstream of the ribbon, they observed that the amplitude of the three-dimensional component intensified gradually until transition to turbulence occurred.

The gradual intensification of the three-dimensional component does not corroborate the presence of an explosive instability in which the amplitudes become infinite in a finite time. Moreover, Craik found that the conditions of triad resonance are satisfied only for the frequency 145 c/s but not for the frequency 65 c/s. Therefore, Craik concluded that the triad-resonance model is hardly adequate for the experimental situation of Klebanoff, Tidstrom, and Sargent. Even though the condition of triad resonance is satisfied for the frequency 145 c/s, it is satisfied only at a single location, namely the ribbon location. Thus, if an explosive instability exists, it would have manifested itself at the ribbon location, and the three-dimensionality would not have intensified gradually. This led Nayfeh and Bozatli [33] to suggest that, in a growing boundary layer, the explosive instability suggested by Craik would not occur due to the spatial detuning and that a nonexplosive instabilty mechanism is responsible for the gradual intensification of the three-dimensional waves. To explain the aligned peak-valley splitting observed by Klebanoff and co-workers [2], Nayfeh and Bozatli [33] proposed a four-wave interaction model consisting of two two-dimensional waves with the frequencies ω and 2ω and two three-dimensional waves with the frequency ω.

The parametric approach was used by Kelly [34] and Pierrehumbert and Widnall [35] to analyze the stability of a finite-amplitude periodic flow and an array of finite-amplitude vortices in a shear layer, respectively. Kelly found that vortex pairing in an inviscid shear layer is the result of principal parametric resonances. Pierrehumbert and Widnall found that principal parametric resonances lead to vortex pairing and fundamental parametric resonances lead to three-dimensional modes of instability. Nayfeh and Bozatli [36] investigated the principal parametric instability on flat plates in the form of two-dimensional T-S waves. Maseev [37], Orszag and Patera [15] and Herbert [38] investigated the three-dimensional fundamental parametric instability, whereas Herbert [39,40], Bertolotti [41] and Nayfeh [42] investigated the three-dimensional principal parametric instability. Nayfeh [43] analyzed principal parametric instabilities in the form of two oblique three-dimensional T-S waves when the primary instability is a streamwise vortex, Herbert and Morkovin [44] analyzed fundamental parametric instabilities in the form of Gortler vortices when the primary instability is a T-S wave, and Floryan and Saric [45] analyzed principal parametric instabilities in the form of streamwise vortices when the primary instability is a Gortler vortex. Reed [46] analyzed the influence of the cross-flow instability on three-dimensional T-S waves.

There are a number of methods available for analyzing parametric resonances [47-49]. When the primary wave is periodic and its amplitude is constant, one can use Floquet theory and the method of harmonic balance to determine the characteristic exponents (growth rates). When the primary wave is not periodic or its amplitude is not constant, one can use either the method of multiple scales or a version of the method of averaging to determine the modulation of the amplitudes and phases of the secondary wave.

In the next section, we use a two-degree-of-freedom example to describe and compare the mutual interaction and parametric approaches. In Section 3, we present the mathematical problem governing nonlinear three-dimensional disturbances in growing incompressible boundary layers. In Section 4, we discuss the linear quasi-parallel problem, including the adjoint and bi-orthogonality condition. In Sections 5-8, we discuss application of the mutual interaction and parametric approaches to the problem described in Section 3.

2. Comparison of Mutual Interaction and Parametric Approaches

In this section, we discuss and compare the parametric and mutual interaction approaches using a two-degree-of-freedom system, namely, the swinging spring shown in Figure 2. The equations describing the motion of the mass m are [48]

$$\ddot{x} + \frac{k}{m} x + g(1 - \cos\theta) - (\ell + x)\dot{\theta}^2 = 0 \tag{1}$$

$$\ddot{\theta} + \frac{g}{\ell+x} \sin\theta + \frac{2}{\ell+x} \dot{x}\dot{\theta} = 0 \tag{2}$$

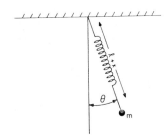

Figure 2. Spring pendulum.

Expanding (1) and (2) for small x and θ yields

$$\ddot{x} + \omega_2^2 x + \frac{1}{2} g\theta^2 - \ell\dot{\theta}^2 - x\dot{\theta}^2 + \ldots = 0 \tag{3}$$

$$\ddot{\theta} + \omega_1^2 \theta - \frac{g}{\ell^2} x\theta + \frac{2}{\ell} \dot{x}\dot{\theta} - \frac{g}{6\ell} \theta^3 + \frac{g}{\ell^3} x^2\theta - \frac{2}{\ell^2} x\dot{x}\dot{\theta} + \ldots = 0 \tag{4}$$

where

$$\omega_1^2 = \frac{g}{\ell}, \quad \omega_2^2 = \frac{k}{m} \tag{5}$$

For small θ, (3) reduces to

$$\ddot{x} + \omega_2^2 x = 0 \tag{6}$$

whose solution is

$$x = a_2 \cos(\omega_2 t + \beta_2) \tag{7}$$

where a_2 and β_2 are constants. This is the response of the spring mode.

Substituting (7) into (4) yields

$$\ddot{\theta} + \omega_1^2 \theta - \frac{ga_2}{\ell^2} \left[\cos(\omega_2 t + \beta_2) + \frac{a_2}{\ell} \cos^2(\omega_2 t + \beta_2) \right] \theta$$
$$- \frac{2a_2\omega_2}{\ell} \sin(\omega_2 t + \beta_2)\left[1 - \frac{a_2}{\ell}\cos(\omega_2 t + \beta_2)\right]\dot{\theta} - \frac{g}{6\ell}\theta^3 = 0 \tag{8}$$

which governs the pendulum mode as a secondary instability. In the parametric approach, one neglects the nonlinear terms in (8) and obtains

$$\ddot{\theta} + \omega_1^2 \theta - \frac{ga_2}{\ell^2} \theta\cos(\omega_2 t + \beta_2) - \frac{2a_2\omega_2}{\ell}\dot{\theta}\sin(\omega_2 t + \beta_2) = 0 \tag{9}$$

which is a linear equation with periodic coefficients. Equation (9) possesses parametric resonances [48] when $2\omega_1/\omega_2 \approx 0, 1, 2, 3, 4, \ldots$. The $a_2 - 2\omega_1/\omega_2$ plane is divided into regions of stable and unstable solutions. In the unstable regions, the pendulum mode (i.e., θ) grows exponentially with t.

The exponential growth predicted by the linear parametric approach is not realistic. As θ grows, the nonlinear terms in (8) limit the growth of θ to a limit cycle [48]. Moreover, the nonlinear terms may produce a subcritical instability as shown in Figure 3. The nonlinear terms in (3) may also limit this growth.

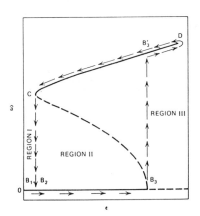

Figure 3. Response curves for a parametrically excited Duffing equation. Whereas the linear model predicts exponential growth in Region III, including the nonlinear terms limits the growth to a limit cycle. Moreover, whereas the linear model predicts exponentially decaying response in Region II, the nonlinear model predicts a subcritical instability and the response decays with time or grows with time to a limit cycle, depending on the initial conditions.

Analyzing the solutions of (3) and (4) when $\omega_2 \approx 2\omega_1$ (corresponding to principal parametric resonance) yields the following expansion [47,48]:

$$x = a_2\cos(\omega_2 t + \beta_2) + \ldots \tag{10}$$

$$\theta = a_1\cos(\omega_1 t + \beta_1) + \ldots \qquad (11)$$

where

$$\dot{a}_2 = -\frac{3g}{8\omega_2} a_1^2 \sin\gamma \qquad (12)$$

$$a\dot{\beta}_2 = \frac{3g}{8\omega_2} a_1^2 \cos\gamma \qquad (13)$$

$$\dot{a}_1 = -\frac{\omega_1 - 2\omega_2}{4\ell} a_1 a_2 \sin\gamma \qquad (14)$$

$$\dot{\beta}_1 = -\frac{\omega_1 - 2\omega_2}{4\ell} a_2 \cos\gamma \qquad (15)$$

$$\gamma = 2\beta_1 - \beta_2 + (2\omega_1 - \omega_2)t \qquad (16)$$

Equations (12)-(16) describe the mutual interaction of the amplitudes and phases of the spring and pendulum modes. Figure 4 shows typical time variations of a_1

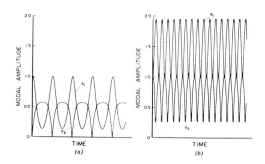

Figure 4. Free-oscillation amplitudes a_1 and a_2 of the spring pendulum when $\omega_2 \simeq 2\omega_1$: (a) $a_1(0) = 1$, $a_2(0) = 0$; (b) $a_1(0) = a_2(0) = 1$.

and a_2. Contrary to the exponential growth predicted by the parametric approach, the mutual-interaction approach predicts that the response consists of a continual exchange of energy between the two modes. Consequently, the parametric approach is limited to infinitesimal amplitudes of the secondary instability and cannot predict any subcritical instability.

3. Disturbance Equations

We consider the nonlinear, nonparallel stability of a three-dimensional, incompressible, growing, boundary-layer flow over a flat surface. Let x, y, z be a Cartesian coordinate system such that the x and z coordinates lie on the surface and the y coordinate is normal the surface. Let us denote the boundary-layer flow variables by $U(x,y,z)$, $V(x,y,z)$, $W(x,y,z)$ and $P(x,z)$, where U and W are slowly varying functions of x and z and V is small compared with U.

To study the stability of such a flow, we superpose on it a disturbance to obtain

$$\tilde{q}(x,y,z,t) = Q(x,y,z) + q(x,y,z,t) \tag{17}$$

where q stands for u, v, w, and p. Substituting (17) into the dimensionless Navier-Stokes equations and subtracting the mean-flow quantities, we obtain the disturbance equations

$$\frac{\partial u}{\partial x} + \frac{\partial v}{\partial y} + \frac{\partial w}{\partial z} = 0 \tag{18}$$

$$\frac{\partial u}{\partial t} + U\frac{\partial u}{\partial x} + W\frac{\partial u}{\partial z} + v\frac{\partial U}{\partial y} + \frac{\partial p}{\partial x} - \frac{1}{R}\nabla^2 u = -u\frac{\partial U}{\partial x} - V\frac{\partial u}{\partial y} - w\frac{\partial U}{\partial z} - u\frac{\partial u}{\partial x} - v\frac{\partial u}{\partial y} - w\frac{\partial u}{\partial z} \tag{19}$$

$$\frac{\partial v}{\partial t} + U\frac{\partial v}{\partial x} + W\frac{\partial v}{\partial z} + \frac{\partial p}{\partial y} - \frac{1}{R}\nabla^2 v = -u\frac{\partial V}{\partial x} - V\frac{\partial v}{\partial y} - v\frac{\partial V}{\partial y} - w\frac{\partial V}{\partial z} - u\frac{\partial v}{\partial x} - v\frac{\partial v}{\partial y} - w\frac{\partial v}{\partial z} \tag{20}$$

$$\frac{\partial w}{\partial t} + U\frac{\partial w}{\partial x} + W\frac{\partial w}{\partial z} + v\frac{\partial W}{\partial y} + \frac{\partial p}{\partial z} - \frac{1}{R}\nabla^2 w = -u\frac{\partial W}{\partial x} - V\frac{\partial w}{\partial y} - w\frac{\partial W}{\partial z} - u\frac{\partial w}{\partial x} - v\frac{\partial w}{\partial y} - w\frac{\partial w}{\partial z} \tag{21}$$

where $R = U_\infty \delta/\nu$, ν is the kinematic viscosity of the fluid, and U_∞ is the free-stream velocity. Distances and time were made dimensionless using a reference boundary-layer displacement thickness δ and δ/U_∞, respectively.

The no-slip and no-penetration conditions require that

$$u = v = w = 0 \quad \text{at} \quad y = 0 \tag{22}$$

Demanding that the disturbances decay away from the surface, we have

$$u, v, w, p \to 0 \quad \text{as} \quad y \to \infty \tag{23}$$

4. Linear Quasi-Parallel Solution

In the case of linear quasi-parallel flow, the right-hand sides of (19)-(21) are taken to be zero and (18)-(23) have solutions of the form

$$u = \zeta_1(x,y,z)e^{i\theta}, \quad v = \zeta_3(x,y,z)e^{i\theta}$$

$$p = \zeta_4(x,y,z)e^{i\theta}, \quad w = \zeta_5(x,y,z)e^{i\theta} \tag{24}$$

where

$$\frac{\partial \theta}{\partial x} = \alpha(x,z), \quad \frac{\partial \theta}{\partial z} = \beta(x,z), \quad \frac{\partial \theta}{\partial t} = -\omega \tag{25}$$

Substituting (24) and (25) into the linear quasi-parallel form of (18)-(23) yields

$$i\alpha\zeta_1 + D\zeta_3 + i\beta\zeta_5 = 0 \tag{26}$$

$$\Lambda R^{-1}\zeta_1 + \zeta_3 DU + i\alpha\zeta_4 - R^{-1}D^2\zeta_1 = 0 \tag{27}$$

$$\Lambda R^{-1}\zeta_3 + D\zeta_4 - R^{-1}D^2\zeta_3 = 0 \tag{28}$$

$$\Lambda R^{-1}\zeta_5 + \zeta_3 DW + i\beta\zeta_4 - R^{-1}D^2\zeta_5 = 0 \tag{29}$$

$$\zeta_1 = \zeta_3 = \zeta_5 = 0 \quad \text{at} \quad y = 0 \tag{30}$$

$$\zeta_n \to 0 \quad \text{as} \quad y \to \infty \tag{31}$$

where

$$\Lambda = iR(\alpha U + \beta W - \omega) + \alpha^2 + \beta^2 \tag{32}$$

For a given mean flow U and W and given five of the seven parameters R, α_r, α_i, β_r, β_i, ω_r, and ω_i, the eigenvalue problem (26)-(31) can be solved numerically to determine the last two parameters. Then, the solution of the linear quasi-parallel problem can be written as

$$u = \sum_n A_n \zeta_{1n} e^{i\theta_n} + cc, \quad v = \sum_n A_n \zeta_{3n} e^{i\theta_n} + cc$$

$$p = \sum_n A_n \zeta_{4n} e^{i\theta_n} + cc, \quad w = \sum_n A_n \zeta_{5n} e^{i\theta_n} + cc \tag{33}$$

where cc stands for the complex conjugate of the preceding terms and the summation is taken over all possible solutions of the eigenvalue problem and the A_n are constants. For the spatial stability case, the ω_n are real and the α_n and β_n are complex.

4.1. Adjoint

Next, we define the adjoint of the linear quasi-parallel problem [49]. To this end, we multiply (26) by ζ_4^*, (27) by ζ_1^*, (28) by ζ_3^*, and (29) by ζ_5^*, add the results, integrate the resulting expression by parts to transfer the derivatives from the ζ_n to the ζ_n^*, and obtain

$$\int_0^\infty \{\zeta_1[i\alpha\zeta_4^* + R^{-1}\Lambda\zeta_1^* - R^{-1}D^2\zeta_1^*] + \zeta_4[i\alpha\zeta_1^* - D\zeta_3^* + i\beta\zeta_5^*]$$
$$+ \zeta_3[-D\zeta_4^* + \zeta_1^*DU + \zeta_5^*DW + R^{-1}\Lambda\zeta_3^* - R^{-1}D^2\zeta_3^*] + \zeta_5[i\beta\zeta_4^* + R^{-1}\Lambda\zeta_5^* - R^{-1}D^2\zeta_5^*]\}dy$$
$$= -[\zeta_4^*\zeta_3 + \zeta_3^*\zeta_4 - R^{-1}(\zeta_1^*D\zeta_1 - \zeta_1D\zeta_1^*) - R^{-1}(\zeta_3^*D\zeta_3 - \zeta_3D\zeta_3^*) - R^{-1}(\zeta_5^*D\zeta_5 - \zeta_5D\zeta_5^*)]\Big|_0^\infty \quad (34)$$

The equations governing the adjoint are obtained by setting each of the coefficients of ζ_1, ζ_3, ζ_4, and ζ_5 in the integrand in (34) equal to zero. The result is

$$i\alpha\zeta_1^* - D\zeta_3^* + i\beta\zeta_5^* = 0 \quad (35)$$

$$R^{-1}\Lambda\zeta_1^* + i\alpha\zeta_4^* - R^{-1}D^2\zeta_1^* = 0 \quad (36)$$

$$R^{-1}\Lambda\zeta_3^* + \zeta_1^*DU + \zeta_5^*DW - D\zeta_4^* - R^{-1}D^2\zeta_3^* = 0 \quad (37)$$

$$R^{-1}\Lambda\zeta_5^* + i\beta\zeta_4^* - R^{-1}D^2\zeta_5^* = 0 \quad (38)$$

Then, (34) reduces to its right-hand side being equal to zero. The value at infinity vanishes if we demand that

$$\zeta_n^* \to 0 \quad \text{as} \quad y \to \infty \quad (39)$$

Then, using the boundary conditions (30), we reduce (34) to

$$\zeta_3^*\zeta_4 - R^{-1}(\zeta_1^*D\zeta_1 + \zeta_3^*D\zeta_3 + \zeta_5^*D\zeta_5) = 0 \quad \text{at} \quad y = 0$$

We define the boundary conditions for the adjoint at $y = 0$ by requiring that each of the coefficients of ζ_4, $D\zeta_1$, $D\zeta_3$, and $D\zeta_5$ vanish independently. The result is

$$\zeta_1^* = \zeta_3^* = \zeta_5^* = 0 \quad \text{at} \quad y = 0 \quad (40)$$

For a given mean flow, the eigenvalue problem (35)-(40) yields the same eigenvalues as the linear quasi-parallel problem.

4.2. Bi-orthogonality Condition

We consider the linear quasi-parallel case, (26)-(31), for the nth mode; that is, $\zeta_k = \zeta_{kn}$, $\alpha = \alpha_n$, $\beta = \beta_n$, and $\omega = \omega_n$. Moreover, we consider the adjoint problem corresponding to the mth mode; that is, $\zeta_k^* = \zeta_{km}^*$, $\alpha = \alpha_m$, $\beta = \beta_m$, and $\omega = \omega_m$. We multiply (26) by ζ_{4m}^*, (27) by ζ_{1m}^*, (28) by ζ_{3m}^*, and (29) by ζ_{5m}^*, add the results, integrate the resulting expression by parts to transfer the derivatives from

ζ_{kn} to ζ_{km}^*, and obtain

$$\int_0^\infty \{i\alpha_n \zeta_{1n}\zeta_{4m}^* - \zeta_{3n}D\zeta_{4m}^* + i\beta_n\zeta_{4m}^*\zeta_{5n} + R^{-1}\Lambda_n\zeta_{1m}^*\zeta_{1n} + \zeta_{1m}^*\zeta_{3n}DU + i\alpha_n\zeta_{1m}^*\zeta_{4n}$$
$$- R^{-1}\zeta_{1n}D^2\zeta_{1m}^* + R^{-1}\Lambda_n\zeta_{3m}^*\zeta_{3n} - \zeta_{4n}D\zeta_{3m}^* - R^{-1}\zeta_{3n}D^2\zeta_{3m}^* + R^{-1}\Lambda_n\zeta_{5m}^*\zeta_{5n}$$
$$+ \zeta_{3n}\zeta_{5m}^*DW + i\beta_n\zeta_{5m}^*\zeta_{4n} - R^{-1}\zeta_{5n}D^2\zeta_{5m}^*\}dy + [\zeta_{4m}^*\zeta_{3n} + \zeta_{3m}^*\zeta_{4n}$$
$$- R^{-1}(\zeta_{1m}^*D\zeta_{1n} - \zeta_{1n}D\zeta_{1m}^* + \zeta_{3m}^*D\zeta_{3n} - \zeta_{3n}D\zeta_{3m}^* + \zeta_{5m}^*D\zeta_{5n} - \zeta_{5n}D\zeta_{5m}^*)]\Big|_0^\infty = 0 \quad (41)$$

where

$$\Lambda_n = iR(\alpha_n U + \beta_n W - \omega_n) + \alpha_n^2 + \beta_n^2 \quad (42)$$

The last term in (41) vanishes on account of the boundary conditions (30), (31), (39), and (40). Using (35)-(38) to eliminate the derivatives of the ζ_{km}^* from the integrand in (41) and rearranging, we obtain

$$(\omega_n - \omega_m)q_{1nm} + (\alpha_n - \alpha_m)q_{2nm} + (\beta_n - \beta_m)q_{3nm} = 0 \quad (43)$$

where

$$q_{1nm} = -\int_0^\infty (\zeta_{1n}\zeta_{1m}^* + \zeta_{3n}\zeta_{3m}^* + \zeta_{5n}\zeta_{5m}^*)dy \quad (44)$$

$$q_{2nm} = \int_0^\infty [\zeta_{1n}\zeta_{4m}^* + \zeta_{4n}\zeta_{1m}^* + (U - i\frac{\alpha_n+\alpha_m}{R})(\zeta_{1n}\zeta_{1m}^* + \zeta_{3n}\zeta_{3m}^* + \zeta_{5n}\zeta_{5m}^*)]dy \quad (45)$$

$$q_{3nm} = \int_0^\infty [\zeta_{5n}\zeta_{4m}^* + \zeta_{4n}\zeta_{5m}^* + (W - i\frac{\beta_n+\beta_m}{R})(\zeta_{1n}\zeta_{1m}^* + \zeta_{3n}\zeta_{3m}^* + \zeta_{5n}\zeta_{5m}^*)]dy \quad (46)$$

It follows from (43) that

(i) if $\omega_n = \omega_m$, $\beta_n = \beta_m$, but $\alpha_n \neq \alpha_m$, then $q_{2nm} = 0$ \quad (47)

(ii) if $\omega_n = \omega_m$, $\alpha_n = \alpha_m$, but $\beta_n \neq \beta_m$, then $q_{3nm} = 0$ \quad (48)

(iii) if $\alpha_n = \alpha_m$, $\beta_n = \beta_m$, but $\omega_n \neq \omega_m$, then $q_{1nm} = 0$ \quad (49)

5. Mutual Interaction Approach

In the nonlinear nonparallel case, we continue to represent the solution as in (33), the linear quasi-parallel case, except that A_n is a function of x, z and t. Then, it follows from (33) that

$$\frac{\partial u}{\partial x} = \sum_n [i\alpha_n A_n \zeta_{1n} + \frac{\partial A_n}{\partial x}\zeta_{1n} + A_n \frac{\partial \zeta_{1n}}{\partial x}]e^{i\theta_n} \quad (50)$$

$$\frac{\partial^2 u}{\partial x^2} = \sum_n [-\alpha_n^2 A_n \zeta_{1n} + 2i\alpha_n \frac{\partial A_n}{\partial x} \zeta_{1n} + 2i\alpha_n A_n \frac{\partial \zeta_{1n}}{\partial x}$$
$$+ 2\frac{\partial A_n}{\partial x}\frac{\partial \zeta_{1n}}{\partial x} + \frac{\partial^2 A_n}{\partial x^2}\zeta_{1n} + A_n \frac{\partial^2 \zeta_{1n}}{\partial x^2} + i\frac{\partial \alpha_n}{\partial x} A_n \zeta_{1n}]e^{i\theta_n} \quad (51a)$$

Since $R \gg 1$ and A_n, α_n, β_n and ζ_{kn} are slowly varying functions of x and z, we have

$$\frac{1}{R}\frac{\partial^2 u}{\partial x^2} \approx -R^{-1} \sum_n \alpha_n^2 A_n \zeta_{1n}, \quad \frac{1}{R}\frac{\partial^2 u}{\partial z^2} \approx -R^{-1} \sum_n \beta_n^2 A_n \zeta_{1n} \quad (51b)$$

with similar expressions for the second derivatives of v and w with respect to x and z. Substituting (33) and these derivatives into (18)-(21) yields

$$\sum_n (\frac{\partial A_n}{\partial x}\zeta_{1n} + \frac{\partial A_n}{\partial z}\zeta_{5n} + A_n \frac{\partial \zeta_{1n}}{\partial x} + A_n \frac{\partial \zeta_{5n}}{\partial z})e^{i\theta_n} + cc = 0 \quad (52)$$

$$\sum_n (\frac{\partial A_n}{\partial t}\zeta_{1n} + U\frac{\partial A_n}{\partial x}\zeta_{1n} + W\frac{\partial A_n}{\partial z}\zeta_{1n} + \frac{\partial A_n}{\partial x}\zeta_{4n})e^{i\theta_n} + cc$$

$$= -\sum_n A_n[U\frac{\partial \zeta_{1n}}{\partial x} + W\frac{\partial \zeta_{1n}}{\partial z} + \frac{\partial \zeta_{4n}}{\partial x} + \frac{\partial U}{\partial x}\zeta_{1n} + VD\zeta_{1n} + \frac{\partial U}{\partial z}\zeta_{5n}]e^{i\theta_n}$$

$$- \sum_{n,k} A_n A_k [i\alpha_k \zeta_{1n}\zeta_{1k} + \zeta_{3n}D\zeta_{1k} + i\beta_k \zeta_{5n}\zeta_{1k}]e^{i(\theta_n+\theta_k)}$$

$$- \sum_{n,k} A_n \bar{A}_k [-i\bar{\alpha}_k \zeta_{1n}\bar{\zeta}_{1k} + \zeta_{3n}D\bar{\zeta}_{1k} - i\bar{\beta}_k \zeta_{5n}\bar{\zeta}_{1k}]e^{i(\theta_n-\bar{\theta}_k)} + cc$$
$$(53)$$

$$\sum_n (\frac{\partial A_n}{\partial t}\zeta_{3n} + U\frac{\partial A_n}{\partial x}\zeta_{3n} + W\frac{\partial A_n}{\partial z}\zeta_{3n})e^{i\theta_n} + cc$$

$$= -\sum_n A_n[U\frac{\partial \zeta_{3n}}{\partial x} + W\frac{\partial \zeta_{3n}}{\partial z} + VD\zeta_{3n} + \zeta_{3n}DV]e^{i\theta_n}$$

$$- \sum_{n,k} A_n A_k [i\alpha_k \zeta_{1n}\zeta_{3k} + \zeta_{3n}D\zeta_{3k} + i\beta_k \zeta_{5n}\zeta_{3k}]e^{i(\theta_n+\theta_k)}$$

$$- \sum_{n,k} A_n \bar{A}_k [-i\bar{\alpha}_k \zeta_{1n}\bar{\zeta}_{3k} + \zeta_{3n}D\bar{\zeta}_{3k} - i\bar{\beta}_k \zeta_{5n}\bar{\zeta}_{3k}]e^{i(\theta_n-\bar{\theta}_k)} + cc$$
$$(54)$$

$$\sum_n (\frac{\partial A_n}{\partial t} \zeta_{5n} + U \frac{\partial A_n}{\partial x} \zeta_{5n} + W \frac{\partial A_n}{\partial z} \zeta_{5n} + \zeta_{4n} \frac{\partial A_n}{\partial z}) e^{i\theta_n} + cc$$

$$= - \sum_n A_n [U \frac{\partial \zeta_{5n}}{\partial x} + W \frac{\partial \zeta_{5n}}{\partial z} + \frac{\partial \zeta_{4n}}{\partial z} + \zeta_{1n} \frac{\partial W}{\partial x} + VD\zeta_{5n} + \zeta_{5n} \frac{\partial W}{\partial z}] e^{i\theta_n}$$

$$- \sum_{n,k} A_n A_k [i\alpha_k \zeta_{1n} \zeta_{5k} + \zeta_{3n} D\zeta_{5k} + i\beta_k \zeta_{5n} \zeta_{5k}] e^{i(\theta_n + \theta_k)}$$

$$- \sum_{n,k} A_n \bar{A}_k [-i\bar{\alpha}_k \zeta_{1n} \bar{\zeta}_{5k} + \zeta_{3n} D\bar{\zeta}_{5k} - i\bar{\beta}_k \zeta_{5n} \bar{\zeta}_{5k}] e^{i(\theta_n - \bar{\theta}_k)} + cc \quad (55)$$

where the overbar stands for the complex conjugate.

Multiplying (52) by ζ^*_{4m}, (53) by ζ^*_{1m}, (54) by ζ^*_{3m}, and (55) by ζ^*_{5m}, adding the results, and using the definitions of the ζ_{kn} and ζ^*_{km}, we obtain

$$\sum_n [q_{1nm} \frac{\partial A_n}{\partial t} + q_{2nm} \frac{\partial A_n}{\partial x} + q_{3nm} \frac{\partial A_n}{\partial z}] e^{i\theta_n} + \sum_n h_{nm} A_n e^{i\theta_n}$$

$$+ \sum_{n,k} g_{nkm} A_n A_k e^{i(\theta_n + \theta_k)} + \sum_{n,k} f_{nkm} A_n \bar{A}_k e^{i(\theta_n - \bar{\theta}_k)} + cc = 0 \quad (56)$$

where the q_{knm} are defined in (44)-(46) and

$$h_{nm} = \int_0^\infty \{\zeta^*_{4m} [\frac{\partial \zeta_{1n}}{\partial x} + \frac{\partial \zeta_{5n}}{\partial z}] + \zeta^*_{1m} [U \frac{\partial \zeta_{1n}}{\partial x} + W \frac{\partial \zeta_{1n}}{\partial z}$$

$$+ \frac{\partial \zeta_{4n}}{\partial x} + \frac{\partial U}{\partial x} \zeta_{1n} + VD\zeta_{1n} + \frac{\partial U}{\partial z} \zeta_{5n}] + \zeta^*_{3m} [U \frac{\partial \zeta_{3n}}{\partial x}$$

$$+ W \frac{\partial \zeta_{3n}}{\partial z} + VD\zeta_{3n} + \zeta_{3n} DV] + \zeta^*_{5m} [U \frac{\partial \zeta_{5n}}{\partial x} + W \frac{\partial \zeta_{5n}}{\partial z}$$

$$+ \frac{\partial \zeta_{4n}}{\partial z} + \zeta_{1n} \frac{\partial W}{\partial x} + VD\zeta_{5n} + \zeta_{5n} \frac{\partial W}{\partial z}]\} dy \quad (57)$$

$$g_{nkm} = \int_0^\infty \{\zeta^*_{1m} [i\alpha_k \zeta_{1n} \zeta_{1k} + \zeta_{3n} D\zeta_{1k} + i\beta_k \zeta_{5n} \zeta_{1k} + i\alpha_n \zeta_{1k} \zeta_{1n} + \zeta_{3k} D\zeta_{1n}$$

$$+ i\beta_n \zeta_{5k} \zeta_{1n}] + \zeta^*_{3m} [i\alpha_k \zeta_{1n} \zeta_{3k} + \zeta_{3n} D\zeta_{3k} + i\beta_k \zeta_{5n} \zeta_{3k}$$

$$+ i\alpha_n \zeta_{1k} \zeta_{3n} + \zeta_{3k} D\zeta_{3n} + i\beta_n \zeta_{5k} \zeta_{3n}] + \zeta^*_{5m} [i\alpha_k \zeta_{1n} \zeta_{5k} + \zeta_{3n} D\zeta_{5k} + i\beta_k \zeta_{5n} \zeta_{5k}$$

$$+ i\alpha_n \zeta_{1k} \zeta_{5n} + \zeta_{3k} D\zeta_{5n} + i\beta_n \zeta_{5k} \zeta_{5n}]\} dy \quad (58)$$

$$f_{nkm} = \int_0^\infty \{\zeta_{1m}^*[-i\bar{\alpha}_k\zeta_{1n}\bar{\zeta}_{1k} + \zeta_{3n}D\bar{\zeta}_{1k} - i\bar{\beta}_k\zeta_{5n}\bar{\zeta}_{1k} + i\alpha_n\bar{\zeta}_{1k}\zeta_{1n} + \bar{\zeta}_{3k}D\zeta_{1n} + i\beta_n\bar{\zeta}_{5k}\zeta_{1n}]$$

$$+ \zeta_{3m}^*[-i\bar{\alpha}_k\zeta_{1n}\bar{\zeta}_{3k} + \zeta_{3n}D\bar{\zeta}_{3k} - i\bar{\beta}_k\zeta_{5n}\bar{\zeta}_{3k} + i\alpha_n\bar{\zeta}_{1k}\zeta_{3n} + \bar{\zeta}_{3k}D\zeta_{3n} + i\beta_n\bar{\zeta}_{5k}\zeta_{3n}]$$

$$+ \zeta_{5m}^*[-i\bar{\alpha}_k\zeta_{1n}\bar{\zeta}_{5k} + \zeta_{3n}D\bar{\zeta}_{5k} - i\bar{\beta}_k\zeta_{5n}\bar{\zeta}_{5k} + i\alpha_n\bar{\zeta}_{1k}\zeta_{5n} + \bar{\zeta}_{3k}D\zeta_{5n} + i\beta_n\bar{\zeta}_{5k}\zeta_{3n}]\}dy$$

(59)

Equation (56) is valid for all possible resonances to first order and any specific first-order resonance configuration can be obtained as a special case. The case of subharmonic resonances (two-to-one resonances) is discussed next.

6. Two-Dimensional Mean Flows

Let us consider the case of subharmonic resonances in a two-dimensional mean flow. Let us consider the interaction of a two-dimensional Tollmien-Schlichting wave having the frequency ω with three-dimensional waves having the frequency $\frac{1}{2}\omega$ and the spanwise wavenumber β. Let us denote the phase of the two-dimensional wave by

$$\theta_s = \int \alpha dx - \omega t \qquad (60a)$$

and its amplitude by $A_s = A$. It follows from symmetry that if the phase of a three-dimensional wave with $\beta_m = \beta$ is given by

$$\int \alpha_m dx + \beta z - \frac{1}{2}\omega t \qquad (60b)$$

then the phase of the three-dimensional wave with $\beta_m = -\beta$ is

$$\int \alpha_m dx - \beta z - \frac{1}{2}\omega t \qquad (60c)$$

Let us denote the amplitudes of the right-running and left-running waves by A_m and B_m, respectively. Using these waves in (56) and separating the time and spanwise variations, we have

$$q_{200}\frac{dA}{dx} + h_{oo}A + \sum_n g_{nno}A_n B_n e^{i\int(2\alpha_n - \alpha)dx} = 0 \qquad (61)$$

$$q_{2mn}\frac{dA_m}{dx} + \sum_n h_{nm}A_n e^{i\int(\alpha_n - \alpha_m)dx} + \sum_n f_{onm}AB_n e^{-i\int(\alpha - \bar{\alpha}_n - \alpha_m)dx} = 0 \qquad (62)$$

$$q_{2mm}\frac{dB_m}{dx} + \sum_n h_{nm}B_n e^{i\int(\alpha_n - \alpha_m)dx} + \sum_n f_{onm}AA_n e^{-i\int(\alpha - \bar{\alpha}_n - \alpha_m)dx} = 0 \qquad (63)$$

For the case of temporal stability, the phase of the primary wave is

$$\theta_s = \alpha x - \omega t \tag{64a}$$

whereas the phases of the oblique waves are

$$\tfrac{1}{2}\alpha x + \beta z - \omega_m t \quad \text{and} \quad \tfrac{1}{2}\alpha x - \beta z - \omega_m t \tag{64b}$$

Using these waves in (56) and separating the space variations, we obtain

$$q_{100}\frac{dA}{dt} + \sum_n g_{mo} A_n B_n e^{i(\omega - 2\omega_n)t} = 0 \tag{65}$$

$$q_{1mm}\frac{dA_m}{dt} + \sum_n f_{onm} \overline{AB}_n e^{-i(\omega_m + \bar{\omega}_n - \omega)t} = 0 \tag{66}$$

$$q_{1mm}\frac{dB_m}{dt} + \sum_n f_{onm} \overline{AA}_n e^{-i(\omega_m + \bar{\omega}_n - \omega)t} = 0 \tag{67}$$

Equations (61)-(63) and (65)-(67) reduce to those of Craik [27,28] and Lekoudis [29] if the nonparallel terms are neglected and the summation is restricted to one mode, namely, two obliques T-S waves. However, Nayfeh [42] showed that the Squire mode is more important than the T-S wave for the subharmonic instability. Moreover, recent work by Nayfeh and Masad [50] show that more than one T-S mode and more than one Squire mode are participating in the instability.

7. Parametric Approach

To simplify the algebra, we consider the case of two-dimensional mean flows and a primary instability consisting of a two-dimensional Tollmien-Schlichting wave. Thus the basic state is given by

$$U_s = U(x,y) + \left[A(x)\zeta_{10}(x,y)e^{i\theta_s} + cc\right] \tag{68a}$$

$$V_s = V(x,y) + \left[A(x)\zeta_{30}(x,y)e^{i\theta_s} + cc\right] \tag{68b}$$

$$P_s = P(x) + \left[A(x)\zeta_{40}(x,y)e^{i\theta_s} + cc\right] \tag{68c}$$

where θ_s is defined in (60a), ω is the frequency, the ζ_{no} are the eigenvectors corresponding to the eigenvalue k, and

$$q_{200}\frac{dA}{dx} + h_{oo}A = 0 \tag{69}$$

Equation (69) describes the modulation of the amplitude and phase due to nonparallelism; it is obtained from (61) by dropping the nonlinear term (i.e., $A_n = B_n = 0$). To study the secondary instability of this basic state using the parametric approach, we superpose on it a three-dimensional disturbance to obtain

$$\tilde{u} = U(x,y) + A(x)\zeta_{10}(x,y)e^{i\theta_s} + \bar{A}(x)\bar{\zeta}_{10}(x,y)e^{-i\theta_s} + u(x,y,z,t) \tag{70a}$$

$$\tilde{v} = V(x,y) + A(x)\zeta_{30}(x,y)e^{i\theta_s} + \bar{A}(x)\bar{\zeta}_{30}(x,y)e^{-i\bar{\theta}_s} + v(x,y,z,t) \tag{70b}$$

$$\tilde{p} = P(x) + A(x)\zeta_{40}(x,y)e^{i\theta_s} + \bar{A}(x)\bar{\zeta}_{40}(x,y)e^{i\bar{\theta}_s} + p(x,y,z,t) \tag{70c}$$

$$\tilde{w} = 0 + 0 + 0 + w(x,y,z,t) \tag{70d}$$

In the parametric approach, we note that the primary waves influence but they are not influenced by the secondary wave, u, v, w, and p, in contrast with the mutual interaction approach. Substituting (68)-(71) into the dimensionless Navier-Stokes equations, subtracting the mean-flow and the two-dimensonal T-S quantities, and linearizing the resulting equations, we obtain

$$\frac{\partial u}{\partial x} + \frac{\partial v}{\partial y} + \frac{\partial w}{\partial z} = 0 \tag{71}$$

$$\frac{\partial u}{\partial t} + [U + (A\zeta_{10}e^{i\theta_s} + cc)]\frac{\partial u}{\partial x} + [\frac{\partial U}{\partial y} + (A\frac{\partial \zeta_{10}}{\partial y}e^{i\theta_s} + cc)]v$$

$$+ \{\frac{\partial U}{\partial x} + [(\frac{\partial A}{\partial x}\zeta_{10} + A\frac{\partial \zeta_{10}}{\partial x} + i\alpha A\zeta_{10})e^{i\theta_s} + cc]\}u$$

$$+ [V + (A\zeta_{30}e^{i\theta_s} + cc)]\frac{\partial u}{\partial y} + \frac{\partial p}{\partial x} - \frac{1}{R}\nabla^2 u = 0 \tag{72}$$

$$\frac{\partial v}{\partial t} + [U + (A\zeta_{10}e^{i\theta_s} + cc)]\frac{\partial v}{\partial x} + \{\frac{\partial V}{\partial x} + [(\frac{\partial A}{\partial x}\zeta_{30} + A\frac{\partial \zeta_{30}}{\partial x}$$

$$+ i\alpha A\zeta_{30})e^{i\theta_s} + cc]\}u + [V + (A\zeta_{30}e^{i\theta_s} + cc)]\frac{\partial v}{\partial y}$$

$$+ [\frac{\partial V}{\partial y} + (A\frac{\partial \zeta_{30}}{\partial y}e^{i\theta_s} + cc)]v + \frac{\partial p}{\partial y} - \frac{1}{R}\nabla^2 v = 0 \tag{73}$$

$$\frac{\partial w}{\partial t} + [U + (A\zeta_{10}e^{i\theta_s} + cc)]\frac{\partial w}{\partial x} + [V + (A\zeta_{30}e^{i\theta_s} + cc)]\frac{\partial w}{\partial y}$$

$$+ \frac{\partial p}{\partial z} - \frac{1}{R}\nabla^2 w = 0 \tag{74}$$

The boundary conditions are the same as (22) and (23).

Equations (71)-(74) are coupled four partial-differential equations whose coefficients are independent of z, periodic in t, quasi-periodic and slowly varying in x, and dependent in a complicated manner on y. Since the boundary conditions (22) and (23) are homogeneous, the z variation can be separated as follows:

$$u(x,y,z,t) = \bar{u}(x,y,t)e^{i\beta z}, \quad v(x,y,z,t) = \bar{v}(x,y,t)e^{i\beta z} \tag{75a}$$

$$w(x,y,z,t) = \bar{w}(x,y,t)e^{i\beta z}, \quad p(x,y,z,t) = \bar{p}(x,y,t)e^{i\beta z} \tag{75b}$$

Substituting (75a) and (75b) into (71)-(74) yields

$$\frac{\partial u}{\partial x} + \frac{\partial v}{\partial y} + i\beta w = 0 \tag{76}$$

$$\frac{\partial u}{\partial t} + [U + (A\zeta_{10}e^{i\theta_s} + cc)]\frac{\partial u}{\partial x} + [\frac{\partial U}{\partial y} + (A\frac{\partial \zeta_{10}}{\partial y}e^{i\theta_s} + cc)]v$$

$$+ \{\frac{\partial U}{\partial x} + [(\frac{\partial A}{\partial x}\zeta_{10} + A\frac{\partial \zeta_{10}}{\partial x} + i\alpha A\zeta_{10})e^{i\theta_s} + cc]\}u$$

$$+ [V + (A\zeta_{30}e^{i\theta_s} + cc)]\frac{\partial u}{\partial y} + \frac{\partial p}{\partial x} - \frac{1}{R}[\frac{\partial^2 u}{\partial x^2} + \frac{\partial^2 u}{\partial y^2} - \beta^2 u] = 0 \quad (77)$$

$$\frac{\partial v}{\partial t} + [U + (A\zeta_{10}e^{i\theta_s} + cc)]\frac{\partial v}{\partial x} + \{\frac{\partial V}{\partial x} + [(\frac{\partial A}{\partial x}\zeta_{30} + A\frac{\partial \zeta_{30}}{\partial x}$$

$$+ i\alpha A\zeta_{30})e^{i\theta_s} + cc]\}u + [V + (A\zeta_{30}e^{i\theta_s} + cc)]\frac{\partial v}{\partial y}$$

$$+ [\frac{\partial V}{\partial y} + (A\frac{\partial \zeta_{30}}{\partial y}e^{i\theta_s} + cc)]v + \frac{\partial p}{\partial y} - \frac{1}{R}[\frac{\partial^2 v}{\partial x^2} + \frac{\partial^2 v}{\partial y^2} - \beta^2 v] = 0 \quad (78)$$

$$\frac{\partial w}{\partial t} + [U + (A\zeta_{10}e^{i\theta_s} + cc)]\frac{\partial w}{\partial x} + [V + (A\zeta_{30}e^{i\theta_s} + cc)]\frac{\partial w}{\partial y}$$

$$+ i\beta p - \frac{1}{R}[\frac{\partial^2 w}{\partial x^2} + \frac{\partial^2 w}{\partial y^2} - \beta^2 w] = 0 \quad (79)$$

where the tilde was dropped for convenience of notation.

For the case of parallel flow (i.e., $V = 0$ and $\partial U/\partial x$), $A = A_0$ is a constant. Moreover, α is also a constant and θ_s of (60a) becomes

$$\theta_s = \alpha x - \omega t + \tau = \alpha_r x - \omega t + i\alpha_i x + \tau$$

where τ is a constant phase. Thus, (68a)-(68c) becomes

$$U_s = U(y) + [A_0 e^{-\alpha_i x}\zeta_{10}(y)e^{i(\alpha_r x - \omega t + \tau)} + cc] \quad (80a)$$

$$V_s = 0 + [A_0 e^{-\alpha_i x}\zeta_{30}(y)e^{i(\alpha_r x - \omega t + \tau)} + cc] \quad (80b)$$

$$P_s = P + [A_0 e^{-\alpha_i x}\zeta_{40}(y)e^{i(\alpha_r x - \omega t + \tau)} + cc] \quad (80c)$$

If we assume that

$$A_0 e^{-\alpha_i x} = a \quad (80d)$$

is a constant, then (76)-(79) constitute a system of four linear coupled partial differential equations whose coefficients are periodic functions of t and x. Therefore, it follows from Floquet theory that (76)-(79), (22), and (23) have solutions of the form

$$[u,v,w,p] = e^{\gamma_1 x + \gamma_2 t}[\phi_1(x,y,t), \phi_2(x,y,t), \phi_3(x,y,t), \phi_4(x,y,t)] \quad (81)$$

where the γ_n are called characteristic exponents and the ϕ_n are periodic functions of $x - \omega t/\alpha_r$. Consequently, one can expand the ϕ_n in Fourier series in $x - \omega t/\alpha_r$ and rewrite (81) as

$$[u,v,w,p] = e^{\gamma_1 x + \gamma_2 t} \sum_{n=-\infty}^{n=\infty} [\eta_{1n}(y), \eta_{2n}(y), \eta_{3n}(y), \eta_{4n}(y)] e^{\frac{1}{2}in(\alpha_r x - \omega t)} \qquad (82)$$

For temporal stability $\gamma_1 = 0$, whereas for spatial stability $\gamma_2 = 0$. Substituting (82) into (76)-(79), (22) and (23), using (80), and separating the different harmonics, we obtain an infinite system of coupled equations for the η_{mn}. To solve for the η_{mn} and the characteristic exponent γ_1 or γ_2, one truncates the Fourier series to N terms and numerically solves the resulting two-point boundary value problem using a shooting, a finite difference, or a collocation technique. With the last two techniques, the problem is reduced to an algebraic eigenvalue problem. The eigenvalue γ_2 appears linearly whereas the eigenvalue γ_1 appears nonlinearly in these equations. Consequently, the temporal stability problem is usually treated using this approach.

When the amplitude of the primary wave is not constant, Floquet theory does not apply and one needs to use another method of solution. For the case of spatial subharmonic secondary instability, we seek a solution in the form

$$u = \sum_n A_n(x) \zeta_{1n}(x,y) e^{i\theta_n} + cc, \quad v = \sum_n A_n(x) \zeta_{3n}(x,y) e^{i\theta_n} + cc$$

$$\qquad (83)$$

$$p = \sum_n A_n(x) \zeta_{4n}(x,y) e^{i\theta_n} + cc, \quad w = \sum_n A_n(x) \zeta_{5n}(x,y) e^{i\theta_n} + cc$$

where

$$\theta_n = \int \alpha_n(x) dx - \frac{1}{2} \omega t \qquad (84)$$

and the ζ_{mn} and α_n are given by (26)-(32). Substituting (83) and (84) into (76)-(79) and using (50) and (51), we obtain

$$\sum_n \left(\frac{dA_n}{dx} \zeta_{1n} + A_n \frac{\partial \zeta_{1n}}{\partial x} \right) e^{i\theta_n} + cc = 0 \qquad (85)$$

$$\sum_n \left(U \frac{dA_n}{dx} \zeta_{1n} + \frac{dA_n}{dx} \zeta_{4n} \right) e^{i\theta_n} + cc$$

$$= -\sum_n A_n \left[U \frac{\partial \zeta_{1n}}{\partial x} + \frac{\partial \zeta_{4n}}{\partial x} + \frac{\partial U}{\partial x} \zeta_{1n} + V D\zeta_{1n} \right] e^{i\theta_n}$$

$$- \sum_n A_n A [i(\alpha + \alpha_n) \zeta_{10} \zeta_{1n} + \zeta_{30} D\zeta_{1n} + \zeta_{3n} D\zeta_{10}] e^{i(\theta_s + \theta_n)}$$

$$- \sum_n A \bar{A}_n [i(\alpha - \bar{\alpha}_n) \zeta_{10} \bar{\zeta}_{1n} + \zeta_{30} D\bar{\zeta}_{1n} + \bar{\zeta}_{3n} D\zeta_{10}] e^{i(\theta_s - \bar{\theta}_n)} + cc \qquad (86)$$

$$\sum_n (U \frac{dA_n}{dx} \zeta_{3n}) e^{i\theta_n} + cc = - \sum_n A_n [U \frac{\partial \zeta_{3n}}{\partial x} + V D\zeta_{3n} + \zeta_{3n} D V] e^{i\theta_n}$$

$$- \sum_n A_n A[i\alpha\zeta_{1n}\zeta_{30} + i\alpha_n\zeta_{10}\zeta_{3n} + \zeta_{3n} D\zeta_{30} + \zeta_{30} D\zeta_{3n}] e^{i(\theta_s + \theta_n)}$$

$$- \sum_n A\bar{A}_n [i\alpha\zeta_{30}\bar{\zeta}_{1n} - i\bar{\alpha}_n\zeta_{10}\bar{\zeta}_{3n} + \zeta_{30} D\bar{\zeta}_{3n} + \bar{\zeta}_{3n} D\zeta_{30}] e^{i(\theta_s - \bar{\theta}_n)} + cc \quad (87)$$

$$\sum_n (U \frac{dA_n}{dx} \zeta_{5n}) e^{i\theta_n} + cc = - \sum_n A_n [U \frac{\partial \zeta_{5n}}{\partial x} + V D\zeta_{5n}] e^{i\theta_n}$$

$$- \sum_n A_n A[i\alpha\zeta_{1n}\zeta_{50} + i\alpha_n\zeta_{10}\zeta_{5n} + \zeta_{3n} D\zeta_{50} + \zeta_{30} D\zeta_{5n}] e^{i(\theta_s + \theta_n)}$$

$$- \sum_n A\bar{A}_n [i\alpha\zeta_{50}\bar{\zeta}_{1n} - i\bar{\alpha}_n\zeta_{10}\bar{\zeta}_{5n} + \zeta_{30} D\bar{\zeta}_{5n} + \bar{\zeta}_{3n} D\zeta_{50}] e^{i(\theta_s - \bar{\theta}_n)} + cc \quad (88)$$

Multiplying (85) by ζ^*_{4m}, (86) by ζ^*_{1m}, (87) by ζ^*_{3m}, and (88) by ζ^*_{5m}, adding the results, using the bi-orthogonality conditions of Section 4, and separating the time variations, we obtain

$$q_{2mm} \frac{\partial A_m}{\partial x} + \sum_n h_{nm} A_n e^{i\int (\alpha_n - \alpha_m) dx} + \sum_n f_{onm} A\bar{A}_n e^{i\int (\alpha - \bar{\alpha}_n - \alpha_m) dx} = 0 \quad (89)$$

where q_{2mm}, h_{nm} and f_{onm} are defined in (45), (57) and (59).

Comparing (89) with (61)-(63) obtained using the mutual interaction approach, we note that the latter reduce to (89) when the amplitudes A_n and B_n of the two oblique waves are the same (i.e., the case of standing waves in the z direction) and when the influence of the subharmonic instability on the primary wave is neglected; that is, the last term in (61) is neglected so that the primary wave is independent of the secondary instability.

ACKNOWLEDGEMENT: This work was supported by the Office of Naval Research under Contract No. N00014-85-K-0011, NR 061-201.

References

1. Klebanoff, P. S. and Tidstrom, K. D., "Evaluation of amplified waves leading to transition in a boundary layer with zero pressure gradient", NASA TN D-195, 1959.

2. Klebanoff, P. S., Tidstrom, K. D. and Sargent, L. M., "The three-dimensional nature of boundary-layer instability", J. Fluid Mech., Vol. 12, 1962, 1-34.

3. Hama, F. R., "Boundary-layer transition induced by a vibrating ribbon on a Flat plate", Proceeding - 1960 Heat Transfer and Fluid Mechanics Institute, Stanford University Press, 1960, 92-105.

4. Kachanov, Yu. S. and Levchenko, V. Ya., "Resonance interactions of disturbances in transition to turbulence in a boundary layer", (in Russian), Preprint No. 10-82, I.T.A.M., USSR Academy of Sciences, Novosibirsk, 1982.

5. Kachanov, Yu, S. and Levchenko, V. Ya., "The resonant interaction of disturbances at laminar-turbulent transition in a boundary layer", J. Fluid Mech., Vol. 138, 1984, 209-247.

6. Saric, W. S. and Thomas, A. S. W., "Experiments on the subharmonic route to turbulence in boundary layers", Proceedings - IUTAM Symposium "Turbulence and Chaotic Phenomena in Fluids", Kyota, Japan, 1983.

7. Nishioka, M., Asai, M. and Iida, S., "An experimental investigation of secondary instabilities", in Laminar-Turbulent Transition (eds. R. Eppler and H. Fasel), Springer-Verlag, 1980, 37-46.

8. Nishioka, M., Iida, S. and Kanbayashi, S., "An experimental investigation of the subcritical instability in plane Poiseuille flow", Proc. 10th Turbulence Symposium, Inst. Space Aeron. Sci., Tokyo Univ., 1978, 55-62.

9. Aihara, Y., "Transition in an incompressible boundary layer along a concave wall", Bull. Aero. Res. Inst., Tokyo Univ., Vol. 3, 1962, 195-240.

10. Tai, I. and Komoda, H., "Boundary-layer transition in the presence of streamwise vortices", J. Aerospace Sci., Vol. 29, 1962, 440-444.

11. Tani, I. and Sakagami, J., "Boundary-layer instability at subsonic speeds", Proc. Int. Council Aero. Sci., Third Congress, Stockholm, 1962, 391-403 (Spartan, Washington, D.C., 1964).

12. Tani, I. and Aihara, Y., "Görtler vortices and boundary-layer transition", ZAMP, Vol. 20, 1969, 609-618.

13. Wortman, F. X., "Visualization of transition", J. Fluid Mech., Vol. 38, 1969, 473-480.

14. Orszag, S. A. and Patera, A. T., "Subcritical transition to turbulence in plane channel flows", Phys. Rev. Lett, Vol. 45, 1980, 989-993.

15. Orszag, S. A. and Patera, A. T., "Secondary instability of wall-bounded shear flows", J. Fluid Mech., Vol. 128, 1983, 347-385.

16. Benney, D. J. and Lin, C. C., "On the secondary motion induced by oscillations in a shear flow", Phys. Fluids, Vol. 3, 1960, 656-657.

17. Benney, D. J., "A nonlinear theory for oscillation in a parallel flow", J. Fluid Mech., Vol. 10, 1961, 209-236.

18. Benney, D. J., "Finite amplitude effects in an unstable laminar boundary layer", Phys. Fluids, Vol. 7, 1964, 319-326.

19. Stuart, J. T., "On three-dimensional non-linear effects in the stability of parallel flows", Adv. Aero. Sci., Vol. 3, 1961, 121-142.

20. Antar, B. N. and Collins, F. G., "Numerical calculation of finite amplitude effects in unstable laminar boundary layers", Phys. Fluids, Vol. 18, 1975, 289-297.

21. Herbert, T., "On finite amplitudes of periodic disturbances of the boundary layer along a flat Plate", Deutsche Forschungs-und Versuchasanstalt fur Luft- und Raumfahrt E. V., Freiburg im Breisgau, Rep. No. DLR-FB 74-53, 1974.

22. Volodin, A. G. and Zel'man, M. B., "Pairwise nonlinear interactions of Tollmien-Schlichting waves in flows of the boundary-layer type", Izv. Akad. Nauk SSR, Mekh. Zh. Gaza, No. 2, 1977, 33-37.

23. Raetz, G. S., "A new theory of the cause of transition in fluid flows", Norair Rep. NOR-59-383, 1959, Hawthorne, Calif.

24. Raetz, G. S., "Current status of resonance theory of transition", Norair Rep. NOR-64-111, 1964, Hawthorne, Calif.

25. Stuart, J. T., "Nonlinear effects in hydrodynamic stability", Proc. 10th Int. Cong. Appl. Mech. (Stressa 1960), Elsevier, 63-67.

26. Lekoudis, S. G., "Resonant wave interactions on a swept wing", AIAA J., Vol. 18, 1980, 122-124.

27. Craik, A. D., "Nonlinear resonant instability in boundary layers", J. Fluid Mech., Vol. 50, 1971, 393-413.

28. Craik, A. D. D., "Second order resonance and subcritical instability", Proc. Roy. Soc. London, Vol. A343, 1975, 351-362.

29. Lekoudis, S. G., "On the triad resonance in the boundary layer", Lockheed-Georgia Company Rep. No. LG77ERO152, July 1977.

30. Craik, A. D. D. and Adam, J. A., "Evolution in space and time of resonant wave triads. I. The pump-wave approximation", Proc. Roy. Soc. London, Vol. A363, 1978, 243-255.

31. Craik, A. D. D., "Evolution in space and time of resonant wave triads. II. A class of exact solutions", Proc. Roy. Soc. London, Vol. A363, 1978, 257-269.

32. Benney, D. J. and Gustavsson, L. H., "A new mechanism for linear and nonlinear hydrodynamic instability", Stud. Appl. Math., Vol. 64, 1981, 185-209.

33. Nayfeh, A. H. and A. N. Bozatli, "Nonlinear wave interactions in boundary layers", AIAA Paper No. 79-1496, 1979.

34. Kelly, R. E., "On the stability of an inviscid shear layer which is periodic in space and time", J. Fluid Mech., Vol. 27, 1967, 657-689.

35. Pierrehumbert R. T. and Widnall, S. E., "The two- and three-dimensional instabilities of a spatially periodic shear layer", J. Fluid Mech., Vol. 114, 1982, 59-82.

36. Nayfeh, A. H. and Bozatli, A. N., "Secondary instability in boundary-layer flows", Phys. Fluids, Vol. 22, 1979, 805-813.

37. Maseev, L. M., "Occurrence of three-dimensional perturbations in a boundary layer", Fluid Dyn., Vol. 3, 1968, 23-24.

38. Herbert, T., "Three-dimensional phenomena in the transitional flat-plate boundary layer", AIAA Paper-85-0489, 1985.

39. Herbert, T., "Subharmonic three-dimensional disturbances in unstable plane shear flows", AIAA Paper No. 83-1759, 1983.

40. Herbert, T., "Analysis of the subharmonic route to transition in boundary layers", AIAA Paper No. 84-0009, 1984.

41. Bertolotti, F. P., "Temporal and spatial growth of subharmonic disturbances in Falkner-Skan flows", VPI & SU, M.S. thesis, 1985.

42. Nayfeh, A. H., "Three-dimensional spatial secondary instability in boundary-layer flows", AIAA Paper No. 85-1697, 1985.

43. Nayfeh, A. H., "Effect of streamwise vortices on Tollmien-Schlichting waves", J. Fluid Mech., Vol. 107, 1981, 441-453.

44. Herbert, T. and Morkovin, M. V., "Dialogue on bridging some gaps in stability and transition research", in Laminar-Turbulent Transition, R. Eppler and H. Fasel (eds.), IUTAM Symposium, Stuttgart, Germany, Sept. 16-22, 1979, Springer-Verlag, New York, 47-72.

45. Floryan, J. M. and Saric, W. S., "Wavelength selection and growth of Görtler vortices", AIAA Paper No. 80-1376, 1980.

46. Reed, H., "Disturbance-wave interactions in flows with crossflow", AIAA Paper No. 85-0494, 1985.

47. Nayfeh, A. H., Perturbation Methods, Wiley-Interscience, New York, 1973.

48. Nayfeh, A. H. and Mook, D. T., Nonlinear Oscillations, Wiley-Interscience, New York, 1979.

49. Nayfeh, A. H., Introduction to Perturbation Techniques, Wiley-Interscience, New York, 1981.

50. Nayfeh, A. H. and Masad, J., "Subharmonic instability in the Blasius boundary layer", in preparation.

Floquet Analysis of Secondary Instability in Shear Flows

THORWALD HERBERT, FABIO P. BERTOLOTTI
AND GERMAN R. SANTOS

ABSTRACT

In previous work, the parametric excitation of secondary, three-dimensional disturbances in the Blasius boundary layer was investigated using the concept of temporal growth. The analysis was restricted to fundamental (peak-valley splitting) modes and subharmonic modes, i.e. to disturbances with the same or twice the streamwise wavelength of the primary TS wave. Here, we generalize these studies in two directions. First, we analyze the spatial growth of secondary disturbances and develop an analogue of Gaster's transformation. Second, we study the secondary instability with respect to detuned modes, i.e. disturbances with arbitrary streamwise wavelength, and relate the results to observations of subharmonic and combination resonance in the flat-plate boundary layer. As a by-product, we show the rapid convergence of the Fourier series that governs the streamwise structure of the disturbances.

1. Introduction

Ever since Klebanoff, Tidstrom & Sargent (1962) reported their observations on *the three-dimensional nature of boundary-layer instability*, explanation of the early stages of transition in wall-bounded shear flows has challenged theoretical analysis. Perhaps early attempts have been biased by the association of secondary instability with high-frequency *spikes* immediately preceding breakdown, while the large-scale peak-valley splitting has been attributed to spanwise differential amplification of TS waves in a slightly nonuniform basic flow. Nevertheless, a variety of weakly nonlinear models has been developed in order to explain the origin of peak-valley splitting. These models differ in the intuitive choice of interacting Orr-Sommerfeld modes. In some of these models (Benney & Lin 1960, Herbert & Morkovin 1980), the spanwise wavelength of the peak-valley structure appears as a model parameter. This disregards the observation of a rather specific wavelength in the experiments. Other models (Raetz 1959, 1964, Craik 1971, Nayfeh & Bozatli 1979) therefore rest on the concept of resonant wave interaction at specific spanwise wavelength. However, neither of these weakly nonlinear theories provides a satisfactory explanation or quantitative characterization of the experimental facts. The shortcoming of the weakly nonlinear theory seems to be twofold. First, there is virtually no rational way of selecting an appropriate and complete set amongst the eigenmodes of the linear disturbance equations. Second, rapid convergence of the perturbation series and accuracy of the results cannot be verified at low order of truncation.

Blackwelder (1979) suggested that peak-valley splitting and the associated longitudinal vortex system originate from secondary instability in the boundary layer. This point of view was shared by Herbert & Morkovin (1980) who emphasized the possibility of parametric excitation of three-dimensional disturbances in the presence of a finite-amplitude TS wave. In fact, the growth of primary TS waves on a slow viscous time scale leads to a streamwise (almost) periodic flow. Linear stability analysis of this flow rests on differential equations with periodic coefficients. Therefore, guidance on classes and form of solutions can be obtained from Floquet theory.

The use of Floquet systems in the stability analysis of fluid flows has some history. Applications were primarily for oscillatory, i.e. purely time-periodic flows (Davis 1976). However, in this case the streamwise vorticity distribution is uniform at any instant, and the effect of time-periodicity is weak (von Kerczek 1985). Spatially periodic flows have been analyzed by Kelly (1967) for the mixing layer where secondary instability occurs as vortex pairing. Maseev (1968) studied the origin of three-dimensional disturbances in the boundary layer by means of a Floquet system, but neither his method nor his results received particular attention.

The development of this approach has been fostered by new observations in experiments and computer simulations. Hot-wire data and flow visualizations in the Blasius boundary layer (Kachanov & Levchenko 1982, 1984, Saric & Thomas 1983) have shown that the route to transition is non-unique and sensitively depends on experimental conditions. The difference is in the nature and scale of the three-dimensional disturbances that generate different characteristic patterns of Λ-shaped vortex loops in flow visualizations as shown in Fig. 1. In the K-breakdown associated with peak-valley splitting, the Λ vortices are aligned along the peaks (Fig. 1a) and repeat with the wavelength λ_x of the TS wave. At lower levels of the TS wave, a staggered arrangement of the Λ vortices appears (Fig. 1b). This pattern repeats itself with wavelength $2\lambda_x$ and a stationary hot-wire records subharmonic signals. The spanwise wavelength λ_z may be larger or less than λ_x depending on the TS amplitude. The same types of patterns have been observed in plane Poiseuille flow (Kozlov & Ramasanov 1983), and may therefore be considered a generic phenomenon in a wide class of plane shear flows.

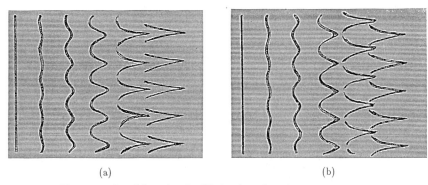

(a)   (b)

Figure 1. Streaklines in the Blasius boundary layer.
(See flow visualizations by Saric & Thomas, 1984). The flow is from left to right. (a) Ordered (peak-valley) arrangement of Λ vortices. Fundamental mode. (b) Staggered arrangement of Λ vortices. Subharmonic mode.

In numerical simulations of transition in plane Poiseuille flow, Orszag & Kells (1980) observed exponential growth of small three-dimensional disturbances in large-amplitude two-dimensional equilibrium states. This result is indicative of a linear instability mechanism in a streamwise periodic flow and led Orszag & Patera (1980, 1981) to a study of secondary instability based on a Floquet system. Pierrehumbert & Widnall (1982) performed a linear stability analysis of an array of Stuart vortices in a mixing layer with respect to fundamental and subharmonic disturbances. Beyond the two-dimensional vortex pairing, they found three-dimensional modes that result in the patterns of figure 1. Secondary (and tertiary) instability of a cubic velocity profile in an inclined fluid layer heated from above has been analyzed by Nagata & Busse (1983). However, the concepts of the stability analysis are difficult to discern from the semi-analytical method of solution.

In our work, we aimed at developing the Floquet analysis of secondary instability as close as possible along the lines of the classical theory of primary instability, and as a self-supporting tool for a systematic identification of classes and modes of disturbances. Earlier applications to plane Poiseuille flow (Herbert 1981a,b) verified the parametric excitation of peak-valley splitting in equilibrium states of small but finite amplitude. The results are consistent with experiments (Nishioka, Iida & Ichikawa 1975, Nishioka & Asai 1985) and computer simulations of transition (Kleiser 1982). Beyond the peak-valley splitting modes, subharmonic modes of secondary instability were found (Herbert 1983a). The limit of vanishing amplitude was studied in order to evaluate existing weakly nonlinear models. This study emphasized the important role of Squire modes (eigenmodes of the homogeneous equation governing the vorticity normal to the wall) in models of three-dimensional phenomena. Approximations were introduced in order to obtain a simpler formulation and to broaden the range of applicability (Herbert 1983b). Analysis of peak-valley splitting in the Blasius boundary layer (Herbert 1985a) provided results consistent with the observations of Klebanoff et al. (1962). Theoretical predictions of growth rates and disturbance velocity profiles for subharmonic modes (Herbert 1984) are in good agreement with the hot-wire data of Kachanov & Levchenko (1982).

The earlier phase of this work established the basic formulation of the theory as well as an initial data base for temporally growing disturbances. This work has been summarized by Herbert (1984b, 1985b). Here, we present a more general formulation of the theory that incorporates the spatial growth concept. The theory is applied to the Blasius boundary layer. Beyond the special cases of subharmonic and fundamental (peak-valley splitting) modes, we consider detuned modes which relate to combination resonance. While earlier results were obtained at lowest order of truncation, we show the effect of higher Fourier modes on growth rates and disturbance velocity profiles.

2. Governing equations

We consider the flow of an incompressible fluid that is governed by the Navier-Stokes equations

$$\frac{\partial \mathbf{v}}{\partial t} + (\mathbf{v}\cdot\nabla)\mathbf{v} = -\nabla p + \frac{1}{R}\nabla^2 \mathbf{v}, \quad \nabla\cdot\mathbf{v} = 0 \tag{1}$$

with appropriate, possibly inhomogeneous boundary conditions. All quantities are nondimensional, and R is the Reynolds number. $\mathbf{v}(x',y,z,t) = (u,v,w)$ and $p(x',y,z,t)$ are velocity and pressure field, respectively, where x' is streamwise, y is normal to the wall(s), z is normal to the x',y plane an t is time. From (1) we derive the transport equation for the vorticity $\omega = \nabla\times\mathbf{v} = (\xi,\eta,\varsigma)$,

$$\frac{\partial \omega}{\partial t} + (\mathbf{v}\cdot\nabla)\omega - (\omega\cdot\nabla)\mathbf{v} = \frac{1}{R}\nabla^2\omega \tag{2}$$

in order to eliminate the pressure. We consider a two-dimensional flow $\mathbf{v}_2 = (u_2,v_2,0)$ subject to small three-dimensional disturbances $\mathbf{v}_3 = (u_3,v_3,w_3)$ according to

$$\mathbf{v}(x',y,z,t) = \mathbf{v}_2(x',y,t) + B\mathbf{v}_3(x',y,z,t) \tag{3}$$

where B is sufficiently small for linearization. Substitution into (2) and comparison in like powers of B provides two sets of equations for the basic flow and the disturbances, respectively, while terms of order $O(B^2)$ have been neglected.

2.1 The periodic basic flow

We assume the basic flow in the form

$$\mathbf{v}_2(x',y,t) = \mathbf{v}_0(y) + A\mathbf{v}_1(x',y,t) \tag{4}$$

where $\mathbf{v}_0 = (u_0, 0, 0)$ is the steady flow solution to equations (1), e.g. plane Poiseuille flow. The component $\mathbf{v}_1$ is considered in the form of a wave, i.e. periodic in t, periodic in x' with wavelength $\lambda_x = 2\pi/\alpha$, and traveling with phase velocity c_r in the x' direction. In a coordinate system x,y,z traveling with the wave, we have

$$\mathbf{v}_1(x',y,t) = \mathbf{v}_1(x,y) = \mathbf{v}_1(x+\lambda_x,y), \quad x = x' - c_r t, \tag{5}$$

and consequently, $\mathbf{v}_1$ and $\mathbf{v}_2$ are steady and streamwise periodic. By proper normalization of $\mathbf{v}_1$, the amplitude A directly measures the maximum streamwise r.m.s. fluctuation (usually denoted as u'_m)

and the streamwise fluctuation assumes this maximum at $x = 0$.

We express $\mathbf{v}_1 = (u_1, v_1, 0)$ and the vorticity $\nabla \times \mathbf{v}_1 = (0, 0, \varsigma)$ in terms of the stream function ψ_1 such that

$$u_1 = \partial \psi_1/\partial y, \quad v_1 = -\partial \psi_1/\partial x, \quad \varsigma_1 = -\nabla^2 \psi_1. \tag{6}$$

The basic flow is then governed by the nonlinear partial differential equation

$$[\frac{1}{R}\nabla^2 - (u_o - c_r)\frac{\partial}{\partial x}]\varsigma_1 + \frac{\partial \varsigma_0}{\partial y}\frac{\partial \psi_1}{\partial x} = A[\frac{\partial \psi_1}{\partial y}\frac{\partial}{\partial x} - \frac{\partial \psi_1}{\partial x}\frac{\partial}{\partial y}]\varsigma_1 \tag{7}$$

with homogeneous boundary conditions on $\partial \psi_1/\partial x$ and $\partial \psi_1/\partial y$. Since ψ_1 is periodic in x, it can be represented by the Fourier series

$$\psi_1(x,y) = \sum_{n=-\infty}^{\infty} \phi_n(y) e^{in\alpha x}, \tag{8}$$

where $\phi_{-n} = \phi_n^\dagger$ is necessary for a real solution, and $\dagger$ denotes the complex conjugate. The functions $\phi_n(y)$, $n \geq 0$, are governed by a nonlinear system of ordinary differential equations.

In some cases like plane Poiseuille flow, approximations to a strictly periodic solution of the Navier-Stokes equations can be found by numerically solving a truncated version of this system for ϕ_n, $0 \leq n \leq N$ (Herbert 1977). Such solutions exist for points on the neutral surface of the flow, i.e. for parameter combinations that satisfy a nonlinear dispersion relation $F(R, \alpha, A, c_r) = 0$. These equilibrium states of constant amplitude allow a clean mathematics of the secondary instability theory. On the other hand, their use severely restricts the scope of the theory. First, three-dimensional motion may branch off from two-dimensional motion with growing or decaying amplitude (Nishioka, Iida & Ichikawa 1975, Orszag & Patera 1983). Second, three-dimensional phenomena appear in flows like the Blasius boundary layer where such equilibrium states are not yet established.

An alternative way of constructing the periodic basic flow rests on the shape assumption (Stuart 1958) and a generalization of the parallel-flow assumption frequently used in the classical stability theory. The shape assumption is equivalent to discarding the nonlinear terms on the right hand side of eq. (7) and considering the amplitude A as a parameter. This assumption is justified by the observations that secondary instability occurs at small amplitudes and originates from the redistribution of vorticity, not from the amplitude-sensitive Reynolds stresses. With the shape assumption, the streamfunction is given by

$$\psi_1(x,y) = \phi_1(y)e^{i\alpha x} + \phi_1^\dagger(y)e^{-i\alpha x} \tag{9}$$

where ϕ_1 is the eigenfunction associated with the principal mode of the Orr-Sommerfeld equation for given R and α, and $\beta = 0$:

$$\{\frac{1}{R}D^2 - i\alpha[(u_0 - c)D^2 - u''_0]\}\phi_1 = 0, \tag{10}$$

where $D^2 = d^2/dy^2 - \alpha^2$, and the prime denotes d/dy. With homogeneous boundary conditions, the eigenvalue problem associated with eq. (10) in general provides complex eigenvalues $c = c_r + ic_i$ or $\alpha = \alpha_r + i\alpha_i$ depending on whether the temporal or spatial growth concept is chosen. In either case, the non-vanishing imaginary part describes exponential growth or decay of the amplitude. Generalizing the parallel-flow assumption, we consider the local (or instantaneous) conditions as independent of the relevant variable x (or t). Obviously, this neglect of the slow changes of the amplitude (on a viscous time scale) is justified only if rapid changes, i.e. sufficiently large growth rates of the three-dimensional disturbances are considered.

2.2. Three-dimensional disturbances

The analysis of three-dimensional disturbances $\mathbf{v}_3$ rests on the vorticity equation

$$(\frac{1}{R}\nabla^2 - \frac{\partial}{\partial t})\omega_3 - (\mathbf{v}_2\cdot\nabla)\omega_3 - (\mathbf{v}_3\cdot\nabla)\omega_2 \qquad (11)$$
$$+ (\omega_2\cdot\nabla)\mathbf{v}_3 + (\omega_3\cdot\nabla)\mathbf{v}_2 = 0.$$

This equation is linear in $\mathbf{v}_3$ due to neglecting terms of order $O(B^2)$. We follow Squire (1933) and eliminate the spanwise disturbance velocity w_3 by exploiting continuity. Since $\partial \eta_3/\partial z$ and $\partial \xi_3/\partial z$ can be expressed in terms of u_3, v_3, we obtain two coupled differential equations for u_3 and v_3,

$$[\frac{1}{R}\nabla^2 - (u_0 - c)\frac{\partial}{\partial x} - \frac{\partial}{\partial t}]\frac{\partial \eta_3}{\partial z} + \zeta_0 \frac{\partial^2 v_3}{\partial z^2} + A\{(-\frac{\partial \psi_1}{\partial y}\frac{\partial}{\partial x} + \frac{\partial \psi_1}{\partial x}\frac{\partial}{\partial y} - \frac{\partial^2 \psi_1}{\partial x \partial y})\frac{\partial \eta_3}{\partial z} \qquad (12)$$
$$+ \frac{\partial^2 \psi_1}{\partial x^2}(\frac{\partial^2 u_3}{\partial x \partial y} + \frac{\partial^2 v_3}{\partial y^2}) - \frac{\partial^2 \psi_1}{\partial y^2}\frac{\partial^2 v_3}{\partial z^2}\} = 0,$$

$$[\frac{1}{R}\nabla^2 - (u_0 - c)\frac{\partial}{\partial x} - \frac{\partial}{\partial t}]\nabla^2 v_3 - \frac{d\zeta_0}{dy}\frac{\partial v_3}{\partial x} + A\{(-\frac{\partial \psi_1}{\partial y}\frac{\partial}{\partial x} + \frac{\partial \psi_1}{\partial x}\frac{\partial}{\partial y})\nabla^2 v_3 \qquad (13)$$
$$+ \frac{\partial^2 \psi_1}{\partial x^2}(\frac{\partial \zeta_3}{\partial y} + \frac{\partial \eta_3}{\partial z}) - \frac{\partial^2 \psi_1}{\partial x \partial y}(\frac{\partial \zeta_3}{\partial x} + \frac{\partial \xi_3}{\partial z}) - \frac{\partial \zeta_1}{\partial x}(2\frac{\partial u_3}{\partial x} + \frac{\partial v_3}{\partial y}) - \frac{\partial \zeta_1}{\partial y}\frac{\partial v_3}{\partial x}$$
$$- (u_3 \frac{\partial}{\partial x} + v_3 \frac{\partial}{\partial y})\frac{\partial \zeta_1}{\partial x}\} = 0,$$

with homogeneous boundary conditions on u_3, v_3, and $\partial v_3/\partial y$. In the limit $A \to 0$, the coefficients in (12), (13) are independent of x, z, and t, and normal modes can be introduced in these variables. Equation (12) reduces to Squire's equation for the y-component of vorticity whereas eq. (13) provides the Orr-Sommerfeld equation for oblique waves in the basic flow $\mathbf{v}_0$, written in a moving coordinate system.

For $A \neq 0$, the x-periodic streamfunction ψ_1 appears in the coefficients. The normal mode concept can still be applied with respect to z and t and the disturbances can be written in the form

$$\mathbf{v}_3(x,y,z,t) = e^{\sigma t} e^{i\beta z} \mathbf{V}(x,y). \qquad (14)$$

We consider the spanwise wave number $\beta = 2\pi/\lambda_z$ as real, whereas $\sigma = \sigma_r + i\sigma_i$ is in general complex. Insight into classes of solutions and their streamwise structure can be obtained from the Floquet theory of ordinary differential equations with periodic coefficients, using Hill's equation or the Mathieu equation with damping as a relevant model. Although $\mathbf{v}_3$ in the form (14) is in general complex, it is important to note that our disturbance equations provide real solutions including the complex conjugate $\mathbf{v}_3^\dagger$. In fact, the system of equations can be written in real form. For this system, Floquet theory suggests solutions in the form

$$\mathbf{V}(x,y) = e^{\gamma x} \tilde{\mathbf{V}}(x,y), \quad \tilde{\mathbf{V}}(x + 2\lambda_z, y) = \tilde{\mathbf{V}}(x,y) \qquad (15)$$

where $\gamma = \gamma_r + i\gamma_i$ is a characteristic exponent, and $\tilde{\mathbf{V}}$ is periodic in x with wavelength $2\lambda_z$. Combining (14) and (15), we can write

$$\mathbf{v}_3 = e^{\sigma t} e^{\gamma x} e^{i\beta z} \sum_{m=-\infty}^{\infty} \hat{\mathbf{v}}_m(y) e^{im\hat{\alpha} x}, \quad \hat{\alpha} = \alpha/2. \qquad (16)$$

The functions $\hat{u}_m$, $\hat{v}_m$ are governed by an infinite system of ordinary differential equations. Since the basic flow $\mathbf{v}_2$ with wavenumber $\alpha = 2\hat{\alpha}$ provides coupling only between components $\hat{\mathbf{v}}_m$ and $\hat{\mathbf{v}}_{m\pm 2}$, this system splits into two separate systems for even and odd m that describe two classes of solutions

$$\mathbf{v}_f = e^{\sigma t} e^{\gamma x} e^{i\beta z} \tilde{\mathbf{v}}_f(x,y), \quad \tilde{\mathbf{v}}_f = \sum_{m \text{ even}} \hat{\mathbf{v}}_m(y) e^{im\hat{\alpha} x}, \qquad (17)$$

$$\mathbf{v}_s = e^{\sigma t} e^{\gamma x} e^{i\beta z} \tilde{\mathbf{v}}_s(x,y), \quad \tilde{\mathbf{v}}_s = \sum_{m \text{ odd}} \hat{\mathbf{v}}_m(y) e^{im\hat{\alpha} x}. \qquad (18)$$

The periodic functions $\tilde{\mathbf{v}}_f$ and $\tilde{\mathbf{v}}_s$ obviously satisfy

$$\tilde{\mathbf{v}}_f(x + \lambda_x, y) = \tilde{\mathbf{v}}_f(x, y), \quad \tilde{\mathbf{v}}_s(x + 2\lambda_x, y) = \tilde{\mathbf{v}}_s(x, y). \tag{19}$$

Therefore, we denote $\mathbf{v}_f$ as the fundamental mode, associated with primary resonance, and $\mathbf{v}_s$ as the subharmonic mode, originating from principal parametric resonance.

The occurrence of two complex quantities, σ and γ, in the eigenvalue problem for secondary disturbances leads to an ambiguity similar to that associated with the Orr-Sommerfeld equation. Only two of the four real quantities σ_r, σ_i, γ_r, γ_i can be determined; the other two must be chosen in some reasonable way. We have as yet identified and studied three physically relevant classes of solutions.

(1) Temporally growing tuned modes. In this case, we assume $\gamma_r = \gamma_i = 0$. The temporal growth rate is given by σ_r, while σ_i can be interpreted as frequency shift. Modes with $\sigma_i = 0$ travel synchronous with the basic flow.

(2) Spatially growing tuned modes. For this case, it is important to note that spatial growth is implicitly related to the laboratory frame x', y, z. Rewriting

$$e^{\sigma t} e^{\gamma x} = e^{(\sigma - \gamma c_r)t} e^{\gamma x'}, \tag{20}$$

we choose $\sigma = \gamma c_r$ in order to suppress temporal effects. Hence, γ_r provides the spatial growth rate in the laboratory frame while γ_i is the shift in the streamwise wavenumber.

(3) Temporally growing detuned modes. In this case, we assume $\gamma_r = 0$ while $\gamma_i = \Delta \alpha$. The meaning of σ_r and σ_i is the same as in class (1).

A fourth class of spatially growing detuned modes with $\sigma_r = \gamma_r c_r$, and given frequency shift σ_i is similar to class (3) but more closely related to the observations of combination resonance in the Blasius boundary layer (Kachanov & Levchenko 1982, 1984).

2.3 Numerical and formal aspects

At this point it seems in order to remember the need for numerically solving truncated versions of the disturbance equations. Considered the efforts for solving a single Orr-Sommerfeld equation, the analysis of the coupled systems of Orr-Sommerfeld and Squire equations is a formidable task, and hence, every simplification is highly welcome.

Primary and secondary stability problems are numerically solved using a spectral collocation method with Chebyshev polynomials. This method converts the ordinary differential equations and boundary conditions into systems of algebraic equations. We prefer the direct treatment of the boundary value problem over shooting methods since we maintain access to the spectrum of eigenvalues for temporally growing modes. In the absence of any prior guidance, the spectrum is extremely helpful for reliably identifying the most relevant modes in different regions of the multi-dimensional parameter space and for untangling their analytical connections.

For boundary layers, we obtain a finite domain by an algebraic mapping $Y = y_0/(y + y_0)$ that transforms $y = 0, \infty$, into $Y = 1, 0$, respectively. The parameter y_0 is used to control the density of collocation points in the neighborhood of the wall. Only odd Chebyshev polynomials are used such that the boundary conditions for $y \to \infty$ are automatically satisfied. Typically, $J = 30$ collocation points are used and y_0 is chosen to place half of the points within the displacement thickness of the boundary layer. For every (real or complex) function $\hat{v}_m(y)$ in (17), (18), $2J + 3$ (real or complex) unknowns have to be included into the homogeneous system of algebraic equations. In view of the size of the resulting systems, the truncation of the series (17), (18) is crucial for the numerical work.

For tuned subharmonic modes, the lowest possible truncation is $|m| \leq 1$, including only $\hat{v}_{-1}$ and $\hat{v}_1$. The lowest approximation for tuned fundamental modes is $|m| \leq 2$, including $\hat{v}_{-2}$, $\hat{v}_0$ and $\hat{v}_2$. Although subharmonic and fundamental modes apparently represent two distinct classes of solutions, they are analytically connected through detuned modes. Since no restriction is imposed on the detuning γ_i, it is permissible to consider $\gamma_i \to \hat{\alpha}$. Therefore, the tuned fundamental mode can be obtained from the subharmonic mode by detuning with $\gamma_i = \hat{\alpha}$. Obviously, this analytical connection cannot be found with $|m| \leq 1$ since the tuned fundamental requires at least the three functions mentioned above. It will be necessary, therefore, to carefully study the effect of low truncation.

There are two additional features that can be exploited. Since the physical solution must be real, any complex solution for $\mathbf{v}_3$ implies a second solution $\mathbf{v}_3^\dagger$. Moreover, only the square β^2 of the spanwise wavenumber appears in the equations such that the results for β and $-\beta$ are identical. For the tuned modes of classes (1) and (2), it is useful to rewrite the Fourier series in trigonometric form, i.e. to introduce

$$\mathbf{v}_m^+ = \frac{1}{2}(\hat{\mathbf{v}}_m + \hat{\mathbf{v}}_{-m}), \quad \mathbf{v}_m^- = \frac{1}{2i}(\hat{\mathbf{v}}_m - \hat{\mathbf{v}}_{-m}), \quad m \geq 0. \tag{21}$$

Since $\hat{u}_{-m} = \hat{u}_m^\dagger$, $\hat{v}_{-m} = \hat{v}_m^\dagger$, $\hat{w}_{-m} = -\hat{w}_m^\dagger$, the components of $\mathbf{v}_0^+$ are either real (u_0^+, v_0^+) or purely imaginary (w_0^+), while $\mathbf{v}_0^- \equiv 0$. At the lowest truncation, the modes of class (1) then take the form

$$\mathbf{v}_f = 2e^{\sigma t} e^{i\beta z}[\frac{1}{2}\mathbf{v}_0^+(y) + \mathbf{v}_2^+(y)\cos\alpha x - \mathbf{v}_2^-(y)\sin\alpha x], \tag{22}$$

$$\mathbf{v}_s = 2e^{\sigma t} e^{i\beta z}[\mathbf{v}_1^+(y)\cos\hat{\alpha} x - \mathbf{v}_1^-(y)\sin\hat{\alpha} x]. \tag{23}$$

After proper rearrangement, the new functions $u_m^\pm$, $v_m^\pm$ are governed by a system of equations with real coefficients. The solution however is only real for real σ. Similar conclusions can be derived for spatially growing modes of class (2).

The use of real instead of complex variables is an obvious advantage for the numerical work, especially for obtaining spectra and for tracing single real eigenvalues. The latter procedure is frequently used since the principal (i.e. most amplified) modes in classes (1) and (2) are usually real. For tracing complex eigenvalues, there is no clear advantage of the real formulation. The use of the complex formulation is also recommended for detuned modes of class (3), which are associated with complex eigenvalues. For these modes, the real formulation requires the simultaneous detuning by γ_i and $-\gamma_i$, a fact consistent with the observations of combination resonance by Kachanov & Levchenko (1982, 1984).

Concerning the choice between temporal and spatial growth concept, the situation is analogue to the primary stability analysis. The temporal eigenvalue σ appears linear in the equations. Therefore, spectra and single eigenvalues can be obtained by standard procedures of linear algebra. In the spatial formulation, the eigenvalue γ appears up to the fourth power. Although methods exist to obtain spectra in this case (Bridges & Morris 1984), the required computations are very demanding. Therefore, we have exploited the fact that neutral behavior is independent of the chosen growth concept. Parameter combinations for neutral behavior, $\sigma_r = 0$, have been identified using the temporal concept. Starting from these points, the principal eigenvalue can be traced using the spatial concept. The local procedure for spatial eigenvalues γ rests on Newton iteration. Although this procedure is more costly than tracing temporal eigenvalues, it is more convenient for following the downstream development of disturbances of fixed dimensional frequency and spanwise wavelength, as it occurs in experiments.

3. Some results for the Blasius boundary layer.

We consider the two-dimensional boundary-layer flow of an incompressible fluid of kinematic viscosity ν along a semi-infinite plate in the plane $y = 0$. The edge velocity U_∞ is constant and x' measures the distance from the leading edge. U_∞ and $\delta_r = (\nu x'/U_\infty)^{1/2}$ are used as reference length, and the Reynolds number is $R = (U_\infty x'/\nu)^{1/2}$. In nondimensional form, the steady laminar flow is given by $\mathbf{v}_0 = (u_0, 0, 0)$ where $u_0(y) = f'$ is the Blasius profile defined by

$$f''' + \frac{1}{2}ff'' = 0, \tag{24}$$

$$f = f' = 0 \text{ at } y = 0, \; f' \to 1 \text{ as } y \to \infty. \tag{25}$$

The prime denotes d/dy. The small transverse velocity is neglected, applying locally the parallel-flow assumption. The periodic basic flow $\mathbf{v}_2$ is constructed under the shape assumption with $\psi_1(x,y)$ given by (9) as a solution to the Orr-Sommerfeld equation. The same growth concept is applied for primary as well as secondary disturbances. For temporally growing disturbances, the wavenumber α is real while the frequency $\omega = \alpha c = \omega_r + i\omega_i$ is complex. For spatially growing disturbances, ω is real while $\alpha = \alpha_r + i\alpha_i$ is complex. For convenience, we introduce the nondimensional parameters $F = 10^6\omega/R$, $a = 10^3\alpha/R$, $b = 10^3\beta/R$ that describe fixed dimensional frequency, streamwise wavelength, and spanwise wavelength, respectively, at different streamwise locations.

An analysis of the subharmonic modes of class (1) has been reported by Herbert (1984a), and more detailed results will be given elsewhere. Calculations have been carried out primarily for the experimentally documented conditions ($F = 124$) of Kachanov & Levchenko. In all cases of instability, the principal mode has been found to be associated with a real eigenvalue $\sigma = \sigma_r$, i.e. the most unstable subharmonic disturbance travels synchronous with the TS wave, except at very small amplitudes A. In general, instability occurs as the amplitude A exceeds a threshold value. In the neighborhood of resonant triads (Craik 1971), however, weak subharmonic growth may occur at $A = 0$ since oblique waves may be unstable as primary disturbances. The growth rate σ_r at otherwise fixed parameters increases strongly with the TS amplitude as well as with the Reynolds number. Instability may occur in a wide range of spanwise wavenumbers, with a sharp cutoff at lower values indicating suppression of vortex pairing ($\beta = 0$) in the presence of the wall.

Tracing the principal eigenvalue to the limit $A \to 0$ has shown that for selected spanwise wavenumbers an analytical connection to Craik's resonant triad exists. In a large band of wavenumbers, however, the subharmonic secondary instability is intricately linked to eigensolutions of Squire's equation. This finding has stimulated the investigation of extended weakly nonlinear models (Nayfeh 1985) that include Orr-Sommerfeld modes as well as Squire modes. Tracing the results of the Floquet analysis to the limit $A \to 0$ seems to be the first rational way of identifying the relevant modes for constructing weakly nonlinear models.

The result for fundamental (peak-valley splitting) modes are qualitatively similar except for the occurrence of resonance at $A \to 0$. Peak-valley splitting is a threshold phenomenon that requires finite TS amplitude A. Modes in a broad band of spanwise wavenumbers may be unstable, the more the larger A and R. Results for $F = 59$ (Herbert 1985a) are consistent with observations of Klebanoff et al. (1962) although the qualitative comparison suffers somewhat from the fact that these experiments were not designed to accurately model the behavior of TS waves in accordance with linear stability theory.

The theoretical results on subharmonic instability are consistent and in quantitative agreement with the available experimental data on disturbance velocities and spatial growth. For this comparison, however, we have converted the temporal growth rate σ_r to the spatial growth rate γ_r by the simple transformation $\gamma_r = \sigma_r/c_r$. Since this transformation may be invalid at the large growth rates under consideration (Gaster 1962), we present in figures 2 and 3 comparable results for subharmonic modes of class (2), i.e. for spatially growing disturbances.

Figure 2 shows the normalized y-distribution of the streamwise rms fluctuation u'_3 of the subharmonic mode for $F = 124$, $R = 608$, and $b = 0.33$. There are two notable agreements. First, the theoretical result agrees very well with the experimental data. Second, this figure is almost indistinguishable from earlier results (Herbert 1984a, fig. 14) obtained for the temporal case. The situation is very similar with respect to the streamwise amplitude growth of TS wave and subharmonic disturbance shown in figure 3. The only adjustment we have made is the choice of the initial subharmonic

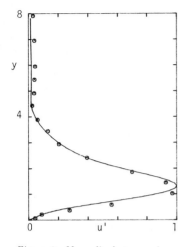

Figure 2. Normalized streamwise rms fluctuation of the spatially growing subharmonic mode at $R = 608$, $F = 124$, $A = 0.0122$, $b = 0.33$. Comparison of theoretical result (———) and experimental data (o) of Kachanov & Levchenko (1984).

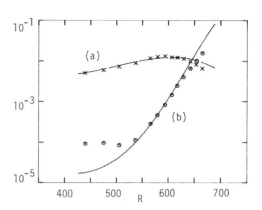

Figure 3. Amplitude variation with the Reynolds number R for (a) the TS wave with initial amplitude $A = 0.0044$ and (b) the subharmonic mode. $F = 124$, $b = 0.33$. Comparison of theoretical results (———) and experimental data (x,o) of Kachanov & Levchenko (1984).

amplitude B_0 such that theoretical and experimental value of B agree at $R = 606$ (branch II). This adjustment is legal in the framework of our linear stability analysis, and was necessary since the experiment does not cover the full range of subharmonic disturbance growth. The data for $R < 500$ are taken shortly downstream of the vibrating ribbon where the small three-dimensional components are not yet fully established.

The good agreement between spatial growth rates and transformed temporal results is unexpected, considered the restriction of Gaster's transformation to the neighborhood of neutral conditions. The growth rates of secondary disturbances at TS amplitudes of $A = 0.01$, say, are much larger than those of TS waves. In view of the reduced computational effort needed for temporal eigenvalue search, we performed a detailed study of transformations relating the two types of growth rates (Bertolotti 1985). At the zeroth order, the spatial growth rate γ_r can be obtained from

$$\gamma_r^{(0)} = \sigma_r / c_r \tag{26}$$

as used in previous work (Herbert 1984a). More accurate transformations for $\gamma_r^{(n)}$ can be obtained by integrating n-term Taylor expansions of $\partial \gamma_r / \partial \sigma_r$ about the temporal result. A comparison of different approximations with the spatial growth rates from direct calculations is shown in figure 4 for $F = 83$, $R = 826$, and $A = 0.01$. The maximum error introduced at zeroth order of less than 5% is surprisingly low and tolerable in view of other approximations incorporated in the analysis. The use of Taylor expansions with $n = 1$ and $n = 2$ improves the agreement on the expense of additional eigenvalue searches for small $\gamma_r \neq 0$. We, therefore, recommend the use of either the temporal concept with the transformation (26) when efficiency of the computation is desired, or the spatial concept whenever the situation requires, e.g. for comparison with experiments.

Considering computational efficiency and the small error introduced by (26), we decided to study detuned modes, the analytical connection between subharmonic and fundamental modes, and the effects of truncation using the temporal growth concept. The effect of truncation is bifold. First, we

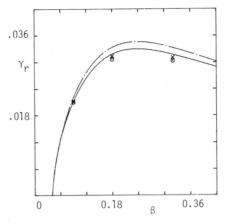

Figure 4. Spatial growth rate γ_r vs. spanwise wavenumber β for $R = 826$, $F = 83$, $A = 0.01$, obtained by direct calculation (———), and by transformation of temporal data: $\gamma_r^{(0)}$ (-- --), $\gamma_r^{(1)}$ (o), $\gamma_r^{(2)}$ (*).

truncate the Chebyshev series, hence reducing the resolution in the y direction. Second, we truncate the Fourier series (17),(18) and reduce the accuracy in resolving the streamwise structure.

Table 1 shows the effect of these truncations on the temporal eigenvalues for subharmonic modes ($\hat{\alpha} + \gamma_i$, $\gamma_i = 0$), detuned modes ($\hat{\alpha} + \gamma_i$, $\gamma_i = \hat{\alpha}/2$), and fundamental modes ($\hat{\alpha} + \gamma_i$, $\gamma_i = \hat{\alpha}$) for different numbers J of collocation points and different numbers of Fourier modes taken into account. A given mode can be obtained by either detuning the subharmonic (17) by γ_i or detuning the fundamental (18) by $\gamma_i - \hat{\alpha}$. In order to remove this ambiguity, in the following we always detune the subharmonic by $0 \leq \gamma_i \leq \hat{\alpha}$ or $0 \leq \epsilon \leq 1$ where $\epsilon = \gamma_i/\hat{\alpha}$. Hence, the fundamental is obtained with $\epsilon = 1$. It is not unexpected that using only two Fourier components provides a reasonable approximation for subharmonic disturbances but is insufficient for representing fundamental modes. For any number ≥ 3 of Fourier modes, however, the modes of the various classes are well described. Convergence of the Fourier series is very rapid. Also, the asymmetry of the model (odd number of Fourier components in (18), even number in (17)) introduces only a small error, clearly indicated by the small imaginary parts σ_i. The number J of collocation points has only a small effect provided $J \geq 20$.

Another interesting aspect of the data in table 1 is the large growth rate of the detuned mode with $\epsilon = 0.5$ The frequency shift σ_i is non-zero indicating a time-periodic modulation of the disturbance amplitude. This behavior is consistent with the observation of a broad spectral peak centered at the subharmonic frequency and the change of phase between subharmonic disturbance and TS wave (Kachanov & Levchenko 1982, 1984). Data similar to those of table 1 are displayed in figure 5 for modes with wavenumbers between $\hat{\alpha}$ and $\alpha = 2\hat{\alpha}$. The result obtained with only two Fourier modes is a good approximation for moderate detuning of the subharmonic disturbance but is useless at larger wavenumbers. The approximations with three and four Fourier modes agree to within 1% over the full range of wavenumbers. The imaginary part σ_i is zero only for integer multiples of $\hat{\alpha}$. The growth rate σ_r exhibits a broad maximum in the neighborhood of $\epsilon = 0$, in agreement with the observed large spectral width of the parametric resonance (Kachanov & Levchenko 1982, 1984). Since $\sigma_i \neq 0$ for $\epsilon \neq 0$, construction of a real disturbance field involves two (complex conjugate) modes with detuning ϵ and $-\epsilon$, and identical growth rates. In the experiments, this fact is reflected by the appearance of two sharp spectral components of (almost) the same amplitude. The growth rate of the

Table 1					
Detuning	Number of Fourier modes	20 collocation points		30 collocation points	
		$10^3\sigma_r$	$10^3\sigma_i$	$10^3\sigma_r$	$10^3\sigma_i$
$\epsilon = 0$ subharmonic mode	2	8.11571	0	8.11789	0
	3	8.15454	0.07349	8.15701	0.01443
	4	8.19404	0	8.19611	0
	5	8.19433	0	8.19625	0
	6	8.19461	0	8.19639	0
	7	8.19461	0	8.19639	0
$\epsilon = 0.5$ detuned mode	2			6.18139	1.64569
	3			6.40703	1.48160
	4			6.41874	1.47093
	5			6.42060	1.47088
	6			6.42048	1.47086
	7			6.42048	1.47087
$\epsilon = 1$ fundamental mode	2	-1.07251	1.96292		
	3	3.63671	0	3.63993	0
	4	3.64732	0.01229	3.65078	0.01015
	5	3.65802	0	3.66205	0
	6	3.65840	0	3.66192	0
	7	3.65878	0	3.66179	0

fundamental mode is relatively small for these parameters that favor subharmonic resonance. This situation may drastically change if the parameters are varied.

Distributions of the streamwise disturbance velocities for $\epsilon = 0, 0.5, 1.0$ are shown in figures 6, 7, and 8, respectively. These results were obtained with 20 collocation points and 7 Fourier modes. Figure 6 shows the components with wavenumbers $\hat{\alpha}$ and $3\hat{\alpha}$ for the subharmonic mode. The maximum amplitude of the high-wavenumber components rapidly decreases. Due to the fast convergence of the Fourier series, the disturbance velocity field is well represented by low-order approximations. While there is only one important component in the subharmonic case, there are two in the other cases. The fundamental mode is characterized by the simultaneous presence of the fluctuating component of wavenumber $\alpha = 2\hat{\alpha}$ and the mean component of zero wavenumber. The latter component causes the different mean velocity distributions across the boundary layer at peak and valley (Klebanoff et al. 1962, fig. 10). The velocities v and w of this mean component generate the observed longitudinal vortex system. The fluctuating component causes the different rms velocity profiles at peak and valley. The peculiar shape of the profile near the wall has been found in results from independent computer programs and for both growth concepts. The little bump near the wall gradually disappears as ϵ decreases (figure 7).

The rms velocity profiles at peak and valley are shown in figure 9. The amplitudes of TS waves and fundamental mode have been chosen to model the experimental conditions at station B (Klebanoff et al. 1962, fig. 5b), although the measurements were made at different frequency and Reynolds number. Figure 9 shows a striking similarity of the theoretical result with the experimental data. In fact, the velocity distributions in figures 6, 7, 8 are typical for the specific amplitude; frequency and

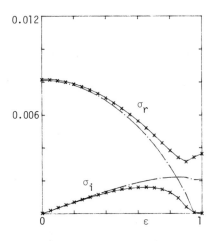

Figure 5. Temporal growth rate σ_r and frequency shift σ_i for detuned modes, obtained with 2 (-- --), 3 (x), and 4 (———) Fourier modes. Subharmonic and fundamental disturbances correspond to $\epsilon = 0$ and $\epsilon = 1$, respectively.

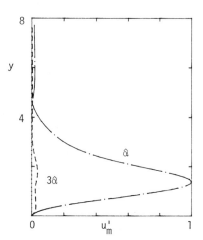

Figure 6. Normalized streamwise velocity components with wavenumbers $\hat{\alpha}$ and $3\hat{\alpha}$ of the temporally growing subharmonic mode ($\epsilon = 0$) at $R = 606$, $\hat{\alpha} = 0.1017$, $A = 0.01$, $\beta = 0.2$.

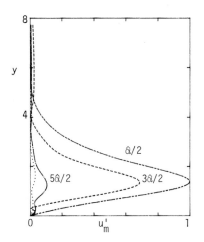

Figure 7. Normalized streamwise velocity components with wavenumbers $\hat{\alpha}/2$, $3\hat{\alpha}/2$, $5\hat{\alpha}/2$, and $7\hat{\alpha}/2$ of the temporally growing detuned mode ($\epsilon = 0.5$) at $R = 606$, $\hat{\alpha} = 0.1017$, $A = 0.01$, $\beta = 0.2$.

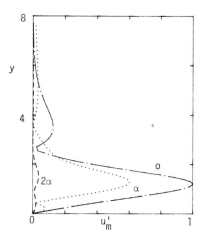

Figure 8. Normalized streamwise velocity components with wavenumbers 0, $2\hat{\alpha}$, and $4\hat{\alpha}$ of the temporally growing fundamental mode ($\epsilon = 1$) at $R = 606$, $\hat{\alpha} = 0.1017$, $A = 0.01$, $\beta = 0.2$.

Reynolds number changes have little effect on their shapes.

Measurements of the subharmonic disturbance velocity have been compared with results of the lowest approximation for temporally (Herbert 1984a, fig. 14) and spatially (fig. 2) growing

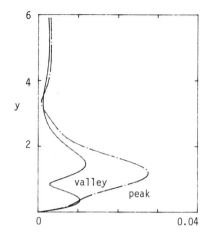
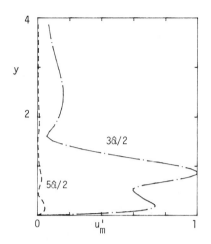

Figure 9. Streamwise rms fluctuation at peak and valley for temporally growing fundamental mode at $R = 606$, $\alpha = 0.2034$, $A = 0.01$, $\beta = 0.2$.

Figure 10. Normalized streamwise velocity components with $3\hat{\alpha}/2$ and $5\hat{\alpha}/2$ of the temporally growing subharmonic mode at $R = 606$, $\hat{\alpha} = 0.1017$, $A = 0.01$, $\beta = 0.2$.

disturbances. Here, we show in figure 10 the distribution of the components with wavenumbers $3\hat{\alpha}$ and $5\hat{\alpha}$. These components are comparable with those of frequencies $3f/2$ and $5f/2$ in the experiments. Although the maximum amplitudes are only 3.7% and 0.2% of the components with $\hat{\alpha}$ (2.5% and 0.4% in the experiments), figure 10 shows essential agreement with the measured distributions (Kachanov & Levchenko 1984, fig. 24). Minor differences may be due to experimental error at very low amplitudes, especially near the wall, as well as due to the approximations incorporated into the theory.

4. Conclusions

Judging from all the results obtained to date, the Floquet analysis of secondary instability appears as a very powerful concept. Although this analysis rests on linearization in the amplitude of the secondary disturbances, it seems superior to the analysis of weakly nonlinear interactions. In fact, Floquet analysis provides rational guidance on the modes interacting in the limit $A \to 0$. Thus, it may help in forming new, complete, and relevant models of weakly nonlinear interactions.

The two distinct classes of subharmonic and fundamental modes with temporal growth can now be considered special cases of a generalized class of secondary disturbances. These disturbances may grow in time or space. To within acceptable error bounds, temporal and spatial growth rates are related by a simple transformation. Inevitable truncations in representing the disturbance structure streamwise and normal to the wall have effects negligible for practical purposes.

Most notable, however, is the consistency of the theoretical result with the scarce experimental data base on secondary instability. While the quantitative comparison with the classical results of Klebanoff, Tidstrom & Sargent (1962) on peak-valley splitting suffers somewhat from irreproducible conditions in these early experiments, the comparison with the more recent measurements of subharmonic and detuned disturbances are very encouraging. The interpretation of three-dimensional phenomena in the early stages of transition as a parametric instability in a streamwise periodic flow appears as a very sound concept.

ACKNOWLEDGEMENT

This work is supported by the Office of Naval Research under Contract N00014-84-K-0093 and by the Air Force Office of Scientific Research under Contract F49620-84-K-0002.

REFERENCES

Benney, D. J. & Lin, C. C. 1960 *On the secondary motion induced by oscillations in a shear flow*, Phys. Fluids 3, 656-657.

Bertolotti, F. P. 1985 *Temporal and spatial growth of subharmonic disturbances in Falkner-Skan flows*, M. S. Thesis, Virginia Polytechnic Institute, Blacksburg, Virginia.

Blackwelder, R. F. 1979 *Boundary-layer transition*, Phys. Fluids 22, 583-584.

Bridges, T. J. & Morris, P. O. 1984 *Differential eigenvalue problems in which the parameter appears nonlinearly*, J. Comp. Physics 55, 437-460.

Craik, A. D. D. 1971 *Nonlinear resonant instability in boundary layers*, J. Fluid Mech. 50, 393-413.

Davis, S. H. 1976 *The stability of time-periodic flows*, Ann. Rev. Fluid Mech. 8, 57-74.

Gaster, M. 1962 *A note on the relation between temporally-increasing and spatially-increasing disturbances in hydrodynamic stability*, J. Fluid Mech. 14, 222-224.

Herbert, Th. 1977 *Finite-amplitude stability of plane parallel flows*, AGARD CP-224, 3/1-10.

Herbert, Th. & Morkovin, M. V. 1980 *Dialogue on bridging some gaps in stability and transition research*, in: *Laminar-Turbulent Transition* (eds. R. Eppler & H. Fasel), 47-72, Springer-Verlag.

Herbert, Th. 1981a *Stability of plane Poiseuille flow - theory and experiment*, VPI-E-81-35, Blacksburg, VA. Published in: Fluid Dyn. Trans. 11, 77-126 (1983).

Herbert, Th. 1981b *A secondary instability mechanism in plane Poiseuille flow*, Bull. Amer. Phys. Soc 26, 1257.

Herbert, Th. 1983a *Secondary instability of plane channel flow to subharmonic three-dimensional disturbances*, Phys. Fluids. 26, 871-874.

Herbert, Th. 1983b *Subharmonic three-dimensional disturbances in unstable shear flows*, AIAA Paper No. 83-1759.

Herbert, Th. 1984a *Analysis of the subharmonic route to transition in boundary layers*, AIAA Paper No. 84-0009.

Herbert, Th. 1984b *Modes of secondary instability in plane Poiseuille flow*, in: *Turbulence and Chaotic Phenomena in Fluids* (ed T. Tatsumi), 53-58, North-Holland.

Herbert, Th. 1985a *Three-dimensional phenomena in the transitional flat-plate boundary layer*, AIAA Paper No. 85-0489.

Herbert, Th. 1985b *Secondary instability of plane shear flows - theory and applications*, in: *Laminar-Turbulent Transition* (ed. V. V. Kozlov), 9-20, Springer-Verlag.

Kachanov, Yu. S. & Levchenko, V. Ya 1982 *Resonant interactions of disturbances in transition to turbulence in a boundary layer* (in Russian), Preprint No. 10-82, I.T.A.M., USSR Academy of Sciences, Novosibirsk.

Kachanov, Yu. S. & Levchenko, V. Ya. 1984 *The resonant interaction of disturbances at laminar-turbulent transition in a boundary layer*, J. Fluid Mech. 138, 209-247.

Kelly, R. E. 1967 *On the stability of an inviscid shear layer which is periodic in space and time*, J. Fluid Mech. 27, 657-689.

Kerczek, C. von 1985 *Stability characteristics of some oscillatory flows - Poiseuille, Ekman and Films*, in these Proceedings.

Klebanoff, P. S., Tidstrom, K. D. & Sargent, L. M. 1962 *The three-dimensional nature of boundary-layer instability*, J. Fluid Mech. 12, 1-34.

Kleiser, L. 1982 *Numerische Simulationen zum laminar-turbulenten Umschlagsprozess der ebenen Poiseuille-Strömung*, Dissertation, Universität Karlsruhe. See also: *Spectral simulations of laminar-turbulent transition in plane Poiseuille flow and comparison with experiments*, Springer Lecture Notes in Physics 170, 280-287 (1982).

Kozlov, V. V. & Ramasanov, M. P. 1983 *Development of finite-amplitude disturbance in Poiseuille flow*, Izv. Akad. Nauk SSSR, Mekh. Zhidk. i Gaza, 43-47.

Maseev, L. M. 1968 *Occurrence of three-dimensional perturbations in a boundary layer*, Fluid Dyn. 3, 23-24.

Nagata, M. & Busse, F. H. 1983 *Three-dimensional tertiary motions in a plane shear layer*, J. Fluid Mech. 135, 1-28.

Nayfeh, A. H. 1985 *Three-dimensional spatial secondary instability in boundary-layer flows*, AIAA Paper No. 85-1697.

Nayfeh, A. H. & Bozatli, A. N. 1979 *Secondary instability in boundary-layer flows*, Phys. Fluids 22, 805-813.

Nishioka, M., Iida, S. & Ichikawa, J. 1975 *An experimental investigation on the stability of plane Poiseuille flow*, J. Fluid Mech. 72, 731-751.

Nishioka, M. & Asai, M. 1983 *Evolution of Tollmien-Schlichting waves into wall turbulence*, in: *Turbulence and Chaotic Phenomena in Fluids* (ed. T. Tatsumi), 87-92, North-Holland.

Nishioka, M. & Asai, M. 1985 *3-D wave disturbances in plane Poiseuille flow*, in: *Laminar-Turbulent Transition* (ed. V. V. Kozlov), 173-182, Springer-Verlag.

Orszag, S. A. & Kells, L. C. 1980 *Transition to turbulence in plane Poiseuille flow and plane Couette flow*, J. Fluid Mech. 96, 159-206.

Orszag, S. A. & Patera, A. T. 1980 *Subcritical transition to turbulence in plane channel flows*, Phys. Rev. Lett. 45, 989-993.

Orszag, S. A. & Patera, A. T. 1981 *Subcritical transition to turbulence in plane shear flows*, in: *Transition and Turbulence* (ed. R. E. Meyer), 127-146, Academic Press.

Orszag, S. A. & Patera, A. T. 1983 *Secondary instability of wall-bounded shear flows*, J. Fluid Mech. 128, 347-385.

Pierrehumbert, R. T. & Widnall, S. E. 1982 *The two- and three-dimensional instabilities of a spatially periodic shear layer*, J. Fluid Mech. 114, 59-82.

Raetz, G. S. 1959 *A new theory of the cause of transition in fluid flows*, NORAIR Report NOR-59-383, Hawthorne, CA.

Raetz, G. S. 1964 *Current status of resonance theory of transition*, NORAIR Report NOR-64-111, Hawthorne, CA.

Saric, W. S. & Thomas, A. S. W. 1984 *Experiments on the subharmonic route to turbulence in boundary layers*, in: *Turbulence and Chaotic Phenomena in Fluids* (ed. T. Tatsumi), 117-122, North-Holland.

Stuart, J. T. 1958 *On the non-linear mechanics of hydrodynamic stability*, J. Fluid Mech. 4, 1-21.

The Generation of Tollmien–Schlichting Waves by Long Wavelength Free Stream Disturbances

M.E. GOLDSTEIN

ABSTRACT

The paper is primarily concerned with explaining how very long wavelength free stream disturbances are able to generate very short wavelength Tollmien-Schlichting waves in laminar boundary layers. We consider the case where the disturbances are of small amplitude and have harmonic time dependence and where the Mach number is effectively zero. It is shown that the free stream wavelength reduction occurs as a result of nonparallel flow effects which can arise from: (1) the slow viscous growth of the boundary layer, and (2) small but abrupt changes in surface geometry that produce only very weak static pressure variations. Analyses of these two mechanisms are carried out by linearizing the unsteady motion about an appropriate steady flow and asymptotically expanding the result in inverse powers of an appropriate Reynolds number. The analyses are compared with each other and with available experimental data, and
they are used to explain the physics of the two mechanisms.

1. Introduction

We now know that transition to turbulence in boundary layers is often the result of a chain of events that begins with the excitation of Tollmien-Schlichting waves in the laminar portion of the boundary layer by very weak free stream disturbances. A number of investigators have attempted to study this phenomena--referred to as the receptivity problem by Morkovin (1969)--by imposing controlled small amplitude, harmonic disturbances (i.e., disturbances of a single frequency) on their flows which then generate spatially amplifying Tollmien-Schlichting waves that are intimately connected with the more familiar temporally growing Tollmien-Schlichting instability waves and are, in fact, themselves a kind of spatial instability (Briggs, 1964).

Even the original experiment of Schubauer and Skramstad (1943), which first demonstrated the existence of Tollmien-Schlichting waves and, to my knowledge was conducted soley for that purpose, was a kind of receptivity experiment that generated spatially amplifying waves of this type. The more recent experiments directed at understanding receptivity phenomenon can usually be grouped into two categories (see figure 1): (1) those in which the disturbance is introduced locally within the boundary layer (or perhaps just outside it), which includes the Schubauer and Skramstad experiment, and (2) those in which the disturbance is introduced at some distance from the boundary layer and, therefore, generates an intermediate free stream disturbance (usually of the acoustic or vortical type), which in turn generates the Tollmien-Schlichting wave. The latter experiments are generally much more difficult to control than the former and have often led to conflicting interpretations of the results (Nishioka and Morkovin, 1985).

This paper is primarily concerned with this latter group of experiments, but it is appropriate to briefly consider the former, since it raises an issue that is relevant to the distant source problem. For this purpose, we can treat the boundary layer as a locally parallel flow, and the unsteady solution, outside the source region (see figure 2), can be treated as a superposition of so-called "normal modes" which, in general, includes both a discrete and continuous spectra. Since the boundary layer is an active system, i.e., it contains an unlimited source of energy, the former involves both amplifying and evanescent waves. In a passive system, such as a wave guide, which involves only propagating and evanescent waves, the solution can be made unique by requiring that it contain only outward propagating waves or waves that decay away from the source (Briggs, 1964). This amounts to determining the side of the source on which any given mode will appear in the solution. But it is not always that easy to distinguish between the spatially amplifying and evanescent waves that occur in active systems and, therefore, to determine on which side of the source these waves will lie.

Gaster (1965) resolved this issue by finding the long time limit of the solution to an initial value problem with the source turned off for the time--$t < 0$, i.e., by imposing a causality condition on the solution. This insured that each mode would propagate outward from its source and, therefore, unambiguously determined the side of the source on which the mode appeared. Fortunately, it turns out that it is possible to obtain the steady state limit of the solution to the initial value problem, without first obtaining the complete solution to this problem, by using a procedure such as the one developed by Briggs (1964).

With that as background, we turn to the case where the disturbance is introduced at some distance from the boundary layer. The instability waves can only be generated if the streamwise component of the free stream disturbance wavelength can be made to match the Tollmien-Schlichting wavelength. But with the exception of Mack's (1975) pioneering supersonic work, most receptivity experiments were carried out at very low Mach numbers where the free stream disturbance wavelengths are typically quite large compared to the Tollmien-Schlichting wavelength. In fact, acoustic disturbances, which are especially effective Tollmien-Schlichting wave generators (primarily because of their large spanwise coherence), have effectively infinite wavelengths in the zero Mach number limit to which we now restrict the discussion.

It is appropriate to approach this problem analytically by seeking asymptotic expansions in inverse powers of some appropriate Reynolds number, since the boundary layer on which the Tollmien-Schlichting waves ride is only defined in the infinite Reynolds number limit. The discussion will also be restricted to this case.

Figure 3 is a schematic representation of the stability diagram for Tollmien-Schlichting waves on a Blasius-like boundary layer. Here, ω denotes the frequency, ν the kinematic viscosity, U_∞ the free stream velocity, and Re_x the length Reynolds number. A spatially "amplifying" wave of constant frequency moves downstream along a horizontal line, first decaying until it encounters the lower branch of the neutral stability curve, and then, if nonlinear effects do not first intervene, growing until it encounters the upper branch--at which point it again begins to decay.

An appropriate expansion Reynolds number for the asymptotic solution is the critical Reynolds number Re_c, at which the flow becomes unstable at the imposed frequency ω. Since the classical high Reynolds number-long wavelength solutions of Heisenberg (1924), Lin (1945, 1946), Tollmien (1929), etc. show that $F = \omega\nu/U_\infty^2 \propto Re_c^{-3/4}$ (See Reid, 1965.), we can also carry out the expansion in terms of F. For reasons that will become clear subsequently, it is convenient to take the actual expansion parameter, say ϵ, to be $F^{1/6}$ or, equivalently, $Re_c^{-1/8}$ (since an O(1) constant is irrelevant in defining an expansion parameter). Then, at all streamwise distances greater than or equal to a convective wavelength U_∞/ω, the local Reynolds number Re_x (which is equal to the streamwise coordinate x normalized by $U_\infty/\omega\epsilon^6$) will always be large.

The asymptotic analysis frequently provides a "rational" method for lumping the receptivity effects into a coupling coefficient that multiplies the Tollmien-Schlichting wave. This tends to make the result of more universal applicability, since the Tollmien-Schlichting wave usually depends on such things as externally imposed pressure gradients that vary on a scale that is large compared to the Tollmien-Schlichting wavelength and do not have an important influence on the receptivity phenomenon, while the coupling coefficient usually depends on local conditions on the scale of the Tollmien-Schlichting wavelength, which do not have an important effect on the subsequent growth of the Tollmien-Schlichting wave.

It is appropriate to require that the spatial scale of the free stream disturbance be $O(U_\infty/\omega)$ or larger in the limit $\epsilon \to 0$. We suppose that it is larger in order to avoid some additional complexities. Then (Goldstein, 1983a) the ratio of the Tollmien-Schlichting wavelength to the free stream disturbance wavelength will be less than $O(\epsilon)$ in this limit, i.e., the free stream wavelength will be very long compared to the Tollmien-Schlichting wavelength. In 1976, Reshotko indicated that the main unresolved issue in the receptivity problem is to explain how the very long wavelength of the free stream disturbance gets reduced down to the Tollmien wavelength. It is now clear that this can usually be attributed to nonparallel flow effects, which can, in turn, be attributed to: (A) slow viscous boundary layer growth and (B) small but abrupt changes in static pressure which can in turn result from small changes in static pressure or can occur naturally at the minimum skin friction point for boundary layers on the verge of separation.

2. Tollmien-Schlichting wave generation by viscous boundary layer growth

The generic problem for the first category of nonparallel flow effects is that of an infinitely thin flat plate with the unsteady flow produced by a small amplitude fluctuation of an otherwise uniform velocity stream. The deviation from parallesion can be treated as a small perturbation at large distances from the leading edge, and the Tollmien-Schlichting waves appear to be generated by a source term which is the product of a Stokes shear wave and the streamwise gradient U_x of the mean flow velocity in this region. Since the dominant contribution to the resulting coupling coefficient seems to come from the singularity in U_x at the leading edge, it is appropriate to begin with the region near the leading edge, and the smallest region that one needs to consider is that in which $X \equiv \omega x/U_\infty = 0\,(1)$.

The unsteady flow is then governed by the linearized unsteady boundary layer equation, which implies that the nonparallel mean flow effects are no longer small but influence the unsteady solution to lowest order of approximation. The cross stream pressure fluctuations can, however, now be neglected (Goldstein, 1983a; Lam and Rott, 1960; Ackerberg and Phillips, 1972).

The solution to this problem must, in general, be obtained numerically, but it develops a two-layer structure asymptotically far downstream, and consists of a Stokes shear wave solution plus a linear combination of "asymptotic eigenfunctions" (Goldstein, 1983a; Lam and Rott, 1960; Ackerberg and Phillips, 1972). The Stokes solution, which is the solution for a doubly infinite wall oncillating parallel to itself in a medium at rest, is independent of the upstream conditions. That it arises as a solution to the present problem should come as no surprise, since the unsteady boundary layer equation is invariant under a Gallean transform even into an accelerating reference frame. The moving stream and moving wall problems should then be equivalent at sufficiently large distances from the leading edge. The Stokes wave is independent of the streamwise coordinate and adjusts to the free stream velocity fluctuation across a Stokes layer of thickness

$$\delta_{st.} = \sqrt{\nu/\omega} \qquad [1]$$

which coincides asymptotically with the thickness of the "inner region" of the two-layer structure.

The asymptotic eigenfunctions account for the upstream conditions in the unsteady boundary layer and are linearly independent of each other and the Stokes shear wave. They decay exponentially fast in the downstream direction, and, in fact, they behave like

$$\exp.(\lambda x^{3/2} - i\omega t)$$

where t is the time and λ is a complex constant with $\mathrm{Re}\,\lambda > 0$. Hence their "wave lengths" decrease like $x^{-1/2}$ as they propagate downstream (Ackerberg and Phillips, 1972; Lam and Rott, 1960). This occurs because the asymptotic eigensolutions, being homogeneous solutions of the unsteady boundary layer equations, can produce no pressure fluctuations and must, therefore, behave somewhat like convected disturbances propagating into a region of decreasing streamwise velocity. Since a convected disturbance is one with zero convective derivative $\partial/\partial t + U \partial/\partial x$, its phase must be $\omega t - \int dx/U_t$ and its wavelength must, therefore, decrease in the streamwise direction if the typical mean velocity U_t does. Near the wall U_t is proportional to the similarity variable $y/\sqrt{x}$ for a Blasius boundary layer so that the phase $-\omega t$ must behave like $x^{3/2}$. Thus, the <u>wavelengths of the asymptotic eigensolutions decrease because they must penetrate into a region of decreasing mean velocity without producing pressure fluctuations</u>.

Since the wave lengths of the asymptotic eigensolutions decrease with increasing x while the mean boundary layer thickness increases, cross stream pressure fluctuations, which are neglected in the unsteady boundary layer approximation, must eventually become important, and the asymptotic eigensolutions, which are based on this approximation, must then become invalid. A new solution that applies further downstream can be obtained (Goldstein 1983 a and b) by introducing the scaled, stream-wise coordinate

$$x_1 = \epsilon^2 X \qquad [2]$$

(where $X = \omega x/U_\infty$) and, taking the limit as $\epsilon \to o$ with this coordinate held fixed. The result is essentially the classical large Reynolds number, long wavelength, asymptotic approximations to the Tollmien-Schlichting wave solutions of the Orr-Summerfeld equation--appropriately corrected for the slow growth in boundary layer thickness.

These solutions match, in the matched asymptotic expansion sense, onto the "asymptotic eigensolutions" of the unsteady boundary layer equation and are, therefore, the natural continuations of these solutions into the downstream region (Goldstein, 1983 a and b). There is one Tollmien-Schlichting wave for each "asymptotic eigensolution", but only the lowest order asymptotic eigensolution matches onto a Tollmien-Schlichting wave that ultimately exhibits spatial growth.

The remaining portion of the asymptotic, unsteady boundary layer solution, that is the Stokes-type solution, remains uniformly valid in the downstream region and is, therefore, completely decoupled from the Tollmien-Schlichting waves.

The classical Tollmien-Schlichting wave solutions are basically inviscid except in a thin region near the wall and in a critical layer about the point where the inviscid equation (i.e., the Rayleigh equation) becomes singular. It is well known that the critical and wall layers coincide near the lower branch of the neutral stability curve (Reid, 1965), which lies in the outer region where $x_1 \equiv \epsilon^2 X = O(1)$ (see figure 4). But, there are two inviscid regions outside this wall layer--a main inviscid region where the unsteady velocity is vortical and pressure effects are relatively unimportant and an outer region where the unsteady flow is potential and, consequently, pressure dominated. The relative thicknesses of these regions are of order $\epsilon^5 \ell_c$, $\epsilon^4 \ell_c$, and $\epsilon^3 \ell_c$. This turns out to be (as was pointed out by Smith, 1979) the same "triple deck" structure that was used by Stewartson (1969) and Messiter (1970) to deal with such steady boundary layer problems as the flow over the trailing edge of an infinitely thin flat plate. It gradually merges into a double-layer structure of the unsteady boundary layer region as $x_1 \to 0$ and the Tollmien-Schlichting waves turn into the "asymptotic eigensolutions."

The analysis shows that the coupling coefficient is $C\epsilon^\tau$ where $\tau = 1.384...$, and C is an $O(1)$ constant that has to be computed from the numerical solution of the unsteady boundary layer equation for each type of imposed free stream disturbance (Goldstein, Sockol, Sanz, 1983).

$C = -0.45 + 0.885$ i for a uniform pulsation of the stream and so that its absolute value is close to unity. But for a transverse fluctuation, such as would be produced by an oblique acoustic wave, $C = -7.64 + 2.4$ i, when normalized by the free stream velocity, and its absolute value is therefore an order of magnitude larger than for a uniform pulsation. Thus the coupling coefficient is about 0.1 for a frequency parameter of 0.57×10^{-4} in a uniformly pulsating stream, but it has a magnitude of about unity for a cross flow or transverse disturbance with a somewhat larger frequency parameter of 2×10^{-4}. Needless to say, most of this increase is due to the change in C.

At very small values of F, the amplitude of the Tollmien-Schlichting wave can be very small at the neutral stability curve (relative to the amplitude of the free stream disturbance) because of the large amount of decay that it can undergo before reaching that point--though this will be more than compensated for by the large amount of growth that it can undergo downstream of this curve. In fact, the net logarithmic amplitude reduction ratio is $O(\epsilon^{-3})$ upstream of the lower branch (Goldstein, 1983a) while it can be shown that the net logarithmic amplitude growth ratio between the upper and lower branches is a much larger $O(\varepsilon^{-3}\ln\varepsilon^{-1})$ (Goldstein, 1985; Reid, 1965; Bodonyi and Smith, 1981). The amplitude reduction ratio is about 4×10^{-3} for $F = 0.57\times10^{-4}$. It is (Goldstein, 1985) a much more moderate 0.7×10^{-1} for $F = 2\times10^{-4}$, but the amount of subsequent growth that it can undergo downstream of the neutral curve is then much less than at the lower F.

One should not, however, conclude from this that the mechanism is unimportant, because even very small adverse pressure gradients that can significantly reduce the amount of amplitude reduction that occurs upstream of the lower branch of the neutral stability curve at small F and can open up the "loop" to allow unlimited growth downstream of the lower branch at larger F. For large adverse pressure gradients, such as would occur on a slender body near the critical angle of attack, the unsteady boundary layer region would be replaced by the "marginal separation" region (Stewartson et al., 1982) which moves the generation reqion downstream and reduces the initial wavelength (Goldstein, 1985b). Moreover, the amplitude reduction of three-dimensional waves may be considerably less than for the two-dimensional waves--though the subsequent growth will then be somewhat smaller upstream of the maximum growth line.

A pertinent experiment was carried out by Kachanov, Kozlov, and Levchenko (1978), who generated their unsteady flow with a vibrating ribbon somewhat below and at an unspecified distance upstream of the leading edge of their plate. They observed what appeared to be decaying Tollmien-Schlichting waves downstream of their leading edge region (i.e., where $\omega x/U_\infty = O(1)$--corresponding roughly to 10 mm), but their frequency parameter was rather large, about 3-1/2 times 10^{-4} (probably intentionally chosen to be so in order to minimize the amplitude reduction). Consistent with the predictions of the present theory, they found that it was nearly impossible to generate Tollmien-Schlichting waves unless their free stream disturbance had a significant cross stream velocity component.

3. Tollmien-Schlichting wave generation by changes in wall curvature

The generic problem for the second category of nonparallel flow effects (category B) is that of a relatively thin body with a small local region of large surface curvature (which we refer to as the "interaction region") at some distance, say ℓ downstream from the leading edge (see figure 5). We assume that the ratio ℓ_c/ℓ

(where ℓ_c is the distance to the neutral stability curve at the imposed frequency) remains finite as $\varepsilon \to 0$, so that the ratio of critical to interactive region Reynolds number remains $O(1)$, and that the unsteady motion is again due to a small amplitude harmonic fluctuation of an otherwise constant velocity stream.

We expect that the maximum coupling will occur when the geometry variations in the interaction region take place on the Tollmien-Schlichting wavelength scale, which turns out to be $O(\varepsilon^3 \ell)$ (Goldstein, 1983a, 1985). This can take place without large-scale flow separation if the wall is displaced by an amount $O(\varepsilon^5 \ell)$ over this streamwise distance.

This means that the interaction region geometry can be described by an equation of the form

$$y/\ell = \varepsilon^5 G(x/\ell \varepsilon^3) \qquad [3]$$

$$G(X) \to \begin{cases} \alpha_o X^r; & X \to -\infty, \; 1 < r < 2 \\ 0; & X \to +\infty \end{cases}$$

where y and x are dimensional transverse and streamwise coordinates and α_o is an $O(1)$ constant. There is no slope change but a large change in surface curvature across the interaction region when r is in the range $1 < r < 2$. When $r = 1$ there is a net slope change with the turning angle α related to α_o by

$$\alpha = \varepsilon^2 \alpha_o, \qquad [4]$$

which shows that it is $O(\varepsilon^2)$. The <u>steady</u> flow in this latter configuration, which exhibits the triple deck structure outlined here in figure 6, was analyzed by Stewartson (1970, 1971). He showed that it produces only small, i.e., $O(\varepsilon^2)$, static pressure variations.

The present receptivity problem can, therefore, be analyzed by treating the corresponding unsteady flow as a small, i.e., linear, perturbation of the generalization of Stewartson's solution described above. Needless to say, it will exhibit the same triple deck structure as this solution.

Upstream of the interaction region where the mean flow varies only slowly, i.e., on the scale of the streamwise distance from the leading edge, the unsteady motion is described by the Stokes shear wave solution to lowest approximation (Goldstein, 1985), and is therefore, as we have already indicated, effectively independent of the streamwise coordinate. We also noted that the Stokes layer thickness is $\delta_{st} = \nu/\omega = \nu/U_\infty \varepsilon^3 = \ell_c/\text{Re}_c \varepsilon^3$, which, in view of the asymptotic relation between $F = \varepsilon^6$ and Re_c, can also be written as an order one constant times $\ell_c \varepsilon^5$ and is therefore of the order of the lower deck thickness (since we are assuming that ℓ_c/ℓ is $O(1)$).

Since the wall displacement is also $O(\ell\varepsilon^5)$, it is clear that the latter will produce an $O(1)$ change in the Stokes shear wave as it penetrates downstream into the "<u>lower deck</u>" of the interaction region, i.e., that

$$\Delta u_{st} \equiv u_{st} - u_{st\infty} = O(u_\infty). \qquad [5]$$

where u_∞ is the free stream velocity fluctuation (see figure 7). It is not, however, obvious that such a small wall displacement, relative to the boundary layer thickness, will produce an $O(1)$ change in the velocity fluctuation all the way across the steady boundary layer, which would certainly not be the case if the mean velocity $U(y/\delta)$ were uniform at zero.

But, since the interaction region length $L = \varepsilon^3 \ell$ is long relative to the boundary layer thickness $\delta = \varepsilon^4 \ell$, the velocity perturbation must be a simple streamline displacement, which is the same for every streamline, i.e., independent of the transverse coordinate y. This means that the velocity perturbation is of the form

$$\Delta u \simeq U'(y/\delta) \, a(x/L) \qquad [6]$$

$$v \approx -\varepsilon U(y/\delta) \, a'(x/L) \qquad [7]$$

in the main boundary layer, where U is the mean boundary layer velocity profile, the primes denote differentiation with respect to the indicated arguments, and $a\delta$ denotes the dimensional particle displacement or streamline displacement relative to the mean streamline. In fact, the second equation, being the convection derivative of this quantity, is therefore just the usual relation connecting the transverse velocity and the particle displacement. The factor of ε multiplying v comes about because of the disparity between the streamwise and transverse length scales, i.e., because $\delta/L = O(\varepsilon)$. The first equation is a consequence of the fact that the very small wall protuberance produces only very small (i.e., higher order) pressure perturbations, so that the net Lagrangian velocity perturbation (i.e., the velocity perturbation seen by an observer moving with the fluid) is zero to the lowest order. In other words, the Eularian velocity perturbation Δu is just equal to minus the velocity change $U'a$ due to the particle's moving into a region of different mean velocity. Notice that the latter effect would be absent if the mean flow were uniform, and the streamwise velocity perturbation would then be of higher order.

Since velocity perturbation Δu_{st} in the lower deck must match Δu in the main boundary layer, it is clear that $U'(o)a$ and, consequently, Δu must be $O(u_\infty)$, which is to say that the streamwise velocity fluctuation is $O(1)$ all across the main boundary layer. Then, since Δu depends on the streamwise coordinate on the scale $\varepsilon^3 \ell$ of the Tollmien-Schlichting wavelength, it is natural that the Tollmien-Schlichting wave should appear as part of the solution in the region downstream of the interaction zone, and that the coupling coefficient should be $O(1)$.

Thus, very small wall displacements, lying well within the depths of the steady boundary layer and producing only small, i.e., $O(\varepsilon^2)$, static pressure variations, are able to scatter very long wavelength free stream disturbances into very short wavelength Tollmien-Schlichting waves in a very efficient manner.

The analysis (Goldstein, 1985) shows the coupling coefficient based on the free stream velocity fluctuation at the outer edge of the interaction region is given (with some approximation) by

$$\text{Coupling Coefficient} = \lambda \bar{G}(\kappa_o) \Lambda (F_o/\lambda)/F_o^{4/3} \qquad [8]$$

where λ is the scaled skin friction just upstream of the interaction region (= 0.3321 for the Blasius boundary layer),

$$\bar{G} \equiv \frac{1}{\sqrt{2\pi}} \int_{-\infty}^{\infty} e^{-i\kappa_o X} G(X) \, dX \qquad [9]$$

denotes the Fourier transform of the scaled wall-shape function G defined by equation (3), κ_o denotes the scaled Tollmien-Schlichting wavenumber (i.e., $\kappa_o = (\ell\varepsilon^3) \times$ Tollmien-Schlichting wavenumber) evaluated at the position of the interaction region, and

$$F_o = \varepsilon^6 \, Re_\ell^{3/4} = F \, Re_\ell^{3/4} \propto (\ell/\ell_c)^{3/4} \qquad [10]$$

is the scaled frequency parameter at the location of the interaction region. Λ is a complex function of F_o/λ, which is independent of the wall geometry and is plotted in figure 8.

Kegelman and Mueller (1984) carried out a relevant experiment on an ogive noised axisymmetric body which had a sudden slope change at the juncture between the noise and cylindrical body section (which corresponds to a scaled turning angle α/ε^2 of about 2). The measured static pressure distribution, shown here as figure 9, is very steep in this region, and the photograph, provided by Professor T.J. Mueller, clearly shows the formation of a Tollmien-Schlichting wave downstream of this region.

An even more relevant experiment was carried out by Leehey and Shapiro (1979), who produced their unsteady flow by placing a large acoustic speaker upstream of the leading edge of their very highly damped flat plate, which consisted of a 6-1 semi-ellipse nose section transitioning abruptly to a long flat plate section (see figure 10). The speaker produced a nearly plane acoustic wave upstream at their leading edge, which acted pretty much like a uniform pulsation of the stream, since the Mach number in their experiment was quite low. They found that: (1) the Tollmien-Schlichting wave amplitude increased linearly with the amplitude of the imposed disturbance--indicating that the coupling was linear in their case (as assumed in the theory), and (2) the maximum amplitude of the Tollmien-Schlichting wave at the theoretical location of the lower branch of the neutral stability curve was about equal to their free stream disturbance amplitude, indicating that the coupling coefficient was greater than or equal to unity in their case.

Their frequency parameter was significantly smaller than Kachanov, Kozlov, and Levchenko's (1978)--about 0.57×10^{-4} (Shapiro, 1977)--so that the leading edge generated instability wave would have to have been exceedingly small at the lower branch of the neutral stability curve. Based on Goldstein's (1983a) flat plate analysis, we estimate it to be between 10^{-3} to about 10^{-4} of the free stream velocity (Goldstein, 1985). In fact, we estimate the damping ratio in their experiment, i.e., the ratio of the Tollmien-Schlichting wave amplitude at the lower branch to its amplitude at the leading edge, to be about 4×10^{-3} for a Blasius boundary layer.

Fortunately, Leehey and Shapiro (1979) measured the static pressure variation along the surface of their plate. Their data points are the open circles shown in figure 10. The pressure coefficient is multiplied by a factor of 100 so that the pressure levels really are quite low. The solid curve through the data points is essentially the one drawn by Shapiro but is redrawn in the vicinity of the juncture of the nose and flat plate regions to agree with Stewartson's (1970, 1971) triple deck calculation of the static pressure variation produced by a wall with a sudden slope change with a scaled turning-angle $\alpha_0 = \alpha/\epsilon^2$ of unity. The calculation blends in very nicely with Shapiro's measurements.

Using these results, Goldstein (1985) showed that the absolute value of the coupling coefficient was about 1.07 times the scaled turning-angle, for the conditions of Shapiro's experiment. However, the interaction region was somewhat upstream of the theoretical location of the lower branch of the neutral stability curve in Shapiro's (1977) experiment (The displacement thickness Reynolds number was 363 at the former location and 950 at the latter.), so that the Tollmien-Schlichting wave would have to undergo some amplitude reduction. But that reduction is now much less than that occurring between the leading edge and the lower branch.

The Tollmien-Schlichting wave depends on the streamwise coordinate primarily through a function of the form

$$\exp. \; i\epsilon^{-3} \int_{x_0}^{x_1} \kappa \; dx_1$$

where x_1 is the scaled streamwise coordinate $\epsilon^2 X$ which, as indicated above, is $O(1)$ in the vicinity of the lower branch, and κ is the scaled complex wavenumber of the Tollmien-Schlichting wave introduced above. Its real part is proportional to the reciprocal wavelength, and its imaginary part is a measure of the amount of damping or growth that the Tollmien-Schlichting wave undergoes. The ϵ^{-3} factor accounts for the Reynolds number dependence.

Figure 11 is a plot of the imaginary part of κ as a function of the scaled streamwise distance x_1 from the leading edge, as calculated from the classical Tollmien-Schlichting wave solution (Goldstein 1983 a and b). The dashed curve is the corresponding result for the lowest order asymptotic eigensolution of the unsteady boundary layer equation. It shows that the Tollmien-Schlichting wavenumber approaches this result in the vicinity of the leading edge, as was indicated in section 2.

The location of the juncture between the nose and flat plate regions, i.e., the interaction region, corresponding to the conditions of Shapiro's experiment is also indicated on this plot. Practically all of the damping occurs upstream of this region. Notice that the damping is measured by the area underneath the curve.

Goldstein (1985) showed that the net amplitude reduction between the interaction region and neutral point was about 0.1 in this case, which would imply that the predicted neutrally stable Tollmien-Schlichting wave amplitude is still too small to agree with Shapiro's observation (if we infer from the comparison of Stewartson's 1970, 1971 calculation with Shapiro's measurements that the coupling coefficient should be about unity). But figure 11 also shows that there was a weak adverse pressure gradient between the interaction region and neutral curve. This would, of course, have reduced the amount of damping that actually occurred in the experiment.

The pressure levels appear to scale with ϵ^3, so we expect that the change in the scaled wavenumber κ will also be of this order, and this could produce an order one change in the damping ratio. It is also worth noting that adverse pressure gradients tend to shift the actual location of the neutral stability curve upstream from its theoretical (i.e., flat plate) location. Since Shapiro's (1977) measurements were carried out at the latter location, some of the amplitude reduction was probably recovered before the measuring station was reached. In fact, Shapiro (1977) actually observed some upstream Tollmien-Schlichting wave growth. Thus the effective damping ratio was probably much closer to unity than to 0.1, and the present theory is, therefore, not inconsistent with Shapiro's observations.

The theory applies to any interaction region wall shape that occurs on the appropriate length scale, including small humps and other protuberances. Aizin and Polyakov (1979) conducted an experiment that was similar to Shapiro's but beamed the sound from the downstream direction and triggered their instability wave in a more controlled fashion by placing a very thin (about 0.02 mm thick) strip of Mylar tape over the nose-flat plate region juncture.

It is much easier to compare the theory with Aizin and Polyakov's measurements than with Shapiro's. One such comparison is shown in figure 12. Here the coupling coefficient is plotted against the streamwise dimension d^* of the tape. The agreement is obviously quite good.

Finally, it is worth making a few comments about the range of validity of the theory. Figure 13 is another plot of the neutral stability curve for the Tollmien-Schlichting waves. The analysis of Goldstein (1985) implies that the theory applies to the shaded region portion of this plot which corresponds to $Re_x = O(\epsilon^{-8})$ and $\epsilon \ll 1$. But since (Bodonyi and Smith, 1984) $Re_x \propto \epsilon^{-10}$ along the upper branch, the main unstable region lying between the upper and lower branches corresponds to $Re_x = O(\epsilon^{-10})$ and $\epsilon \ll 1$ and is consequently much larger than the previous region $Re_x = O(\epsilon^{-8})$, $\epsilon \ll 1$. The Tollmien-Schlichting waves no longer exhibit a triple deck structure in this larger region but rather exhibit a four-layer structure, because the critical layer now moves out of the viscous wall layer (Bodonyi and Smith, 1984). Also, the ratio of their wavelength to the downstream distance ℓ is now much smaller in this region, i.e., it is now $O(\epsilon^{9/2})$ rather than $O(\epsilon^3)$ as before.

The streamwise-length scale of the protuberance must therefore be made correspondingly shorter in order to promote efficient coupling. But the corresponding steady flow still possess a triple deck structure, in a certain restricted sense, as long as the protuberance height is not large enough to cause separation (Smith, Brighton, Jackson, and Hunt, 1984, and Smith and Daniels, 1981). This occurs when its height is $O(\ell\epsilon^{13/2})$, which is smaller than the height $O(\ell\epsilon^5)$ of the lower deck.

It is, therefore, rather remarkable that the formula (8) for the coupling coefficient (or more precisely its asymptotic high frequency or short wave limit) still applies to these protuberances. (See also Burggraf and Smith, 1985.) In fact, that result even applies to much thicker protuberances of the full lower deck height $O(\ell\epsilon^5)$, provided the proturberance geometry is replaced by the separated streamline geometry at the appropriate places. But this result is, in a certain sense, only formal since the evaluation of the coupling coefficient would require the solution of the steady flow problem which has not, as yet, been done. Nevertheless, we can deduce by rescaling equation (8) for the smaller size of the protuberance that the coupling coefficient is no longer $O(1)$ as $\epsilon \to 0$ but rather is $O(Re_x^{-3/40})$ for the separated flow case and $O(Re_x^{-1/10})$ for the shallower protuberances where the flow is unseparated (Goldstein, 1985).

References

ACKERBERG, R.C.; PHILLIPS, J.R. 1972 J. Fluid Mech. 51, 137.

AIZIN, L.B.; POLYAKOV, M.F. 1979 Acoustic generation of Tollmien-Schlichting waves over local uneveness of surface immersed in streams (in Russian). Preprint 17, Akad. Nauk USSR, Siberian Div., Inst. Theor. Appl. Mech., Novosibirsk.

BODONYI, R.J.; SMITH, F.T. 1981 Proc. R. Soc. Lond. A 375, 65-92.

BRIGGS, R.J. 1964 Electron-stream interaction with Plasmas. Research Monograph No. 29. The MIT Press, Cambridge Mass. (Chapter 2).

BURGGRAF, O.R.; SMITH, F.T. 1985 On the development of large-sized short-scaled disturbances in boundary layers (leading to the Benjamin-Ono, KdV and Burger's equations). To be published in Proc. R. Soc. Lond. A.

GASTER, M. 1965 J. Fluid Mech. 22, Part 3.

GOLDSTEIN, M.E. 1983a J. Fluid Mech. 127, 59.

GOLDSTEIN, M.E. 1983b Generation of Tollmien-Schlichting waves by free stream disturbances at low Mach numbers. NASA TM 83026.

GOLDSTEIN, M.E.; SOCKOL, P.M.; SANZ, J. 1983 J. Fluid Mech. 129, 443

GOLDSTEIN, M.E. 1985 J. Fluid Mech. 154, pps. 509-530.

GOLDSTEIN, M.E. 1985b Submitted to J. Fluid Mech.

HEISENBERG, W. 1924 Ann. Physic. 74, p. 577; trans as MACA T.M. 1291 (1951)

KACHANOV, YU.S.; KOZLOV, V.V.; LEVCHENKO, V.YA. 1978 Izvestia Soviet Acad. Sci.: Mekhanika Zhidkosti i Gaza, No. 5, pp. 85-94.

KAGELMAN, J.T.; MUELLER, T.J. 1984 Experimental studies of spontaneous and forced transition on an axisymmetric body. McDonnell Douglas Research Laboratories Report MDRL 84-3. To be published in AIAA Journal.

LAM, S.H.; ROTT, N. 1960 Theory of linearized time-dependent boundary layers. Cornell Univ. Grad. School of Aero. Engng Rep. AFOSR TN-60-1100.

LEEHEY, P.; SHAPIRO, P. 1979 In Laminar-Turbulent Transition (ed. R. Eppler & H. Fasel), pp. 321-331. Springer.

LIN, C.C. 1945 Q. Appl. Math. 3, 117.

LIN, C.C. 1946 Q. Appl. Math. 3, 277.

MACK, L.M. 1975 AIAA J. 13, pp. 278-289.

MESSITER, A.F. 1970 SIAM J. Appl. Math 18, 241.

MORKOVIN, M.V. 1969 Air Force Flight Dyn. Lab., Wright-Patterson AFB, Ohio, Rep. AFFDL-TR-68-149.

MURDOCK, J.W. 1980 Proc. R. Soc. Lond A372, 517.

NISHIOKA, M.; MORKOVIN, M.V. 1985 Boundary-layer receptivity to unsteady pressure gradients: experiments and overview, submitted to J. Fluid Mech.

REID, W.H. 1965 In Basic Developments in Fluid Dynamics, vol. I (ed. Maurice olt), pp. 249-307. Academic.

RESHOTKO, E. 1976 Ann. Rev. Fluid Mech. 8, 311.

SCHUBAUER, G.B.; SKRAMSTAD, H.K. 1943 NBS Research Paper 1772.

SHAPIRO, P.J. 1977 The influence of sound upon laminar boundary layer instability. MIT Acoustics and Vibration Lab Rep. 83458-83560-1

SMITH, F.T. 1973 J. Fluid Mech. 57, 803.

SMITH, F.T., BRIGHTON, P.W.M., JACKSON, P.W., & HUNT J.C.R. 1948 J. Fluid Mech. 113. 123-152.

SMITH, F.T. 1979 Proc. R. Soc. Lond A 366, 91.

STEWARTSON, K. 1969 Mathematika 16, 106.

STEWARTSON, K. 1970 Q. J. Mech. Appl. Maths 23, 137.

STEWARTSON, K. 1971 Q. J. Mech. Appl. Maths 24, 387.

STEWARTSON, K.; SMITH, F.T.; KAUPS, K. 1982 Studies in Appl. Maths. 62, 45.

TOLLMIEN, W. 1929 Nachr. Ges. Wiss. Gottingen, Math.-Phy. Kl 21 (Translated as NACA 609 [1931].)

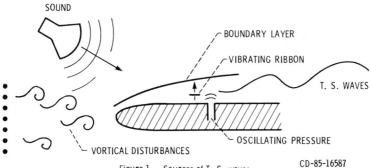

Figure 1. - Sources of T. S. waves.

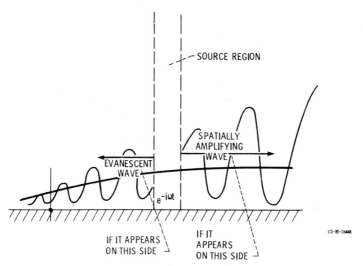

Figure 2. - Amplifying waves.

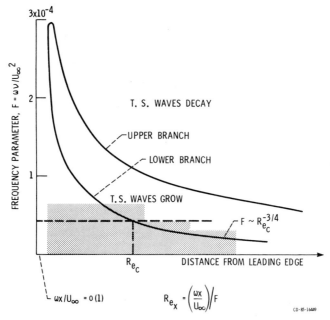

Figure 3. – Stability diagram for Tollmien-Schlichting waves.

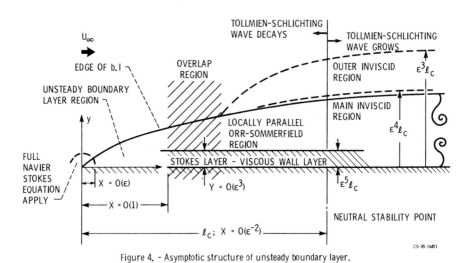

Figure 4. – Asymptotic structure of unsteady boundary layer.

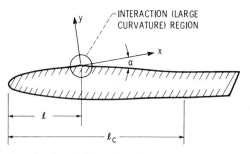

Figure 5. - Type B flow configuration.

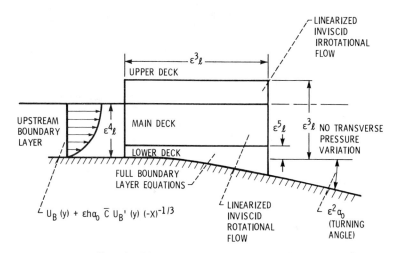

Figure 6. - Triple deck structure of interaction region.

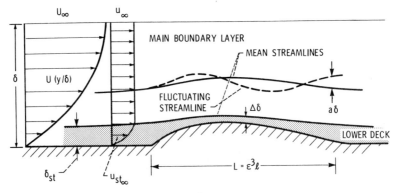

Figure 7. - Change in Stokes shear wave.

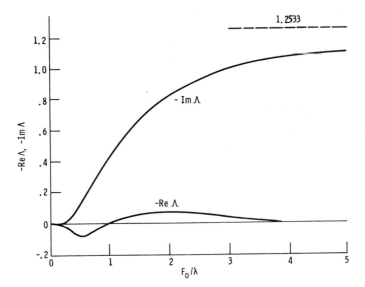

Figure 8. - Real and imaginary parts of $-\Lambda$.

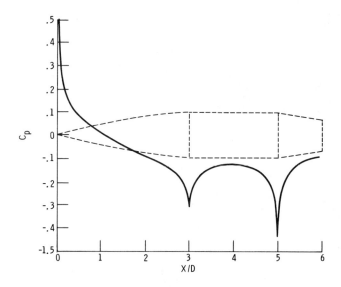

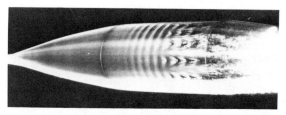

Figure 9. - Sound-induced transition with the sound at 500 Hz, $Re_t = 814\,000$ (courtesy of J. T. Kegelman and T. J. Mueller, University of Notre Dame).

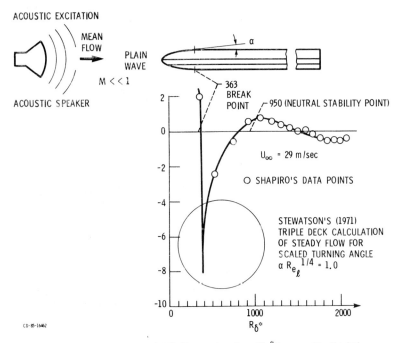

Figure 10. – Static pressure distribution as a function of R_δ^* (measured by Shapiro).

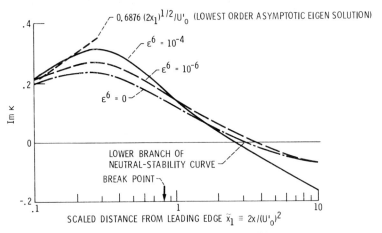

Figure 11. – Variation of scaled complex wavenumber with distance from the leading edge.

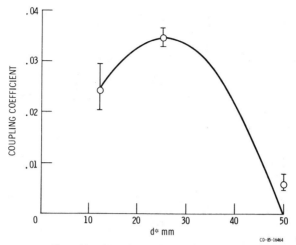

Figure 12. - Comparison with data of Aizin and Polyakov.

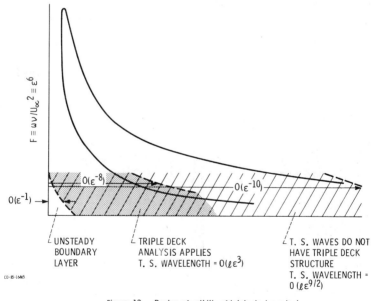

Figure 13. - Region of validity of triple deck analysis.

Numerical Experiments on Boundary-Layer Receptivity

Thomas B. Gatski and Chester E. Grosch

Abstract

The incompressible laminar flow over an infinitely thin flat plate is obtained using a Navier-Stokes code in vorticity-velocity variables. The flow at and near the leading edge of the plate is an integral part of the solution algorithm which requires no special treatment; thus allowing for the flow field in this region to be studied in detail. An incident plane sound wave is imposed in the free-stream flow and the receptivity of the boundary layer is studied with particular emphasis to the flow near and at the leading edge.

I. Introduction

The transition to turbulence in flat-plate boundary layers is the result of the linear and nonlinear development of initially small amplitude Tollmien-Schlichting (T-S) waves into finite-amplitude disturbances which breakdown into a random chaotic motion. The location of the transition point and the initiation of the T-S waves is closely related to the free-stream disturbance; however, the mechanisms responsible for the coupling of the free-stream disturbance and the spatially evolving T-S waves, that is the "receptivity problem," is still a matter of debate.

There have been several experimental, computational and theoretical attempts at studying this problem.[1-5] In the experimental and computational study of Kachanov et al,[1] a vortical disturbance was imposed in the free-stream flow upstream of the leading edge and the receptivity of the boundary layer to this disturbance near the leading edge was examined. The numerical solution involved the linearized Navier-Stokes equations for the disturbance field and a full Navier-Stokes solution for the mean field. In addition to the free-stream vortical disturbance, a cross-flow irradiation of the plate leading edge by a sound wave was also examined. In both cases, the cross-flow disturbance velocity was found to be the dominant triggering mechanism. In another numerical study Murdock[2] has shown that sound waves do feed energy into T-S waves, but only near the leading edge. The study was done in the incompressible limit of infinite sound speed and wavelength and solved a parabolized set of unsteady boundary-layer equations using a spectral method. The results showed that for values of the parameter $s(\equiv \frac{\omega x}{U}) > 1$ no observable interaction between the sound wave and boundary layer occurred; however, for $s < O(1)$ near the leading edge an interaction did occur. Recent analytical work by Goldstein,[3] using the method of matched asymptotic expansions and a "triple deck" decomposition of the boundary layer, has examined the initiation and evolution of the T-S waves near the leading edge, although the effects of the leading edge itself were neglected. In the study it was found that one of the asymptotic eigensolutions of the unsteady boundary-layer equations, whose wavelength decreased with increasing distance downstream, matched the fundamental Tollmien-Schlichting wave further downstream. Thus, the coupling mechanism for wavenumber conversion was established. In both the Murdock[2] and Goldstein[3,4] studies a receptivity factor was established, with values of $O(10^{-4})$ and $O(10^{-2})$ obtained in the respective studies. However, in a recent experimental study by Leehey et al,[5] the receptivity factor was experimentally determined to be of $O(1)$. Thus, deficiencies must still exist in both the computational and theoretical studies performed to date.

The present numerical study is an incompressible two-dimensional Navier-Stokes solution in vorticity-velocity variables of the forced flow over a flat plate. The flow over the leading edge is an integral part of the solution process requiring no modification of the algorithm or the solution process itself. The external forcing is provided by time-dependent irrotational disturbances input outside of the boundary layer in the free stream and upstream of the flat plate. Once again the forcing is a plane sound wave (incompressible limit) as in Reference 2; however, in the present study no _a priori_ assumptions about the nature of the flow _near_ and _at_ the leading edge are made. The intent is to study the receptivity of the boundary layer to such free-stream disturbances in this leading-edge region which will augment the previous computational and theoretical studies of Murdock[2] and Goldstein.[3]

II. Formulation and Methodology

Consider the two-dimensional flow over an infinitely thin flat plate embedded in an otherwise unperturbed flow. Initially the intent is to establish the boundary-layer flow over the flat plate by keeping the flow upstream of the plate uniform and irrotational. The flow variables are scaled so that the flow is subcritical over the streamwise lengh of the computational domain. This scaling is based on parallel theory; if non-parallel theory is invoked then part of the computational domain is in the unstable region. The scaling parameters are as follows: the physical length of the plate, L, is taken as 90 cm and the free-stream velocity, U_∞, is 200 cm/sec; the kinematic viscosity ν is 0.16 cm^2/sec. These values give a boundary-layer thickness, δ, of 1.3416 cm and a displacement thickness of 0.4615 cm; the corresponding Reynolds numbers are Re = 1677 and Re_* = 577. All variables in the numerical code are scaled by the boundary-layer thickness, δ, and the free-stream velocity, U_∞. The computational domain is as shown in Figure 1 with the origin of the coordinate system taken at the plate leading edge. The inflow boundary is taken 30 units upstream of the leading edge of the plate and the length of the plate is taken as 65 units. This length of the plate is slightly less than the non-dimensional length based on 90 cm and was chosen to minimize the region of unstable growth near the outflow boundary. The distance along the plate can also be measured in units of Re_x as well as in dimensionless x. The conversion is simply given by Re_x = xRe, where Re is 1677. In units of Re_x the leading edge is Re_x = 0 and the length of the plate (end of computational domain) is Re_x = 1.09 x 10^5. The flow domain was restricted in the transverse y-direction to the region 0 < y < 6.4. Previous numerical studies using the present numerical algorithm for flows with large scale vortical structures in the boundary layer indicate that the upper limit of y = 6.4 is sufficiently far outside the boundary layer to be taken as free stream.

The numerical algorithm has been used extensively in previous studies for both steady[7] and unsteady flows.[6,8] For completeness, the governing differential equations are given by

$$\frac{\partial u}{\partial x} + \frac{\partial v}{\partial y} = 0 \tag{1}$$

$$\frac{\partial v}{\partial x} - \frac{\partial u}{\partial y} = \omega \tag{2}$$

$$\frac{\partial \omega}{\partial t} + u \frac{\partial \omega}{\partial x} + v \frac{\partial \omega}{\partial y} = \frac{1}{Re} \nabla^2 \omega \tag{3}$$

where u and v are the streamwise and transverse velocities, respectively, ω is the vorticity and Re is the Reynolds number based on boundary-layer thickness. The details of the discretization of these equations are rather cumbersome and will not be presented here. For further details on the algorithm see References 7 and 9. It is important to emphasize that the variables as defined by the numerical algorithm are <u>average values</u> across the edge of a cell; it is for this reason that the leading edge of the plate, which is at the corner of a computational cell, is handled so cleanly by the numerical algorithm. It is certainly not suggested that such an abrupt change of boundary conditions can be handled in a completely non-obtrusive manner with a finite-size grid spacing, however, the nature of the algorithm does smooth the jump over the bounding cell in a manner consistent with the numerics. The grid spacing in the streamwise x-direction is equidistant with $\Delta x = 0.5$ and a total of 190 cells. The grid spacing in the y-direction is non-uniform with cell heights ranging from $\Delta y = 0.01$ to $\Delta y = 0.32$ and a total number of 70 cells. These grid spacings and number of cells were sufficient to obtain the desired accuracy in the parameter range studied.

The detailed requirements for the specification of boundary conditions are described in detail elsewhere,[9] it suffices here to only specify the conditions for the velocities and vorticity on the computational boundaries. At inflow, the streamwise velocity, u, is taken as uniform and equal to unity and the flow is irrotational. At the free-stream boundary, the streamwise velocity is once again taken as unity but the vorticity is determined from the flux condition $\frac{D\omega}{Dt} = 0$. At outflow, flux conditions are used for both the transverse velocity, v, and the vorticity, that is $\frac{Dv}{Dt} = 0$ and $\frac{D\omega}{Dt} = 0$. Previous tests[6,8] have indicated that moderate disturbances are allowed to propagate "cleanly" through the computational boundaries using these conditions. This is not that surprising when it is recognized that the condition $\frac{Dv}{Dt} = 0$ is the finite-amplitude inviscid Rayleigh stability equation. In a recent theoretical study by Halpern,[10] it was shown that this type of transport condition is more accurate than the usual Neumann condition of $\frac{\partial}{\partial x} = 0$. Finally, along the line $y = 0.0$ the v-velocity is taken as zero and the vorticity is obtained from a second-order accurate difference approximation to Equation 2. Note that the v = 0 boundary condition is also used in the region between the inflow boundary and the plate leading edge. Of course, the more general problem of the numerical solution of the flow above and below the plate is more physically relevant; however, in this initial study where the effects of the leading edge are included, the problem formulation presented here will suffice. Future computational efforts in this area will include an examination of the more physically relevant boundary conditions.

In order to have a well-posed physical problem, it would be desirable to first reach a steady-state flow field and then begin the forcing experiments. It was found that this was a rather illusive task due to the fact that the effects of the start-up persisted for long periods of time.

The initial start-up was an impulsive one. It was found that at small times the vorticity was very large in a small region surrounding the leading edge; essentially zero in front of the plate, and of moderate size in a thin shear layer on the plate downstream of the leading edge. The shear layer downstream of the leading edge was uniform in x and was the Stokes boundary layer due to the impulsive start of an infinite plate. This shear layer grew in time and, simultaneously, the vorticity at the leading edge diffused and was advected downstream. This process produced large amplitude shear waves which were the mechanism by which the Stokes layer was changed to a spatially growing boundary layer. This was a lengthy process and accounted for the long start-up time. This process is quite interesting in its own right and will be fully described elsewhere.

The region in the vicinity of the plate leading edge did, however, become time-dependent and only much farther downstream, in the region where Re_c was approached, did transient effects persist. Nevertheless, since the region of interest in the study is in the vicinity of the leading edge, the downstream transients were assumed irrelevant and the forcing of the flow commenced. A plane sound wave disturbance of the form (cf. Ref. 2)

$$u = 1 + \epsilon \sin \beta (t - t_0) \tag{4}$$

was imposed at both the inflow and free stream boundaries at $t = t_0$. The time-dependent forcing was also applied at the free-stream boundary because in the incompressible limit both the sound speed and wavelength approach infinity and application of the transient condition, Eq. (4), at inflow implies that a similar condition be applied at the free-stream boundary at the same time. The forcing was continued for multiple periods to allow for any finite-propagation velocities of disturbances generated upstream of plate by the incident sound wave. A discussion of the results and the choice of forcing parameters are presented in the next section.

III. Results and Discussion

The Reynolds number range covered by the streamwise extent of the plate was always less than the critical Reynolds number based on parallel theory. If non-parallel theory is used as a guide then approximately the last 20 units in the domain are in the unstable regime. The two parameters that need to be determined for the forcing experiments are the amplitude ϵ and the frequency β. The forcing amplitude chosen is 0.1 percent of the free-stream velocity; this level is consistent with previous studies[1,2] and is still within the linear range of forcing amplitude. The frequency β was determined from the neutral stability curve for boundary-layer flow.[11] For a Reynolds number Re based on the boundary-layer thickness of 1677 (Re$_*$ = 577), a frequency β of 0.316 was unstable for both parallel and non-parallel analyses.

Since the flow near the leading edge is of paramount interest in this study, an indicator of the type of flow field in this region is the displacement thickness. In Figure 2 is plotted the square of the non-dimensional displacement thickness, δ_*^2, as a function of x (solid line). The δ_* values are obtained from the unperturbed computed velocity profiles. Recall that the length scale in this problem is a constant; therefore, using non-dimensional values for such quantities as displacement thickness does not alter the functional dependence. Blasius boundary-layer theory predicts that δ_*^2 is a linear function of x; however, the figure shows that for distances from the leading edge less than approximately 9 units the displacement thickness does not exhibit this behavior. Note that the δ_*^2 curve does not go to a x = 0 value in the figure. This is due to the fact that the numerical algorithm defines variables as average values and place them at the midpoints of cell sides; thus at y = 0 the first set of variables defined on the plate will be located one-half grid spacing downstream of the leading edge. The data presented in Figure 2 suggest that invoking results from boundary-layer theory in the region delimited by 0 < x < 9 is precarious and needs to be examined in more detail. Before continuing the analysis of the region near the leading edge, it is of interest to examine the region x > 9 where it appears that boundary-layer theory holds.

The dashed line in Figure 2 represents a linear continuation of the δ_*^2 curve and its intersection with the x-axis yields the value for the virtual origin, x_0, of the downstream boundary-layer flow. The figure shows that the value for x_0, is approximately 2.5. The value is dependent on the accuracy of the linear continuation (dashed line) of the δ_*^2 curve. The computed results yield a slope for this line of 1.74×10^{-3}. Blasius theory predicts that the slope of the displacement thickness squared curve be equal to $2.96\ Re^{-1}$ and for a Reynolds number of 1677 this slope equals 1.76×10^{-3}. This comparison thus confirms the accuracy of the location of the virtual origin and also suggests that the effects of numerical viscosity are minimal. It is misleading to try to compare the location of the virtual origin obtained in this study with experimental results over surfaces with leading edges of finite thickness because the results would be very sensitive to the leading-edge geometry. It does suffice to mention that as the Reynolds number increases this location moves closer to the leading edge and only subsequent studies will show the exact Reynolds number dependence.

Now that the accuracy of the displacement thickness results have been quantified, it is instructive to return to the flow field near the leading edge. Figure 2 suggests a radical departure of the flow field from the downstream similarity results. This is confirmed in Figure 3 where the unperturbed mean velocity profiles immediately downstream of the leading edge are plotted. The figure clearly shows a velocity overshoot on the order of 5 percent followed by a relaxation to the free-stream value. These results and those from Figure 2 indicate that in the region near the leading edge any results for the mean flow variables derivable from boundary-layer theory should be used with caution.

Figures 4a-d are plots of the instantaneous disturbance velocity, u', versus y at different positions on the plate ($x = 0.25, 1.75, 3.25,$ and 4.75, respectively) and at a time equal to two periods after the beginning of forcing. The values of u' are defined by

$$u'(x,y,t) = u(x,y,t) - U(x,y,t_0) \qquad (5)$$

Note that the "mean" field, $U(x,y,t_0)$, at the start of the forced oscillations is not the same as the average of $u(x,y,t)$ over a period because the latter will contain the second-order steady streaming velocity. The streaming velocity is implicit in $u'(x,y,t)$ so that the time average of this velocity over a period is not zero. The results shown in Figure 4 indicate that the oscillations have a significant qualitative variation with downstream distance. The first two profiles (a and b), as well as the ones at the intermediate x positions which are not plotted, have a strong resemblance to Stokes shear waves, but with an x-dependent phase shift. Further downstream this resemblance is less striking. The interpretation of these as Stokes waves or T-S waves is quite difficult because of the large streamwise gradients in the area near the leading edge. The frequency parameter s, equal to βx with the present non-dimensionalization, varies from 0.08 to x = 0.25 to 1.50 at x = 4.75. However, the profiles of u' are neither Stokes waves deeply buried in the steady boundary layer, as is expected for large s (cf. Ref. 2), nor are they in phase with the free-stream oscillations, as is expected for small s (cf. Ref. 2).

As was shown previously, the mean profile at x = 0.25 overshoots the free-stream value and then relaxes back to it. If the boundary-layer thickness is taken to be the inner value of y where U = 0.99, then at x = 0.25, $\delta \approx 0.035$; but if it is taken to be the outer value of y at which U = 0.99, then $\delta \approx 0.3$. The difference, is that at the inner value of y the peak amplitude (see Fig. 4a) is at the top of the boundary layer and at the outer value of y the entire oscillation is embedded in the steady boundary layer. At the other x positions, 1.75, 3.25, and 4.75, the mean profiles are monotonic, although not of the Blasius form. At these locations the boundary-layer thicknesses are 0.12, 0.16, and 0.20, respectively. From Figures 4b and 4c, it can be seen that there is a phase shift across the steady boundary layer and that most, but not all, of the shear wave is contained within the boundary layer. Even further downstream, x > 5.0, the profiles of the disturbance have the same qualitative features as that shown at x = 4.75; however, these have not as yet evolved to an equilibrium state and will not be discussed.

The variation of u' with x at y = 0.13 (Re_y = 217) and a time equal to two periods after the initiation of forcing is shown in Figure 5. This result is typical of others at the same time and different values of y. There appears to be a modulation of the result by waves having at least two different wavelengths; a short wave with a length of about 8 units and a long wave with a length of about 18 units. It is not clear from the available results as to whether or not this pattern is slowly varying with time or has reached equilibrium. The results appear to be quasi-steady in the region shown (x < 30) but not at larger values of x

Interpretation of these results is difficult because the steady boundary layer is quite thin in this region and is growing with downstream distance. However, the short wavelength disturbances may be the result of the presence of the continuum eigenfunctions.

Examination of the flow field ahead of the plate shows the presence of oscillations with wavelengths of 7 to 8 units in the x-direction. In order to resolve the nature of these oscillations, the calculations need to be continued further in time.

IV. Concluding Remarks

The present numerical experiment was an initial attempt at studying the effects of sound wave forcing <u>at</u> and <u>near</u> the leading edge of an infinitely thin flat plate. Previous studies have been unable to closely examine this region due to constraints on the theoretical models of the flow in this region or due to the inability of the numerical algorithms to handle the leading edge region. However, as has been shown, the numerical algorithm utilized in this study has overcome this latter constraint. An analysis of the behavior of the displacement thickness and mean velocity profiles near the leading edge have clearly shown that the usual boundary-layer assumptions do not hold in this region due to the rapid streamwise acceleration of the flow. Plots of streamwise disturbance velocity have shown qualitative similarities to the Stokes-type solutions including a phase-shift behavior across the boundary layer. In addition, further examination of the disturbance velocity at a fixed y station has also identified a phase-shift behavior in the streamwise direction. Attempts to quantify the effects of wavelength variations within the boundary layer proved less successful. Apparent short and long wavelength modulations of the disturbance profiles distorted the results.

The results of this study have shown that some of the essential disturbance dynamics near the leading edge can be successfully resolved and identified; nevertheless, additional forcing times are needed to establish an equilibrium flow over a larger portion of the boundary layer which would allow for a less ambiguous interpretation of results.

References

1. Kachanov, Yu. S., Kozlov, V. V., Levchenko, V. Ya, and Maksimov, V. P.: The Transformation of External Disturbances into the Boundary Layer Waves. Sixth International Conference on Numerical Methods in Fluid Dynamics, Tbilisi, U.S.S.R., H. Cabannes, M. Holt and V. Rusanov (eds.) Springer-Verlag, Berlin, June 21-24, 1978, pp. 299-307.

2. Murdock, John W.: The Generation of a Tollmien-Schlichting Wave by a Sound Wave. Proc. R. Soc. Lond. A., Vol. 372 (1980), pp. 517-534.
3. Goldstein, M. E.: The Evolution of Tollmien-Schlichting Waves near a Leading Edge. J. Fluid Mechanics (1983), Vol. 127, pp. 59-81.
4. Goldstein, M. E., Sockol, P. M., and Sang, J.: The Evolution of Tollmien-Schlichting Waves near a Leading Edge. Part 2, Numerical Determination of Amplitudes. J. Fluid Mechanics (1983), Vol. 129, pp. 443-453.
5. Leehey, C., Gedney, C. J., and Her, J. Y.: The Receptivity of a Laminar Boundary Layer to External Disturbances. IUTAM Symposium on Laminar-Turbulente Transition, Novosibrisk, U.S.S.R., July 9-13, 1984, V. V. Kozlov (ed.), pp. 233-242.
6. Gatski, T. B.: Drag Characteristics of Unsteady, Perturbed Boundary Layer Flows. AIAA Shear Flow Control Conference, Boulder, CO, March 12-14, 1985. Paper No. 85-0551.
7. Gatski, T. B.; and Grosch, C. E.: Embedded Cavity Drag in Steady Laminar Flow. AIAA Journal, Vol. 23, No. 7, July 1985, pp. 1028-1037.
8. McInville, R. M.; Gatski, T. B.; and Hassan, H. A.: Analysis of Large Vortical Structures in Shear Layers. AIAA Journal, Vol. 23, No. 8, August 1985, pp. 1165-1171.
9. Gatski, T. B.; Grosch, C. E.; and Rose, M. E.: A Numerical Study of the Two-Dimensional Navier-Stokes Equations in Vorticity-Velocity Variables. Journal Comp Physics, Vol. 48, No. 1, October 1982, pp. 1-22.
10. Halpern, L.: Artificial Boundary Conditions for the Linear Advection Diffusion Equation. Internal Report No. 118, Centre De Mathematiques Appliques, Equipe de Recherche Associee du C.N.R.S., No. 747, 1985.
11. Drazin, P. G.; and Reid, W. H.: <u>Hydrodynamic Stability</u>, Cambridge University Press, Cambridge, England. G. K. Batchelor and J. W. Miles (eds.), 1981.

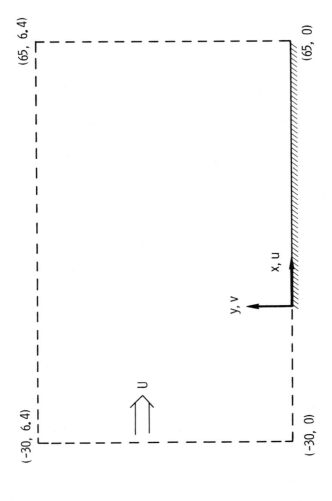

Figure 1. - Computational Domain.

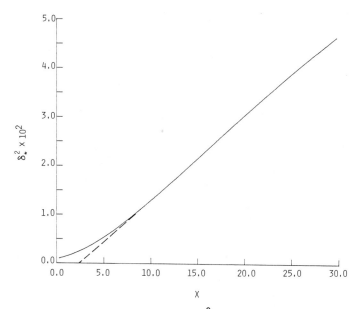

Figure 2. - Variation of δ_*^2 as a Function of Streamwise Location.

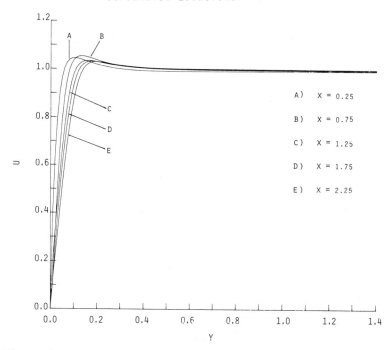

Figure 3. - Mean Velocity Profiles Near the Plate Leading Edge.

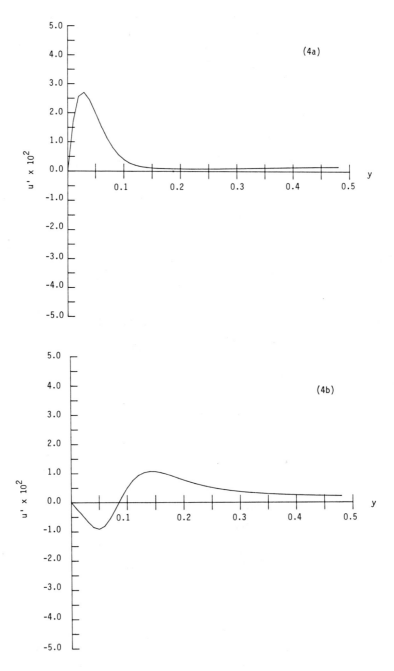

Figure 4 (a-b). For description see next page.

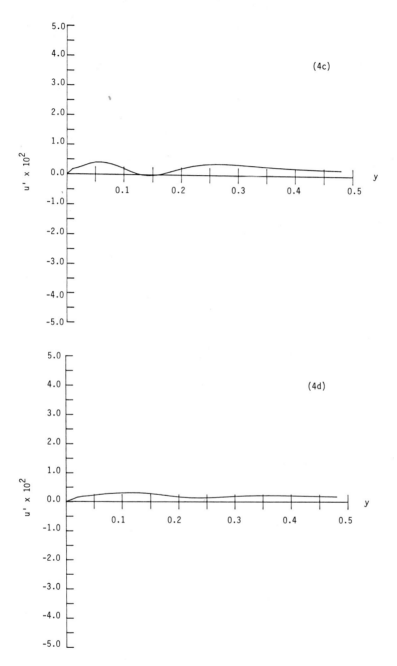

Figure 4. - Disturbance Velocity Profiles: (a) $x = 0.25$; (b) $x = 1.75$; (c) $x = 3.25$; (d) $x = 4.75$.

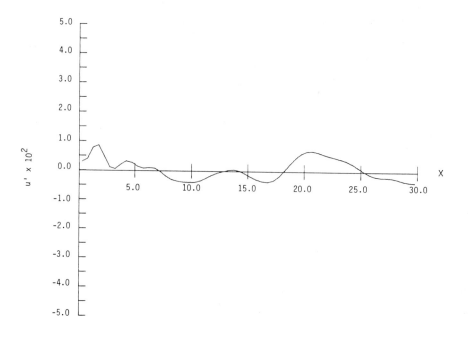

Figure 5. - Streamwise Variation of Disturbance Velocity at Re_y = 217.

The Linear Stability of the Incompressible Boundary Layer on an Ellipsoid at Six Degrees Incidence

S.G. LEKOUDIS

ABSTRACT

The linear stability of the three-dimensional boundary layer on an ellipsoid of revolution at angle of attack is investigated. The appropriate form of the Orr-Sommerfeld equation was used to compute local amplification rates for both streamwise and cross flow instabilities. At six degree incidence, it was found that the spatial growth of unstable waves is essentially governed by two dimensional instability, modified by the presence of cross flow. This behavior explains qualitatively the observed transition patterns on ellipsoids.

INTRODUCTION

Transition prediction methods are needed because aerodynamic loads can largely depend on the location of transition. There are different ways for predicting the location of transition, all of a semi-empirical nature. The most successful and widely used method relies on computing the integral of linear growth rates of unstable disturbances. These rates are based on stability theory and a good description of this theory is in Reference 1.

Because of the operating conditions, it is easier to envision transition control on smooth surfaces of wings. However, future designs show significant wing/fuselage/engine interaction and the boundary layer on fuselage-like bodies becomes more important. Also cylindrical shapes are prominent in underwater vehicles.

It should be emphasized that transition depends on what we usually call nonlinear effects. For the low disturbance environment of high altitude flight, linear theory dictates the trends of the transition location as the pressure gradient and the surface condition varies. Hence, nonlinear effects influence only the later stages of transition.

With the above in mind, we investigated the linear stability characteristics of the boundary layer on an ellipsoid at angle of attack. There are some other reasons for doing these calculations. First, the mean flow has been computed before by different investigators and, therefore, there is confidence in the results. Second, any investigation of the stability of a new flow can benefit from existing experience with results obtained from investigating similar flows. And, finally, there is the good practice of generating a simple mean flow with a sophisticated code and investigating its stability. The ellipsoid boundary layer is a special case that any fuselage code should be able to handle.

A review of the state of the art of linear stability for low speed and transonic boundary layers is given in Reference 1. It should be mentioned that the present investigation is restricted by the assumption of parallel flow. Because, as will be apparent later, significant events occur in regions of adverse pressure gradients, the nonparallel effects (Reference 2) may be very important in the quantitative description of events. However, the qualitative description should remain the same. The procedures used to obtain the boundary layer on the ellipsoid are given in Reference 3. In the next sections of this paper we describe the analytical formulation, the numerical procedures used to obtain the results, and we end with a discussion of the results.

THE ANALYTICAL FORMULATION

Figure 1 shows an ellipsoid of revolution. The coordinate system for the boundary layer is as follows: the x-axis is along the major axis of the ellipsoid. The y-axis is along the circumference and is measured in radians, that are converted to degrees for convenience. The reference quantities are the uniform freestream velocity and the length of the semi-major axis of the ellipsoid. The boundary layer develops along the surface and, in the direction normal to the surface, the usual boundary layer coordinate is used. This coordinate is the length normal to the surface, divided by the square root of the reference Reynolds number.

The above coordinates are used for the boundary layer calculations and more details about them can be found in Reference 3. In the stability calculations reported here the reference velocity is still the freestream velocity but the reference length is the semi-major axis divided by the square root of the freestream Reynolds number. This freestream Reynolds number is based on the freestream velocity, the semi-major axis, and the kinematic viscosity. Therefore, the Reynolds number, R, used in the stability calculations, is the square-root of the freestream Reynolds number.

The equation solved in the stability calculations for this incompressible flow is the Orr-Sommerfeld equation for a three-dimensional mean flow. It is preferred over the primitive variable formulation because it allows the evaluation of group velocities via the eigenfunction and the adjoint eigenfunction, while the system that is being solved is still a fourth order system. Moreover (Reference 1), there is no need for any specific coordinate transformation with the use of the Orr-Sommerfeld equation, and the coordinate system used in the boundary layer calculations is sufficient. This means that the mean flow profiles are used as computed.

The complex group velocity was computed by making use of the solution of the adjoint Orr-Sommerfeld system. The original Orr-Sommerfeld equation for the three-dimensional boundary layer flow was converted into a system of four first order equations. This system was differentiated with respect to the wavenumbers in the x and y-direction and integrated in the direction normal to the wall, after it was multiplied with the solution of the adjoint system. Making use of the properties of the adjoint problem results in the formulas for the components of the group velocity in the x- and y-directions respectively.

The computation of group velocities allows the evaluation of waves that have the maximum local amplification rates. This is done by evaluating waves with a real ratio of the complex group velocity components. In the calculations reported in this paper, the wave frequency is real and the wavenumbers are complex, resulting in growth or decay in space. However, no attempt was made to compute growth factors (sometimes called N-factors) that can be evaluated using several different approaches.

THE NUMERICAL PROCEDURE

The Orr-Sommerfeld system used in this work is stiff because of the value of the Reynolds number. A well tested code SUPORT (Reference 4) was used to overcome this difficulty. Because the flow was new to the authors, no preassigned orthonormalization points were used, and the code decided when and how to orthonormalize with the associated penalty of slower execution.

The eigenvalue search used a simple Newton-Raphson iteration to home on the wavenumber. Convergence was obtained in all the results shown by acquiring three digits in both the real and imaginary parts of the wavenumber components. The direction of growth was kept fixed and equal to the direction of the local potential flow direction. The eigenvalue search was made with fixed direction of the real part of the wavenumber vector (phase direction). The waves with real group velocity component ratio were obtained by a simple Newton-Raphson iteration on the phase direction. This outer iteration converged when the first three digits in the phase direction were obtained. The resulting imaginary part of the ratio of the complex group velocity components became of the order of ten to the minus five.

DISCUSSION OF RESULTS

When the stability of a three-dimensional boundary layer is being investigated, the first issue to be examined is the amount of the existing cross flow. The cross flow profile is defined as the velocity profile in the direction normal to the local potential flow direction. The shape of this profile influences the instability present and is usually responsible for transition on swept wings. A good description of the characteristics of cross flow instability is given in Reference 1.

If one examines the cross flow on the ellipsoid, it is found that the amount of the cross flow for the three degree angle of attack case is small. Thus, the stability characteristics of the boundary layer are only influenced by a small amount by its presence. In Reference 1 there is a good description of the ranges of the influence of magnitude of the cross flow on the stability. However, the situation is different for the case of six degree angle of attack. Here the cross flow cannot be neglected. The stronger cross flow is approximately in the ninety degree area, or about halfway between the lines of symmetry.

Figure 2 shows the spatial amplification rates in the x-direction (negative of the imaginary part of the wavenumber in the x-direction, normalized with the ellipsoid semimajor axis) for different frequencies and several axial locations. The nondimensional frequencies are the usual dimensional frequencies multiplied by the kinematic viscosity and divided by the square of the freestream velocity. Each curve corresponds to a specific circumferential location, and the eigenvalues along it were generated by keeping the direction of growth constant and equal to the local potential flow. It is evident from these figures that the instability is basically the two dimensional instability modified by the presence of cross flow. In the regions of stronger cross flow, the real part of the wavenumber vector (phase) is pointing away from the potential flow direction in the direction opposite to that of the cross flow, for maximum amplification. In regions of weak cross flow the waves with maximum amplification have their real part of the wavenumber vector pointing close to the direction of the local potential flow.

There is another effect that is evident. The amplification rates are higher for locations closer to the lee-line of symmetry of the elliplsoid. This is due to the fact that the boundary layer is thickening in that area, as evidenced by the displacement thickness distribution (see Reference 3).

These curves show that the instability is similar to the instability found in the mid-chord region on swept wings (Ref. 1). It has the characteristics of two-dimensional instability modified by the presence of cross flow.

In order to quantify these effects in a more precise manner, waves with maximum amplification rates (real ratio of their complex group velocity components) were computed. Figures 3 and 4 show the same amplification rate as before, but of maximum amplification; they also show their phase orientation, for two axial stations. Both the effects discussed before, the influence of the cross flow, and the thickening of the boundary layer close to the wind-line of symmetry, are evident from the results shown in these Figures. The maximum amplification occurs in regions beyond the ninety degree area in the circumferential direction, and the phase orientation rapidly changes toward the potential flow direction as we move towards the lee-line of symmetry.

Based on the observations made in the paragraphs above, the following discussion is an attempt to explain the observed transition patterns in the experiments of Meier, et al (Reference 5). In these experiments the following was observed: At zero angle of attack, the transition pattern is symmetric, i.e., transition starts at the same axial location for all circumferential locations, as expected. Between zero and six degree angle of attack the pattern loses its symmetry, with transition starting at earlier axial locations at higher circumferential locations. However, the variation is almost linear and monotonic, i.e., earlier axial locations have higher circumferential values for the beginning of transition. Above six degrees there is a tongue-like shape for the transition pattern, with transition starting earlier in x around the ninety degree area in the circumferential direction.

We propose that the explanation of these patterns is as follows: before approximately six degree angle of attack, the instability is basically dominated by two dimensional instability modified by the presence of cross flow, as examined and discussed in this paper. At higher angles the cross flow becomes stronger and dominates the transition process, which starts in the area of strong cross flow.

REFERENCES

1. Mack, L. M., "Spectral Source on Stability and Transition of Laminar Flow", AGARD Report No. 709, 1985.

2. Reed, H. and Nayfeh, A. H., "Stability of Compressible Three-dimensional Boundary Layer Flows", AIAA Paper 82-1009.

3. Radwan, S. F. and Lekoudis, S. G., "Calculations of the Incompressible Turbulent Boundary Layer on an Ellipsoid in the Inverse Mode", AIAA Paper 85-1654 (to appear in the AIAA J., 1986).

4. Scott, M. R., and Watts, H. A., "Computational Solutions of Two-point Boundary Value Problems via Orthonormalization", SIAM J. of Numerical Analysis, Vol. 14, p. 40, 1977.

5. Meier, H. U., Keplin, H. P. and Vollmers, H., "Development of Boundary Layers and Separation Profiles on a Body of Revolution at Incidence", Proceedings of the 2nd Symposium on Numerical and Physical Aspects of Aerodynamic Flows, California State University, Long Beach, CA, 1983.

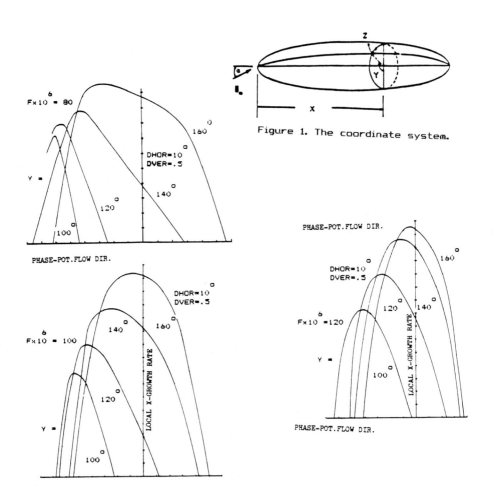

Figure 1. The coordinate system.

Figure 2. Local x-growth rates at x = 0.41, R = 1000.

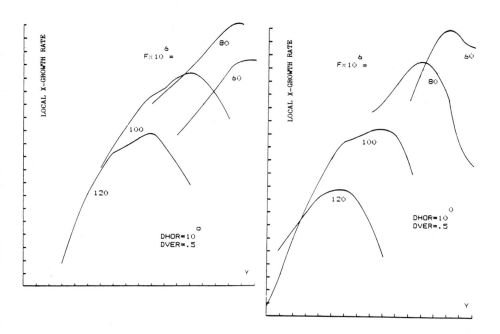

Figure 3. Local x-growth rates for max-amplified waves at x=0.59,0.75,R=1000.

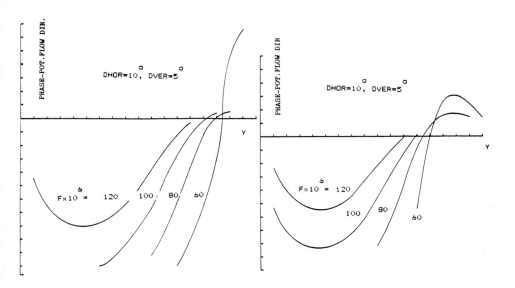

Figure 4. Phase direction for max-amplified waves at x = 0.59,0.75,R=1000.

Non-Linear Effects and Non-Parallel Flows: The Collapse of Separate Motion

F.T. SMITH

SUMMARY

The often-observed collapse or stall of separating flow, due to transition and subsequent abrupt turbulent reattachment of the motion downstream of a laminar separation point, at sufficiently high Reynolds numbers, is considered theoretically here for leading-edge separations and for mid-chord or trailing-edge separations. The former are viewed theoretically as unsteady marginal separations initially and are governed by a nonlinear integro-differential equation for the local unsteady skin friction. Computations and analysis show that in general the flow solution breaks down within a finite time, with a shock-like structure developing then. In the shock there is an abrupt switch from reversed flow just upstream of the shock to forward flow just downstream, i.e., a sudden reattachment. Second, for mid-chord or trailing-edge separations, the unsteady local viscous-inviscid-interactive motion is controlled initially by nonlinear triple-deck theory. Although this can lead on to transition locally for sufficiently large disturbances, it suggests a predominantly inviscid balance emerging beyond separation for smaller disturbances. This yields a prediction for the position of enhanced linear instability downstream of the separation point. On nonlinear grounds, however, the unsteady flow appears to develop a finite-time breakdown with, again, a shock-like collapse arising in the local or global eddy shape. Throughout, the unsteady interactive boundary-layer equations or a composite version would capture many of the flow properties; and emphasis is placed on nonlinear features, which are essential to the theoretical predictions of separating-flow collapse.

SECTION 1 - INTRODUCTION

It is well known that separating flows, nearly separating flows, and the like, are very prone to instability at sufficiently high Reynolds numbers, according both to experimental evidence and to the classical linearized-stability theory associated with infinitesimal disturbances. Such nonparallel basic flows tend to be susceptible to linear instabilities of the Rayleigh, Kelvin-Helmholtz and Görtler types on a local quasi-parallel basis as well as to the Tollmien-Schlichting instabilities which are possible in virtually all realistic boundary layers. The present theoretical contribution is aimed at extending the understanding of the instability of these nonparallel flows to encompass nonlinear features, rather than working further on the well-established existence of linear instabilities. The nonlinear behavior of imposed or free unsteady disturbances in parallel or in nonparallel flows is essential to the process of transition to turbulence in developing boundary layers, in separating flows and in other basic flows of aerodynamic concern. Prime examples of these other basic motions are flows over humps and corners on flat surfaces, modelling trip-wire transition and wall-roughness effects, nonlinear Tollmien-Schlichting flow and other time-dependent motions, trailing-edge motions and three-dimensional flows.

Most of the research described below is concentrated on the nonlinear instability of separating flow, which is a central example theoretically as well as one of much practical interest in aerodynamics, turbo-machinery and elsewhere. In practice (Refs. 1 through 4), it is commonly found that a separating laminar flow collapses abruptly, in spatial terms, with a sudden turbulent reattachment taking place just downstream of the separation point, due to transition and the enhanced entrainment of the resulting turbulent motion. This occurs for airfoil separations near the leading edge, mid-chord and near the trailing edge.

Our aim is to investigate theoretically the possibility of collapse, and this clearly requires a <u>nonlinear</u> approach. The research presented below is largely work in progress, with S. N. Brown, J. W. Elliott and H. K. Cheng, and some of the suggestions involved should be regarded as tentative as yet. The following two sections deal primarily with the nonlinear instability of leading-edge separation (Section 2) and of mid-chord separation (Section 3). The former is controlled by the features of nonlinear unsteady marginal separation, with more details being given in Ref. 8; and the second is viewed initially in terms of the viscous-inviscid triple-deck structure, which describes both the separation process and the nonlinear Tollmien-Schlichting instabilities, although further downstream the entire airfoil properties can be accommodated and Tollmien-Schlichting waves then transform in a sense to inviscid Rayleigh and Kelvin-Helmholtz types. Both of the nonlinear theories of Sections 2 and 3 predict an abrupt breakdown of the unsteady flow, with an effective "shock" forming in the shape of the dividing streamline. This shock-like effect provides the link with the earlier-noted experimental observations of a collapse.

Further comments are presented in Section 4. Nondimensionalized variables are assumed throughout, with Cartesian coordinates (x,y), corresponding velocity components (u,v), pressure p and time t, such that the typical free-stream speed is 1 and, in Section 3, the airfoil chord is 1. In Section 2 characteristic leading-edge coordinates are used. Incompressible-fluid or subsonic motion is studied, with the associated Reynolds number Re taken to be large, and the unsteady flow is assumed to be two-dimensional, as a starting point here, although the important element of three-dimensionality should be included in subsequent work.

SECTION 2 - UNSTEADY BREAKDOWN AND REATTACHMENT, IN LEADING-EDGE SEPARATING FLOW

Leading-edge separations and the possibility of stall arise on certain designs of airfoils and turbine blades at smallish angles of attack α to the oncoming stream $u = 1$, $v = 0$. If the experimental arrangement is "clean", with only small turbulence in the free stream, for example, then laminar flows with small- or larger-scale separations are observed near the leading edge at high Reynolds numbers. When this type of flow is sufficiently disturbed, however, it is found commonly in practice that transition and subsequent turbulent reattachment, due to enhanced entrainment, are produced in an abrupt fashion [Refs. 1-4] within the motion beyond the laminar separation point: see Fig. 1. The phenomenon significantly alters the downstream flow and in particular the lift and drag on the body. A number of turbulence-modelling calculations have been performed [e.g. Ref. 5 and references therein] which tend to tie in with the experimental findings, provided the transition point is chosen carefully in the calculations.

The switch from a laminar separating motion to the turbulent reattaching motion is clearly a nonlinear phenomenon. Here we approach the matter theoretically, from the standpoint of unsteady high-Reynolds-number flow, by taking the laminar separating flow as the starting configuration and studying the development, with time, of a nonlinear disturbance applied locally in the flow. Steady boundary-layer computations and analysis suggest that the starting configuration is an example of marginal separation (Refs. 6, 7) that arises at a critical angle of attack $\alpha = \alpha_c$, with the skin friction then being relatively small near the critical position $x = x_c$, in leading-edge coordinates.

Therefore we consider unsteady marginal separation (Refs. 7, 8, 22), where the unsteadiness is described by the unsteady classical incompressible boundary-layer equations. Then near $x = x_c$, within a length scale which is short, although strictly greater than $O(Re^{-1/5})$ to avoid interaction (see Ref. 8), the unsteady boundary-layer equations can be reduced in the marginal case to a nonlinear integro-differential equation for the scaled skin friction or decrement in displacement, $A(X,T)$, namely

$$A^2 - X^2 + \Gamma = -\int_{-\infty}^{X} \frac{\partial A}{\partial T} \frac{d\xi}{(X-\xi)^{1/4}} . \qquad (2.1)$$

Here X,T are the reduced length and time variables, and the constant Γ is effectively ± 1 or zero. The boundary conditions for (2.1) are of the form $|A| \sim |X|$ as $|X| \to \infty$ in all cases of interest. There are two simple stady states available in principle, for negative values of Γ, $A = \pm(X^2 + |\Gamma|)^{1/2}$, which correspond to totally forward/reversed flow in turn. But the former is found (below) to be attainable only with certain restrictions placed on the initial ($T = 0$) state of the unsteady flow, while the latter is unattainable from virtually any initial state. For, although the general linearized version of (2.1) obtained from a small perturbation of a steady state is difficult to solve because of the nonparallelism present, all the steady or transient flows are found to be very unstable to short-scale disturbances at any station where the local flow is reversed, $A < 0$.

Computations of the nonlinear system (2.1) give results of the kind sketched in Fig. 2. When the unsteady flow solution encounters reversed flow, i.e. $A < 0$, then there is a rapidly growing response in A of a rather localized form. The configurations tend to suggest a nonlinear breakdown of the solution within a finite positive time $T = T_o$, and a corresponding analytical structure can be set up, as follows.

In the nonlinear breakdown of the motion, as $T \to T_o-$, the flow solution focusses around a position $X = X_o$, within a short length scale $O(T_o - T)^{n_1}$, and $|A|$ is large and $O(T_o - T)^{-n_2}$. Here $n_1 = 4(1-n_2)/3$ and n_2 are unknown positive constants. This yields, from (2.1), the nonlinear integral equation

$$[A^*(X^*)]^2 = - \int_{-\infty}^{X^*} \left\{ n_2 A^*(\xi^*) + n_1 \xi^* \frac{dA^*}{d\xi^*} \right\} \frac{d\xi^*}{(X^*-\xi^*)^{1/4}} \qquad (2.2a)$$

for the scaled function $A^*(X^*)$ of the scaled spatial coordinate X^*. The boundary conditions required here are

$$A^*(\pm \infty) = 0 \qquad (2.2b)$$

since locally $|A|$ is large, whereas further away from the $X = X_o$ station $|A|$ remains finite.

Smooth solutions for A^* might be expected at first, but these could not be found from a direct approach, based on shooting from X^* large and negative, because the boundary condition at downstream infinity could never be satisfied. Instead, therefore, a global Newton iteration procedure was adopted, in which guesses were made for the value of n_1 and for $A^*(X^*)$ for all X^*. The Newton-linearized system based on the guesses was then solved in a two-point boundary-value manner, using Gaussian elimination, to satisfy both the end conditions in (2.2b) exactly. The linear increments were then applied to update the guesses and the procedure was repeated until a convergence tolerance was achieved. The resulting converged solution, extrapolated from finite-grid calculations, fixes n_1 and is sketched in Fig. 3. Its most surprising feature is the occurrence of a jump or "shock", in effect, at the origin of X^*. Across the shock the solution for A^* jumps abruptly from $A^*(0-)$ (negative) to $A^*(0+)$ (positive), where

$$A^*(0-) = -A^*(0+), \qquad (2.3)$$

representing an abrupt switch from reversed to forward flow there, or, which is equivalent, an abrupt switch from a much increased displacement just upstream to a much decreased displacement just downstream. The next point to consider is whether the shock is admissible or not, in the sense of being smoothed out within a still shorter region surrounding the origin of X^*.

The inner zone required has extent $O(T_o - T)^{m_1}$ in $(X - X_o)$, where m_1 is an unknown constant exceeding n_1. The amplitude of A is necessarily the same as in the previous zone and so the integral equation controlling the local scaled skin friction, $\hat{A}$ say, is

$$[\hat{A}(\hat{X})]^2 - \sigma^2 = -\int_{-\infty}^{\hat{X}} \frac{d\hat{A}}{dq} \frac{dq}{(\hat{X}-q)^{1/4}}, \qquad (2.4a)$$

where $\hat{X}$ represents the scaled X coordinate. The constant σ is positive and is the value of $-A^*(0-)$ (see (2.3)) derived in the previous calculation. The relevant boundary conditions for (2.4a) are

$$\hat{A}(\pm\infty) = \pm\sigma, \qquad (2.4b)$$

to allow the smoothed jump effect. Two solutions of (2.4a) are simply $\hat{A}(\hat{X}) \equiv \pm\sigma$, but eigensolutions exist also, of exponential-departure form far upstream,

$$\hat{A}(\hat{X}) \sim -\sigma + O(\exp[k\hat{X}]), \qquad (2.4c)$$

with k being a positive constant. It is noteworthy here that the departure solutions exist only for $\hat{A}(-\infty) = -\sigma$, not for $+\sigma$. If we then follow any of these departure solutions numerically and allow it to develop nonlinearly the solution is found to grow towards the downstream asymptote in (2.4b), in the form

$$\hat{A}(\hat{X}) \sim \sigma + O(\hat{X}^{-1/4}), \qquad (2.4d)$$

as shown in Fig. 4. The error term in (2.4d) fits in with the ensuing behavior beyond the shock and serves to determine the value of the constant m_1. Hence, finally, the terminal structure of the solution at its approach to the breakdown time $T = T_0$ is completed. The structure involves an effective shock, necessarily a nonlinear event, and in view of (2.4b,c) the jump across the shock is always from reversed flow to forward flow.

So the theoretical prediction overall, at this stage, is that if the original laminar flow is slightly disturbed or given a small kick, by means of turbulence in the free stream for instance, then (Fig. 5) the effects of the kick become amplified, grow nonlinear, and eventually produce in time an abrupt collapse and short-scale reattachment of the motion, downstream of the laminar separation. The predicted length scale of the process is between $O(1)$ and $O(Re^{-1/5})$ in terms of the leading-edge scaled coordinates.

This seems to tie in well qualitatively with the experimental findings (Refs. 1 through 4) and the computational results (Ref. 5). It should be emphasized here of course that the description is only the beginnings of a full account, the aim being to capture in theoretical terms the important effects of nonlinearity. After the local breakdown just described still shorter scales come into operation. There is indeed a cascade of shortening length scales, including $O(Re^{-2/7})$ (Ref. 7) and $O(Re^{-1/2})$, leading ultimately to the Euler equations holding locally but with a viscous layer inducing eruptions of vorticity (Ref. 10). Probably or possibly, this corresponds eventually to transition to turbulence setting in, if three-dimensional effects are also taken into account.

SECTION 3 - NONLINEAR INSTABILITY, AND BREAKDOWN, IN
SMALL- OR LARGE-SCALE SEPARATIONS AND OTHER DISTURBED FLOWS

Another kind of separation we would like to understand more, with regard to its nonlinear stability and transition, is the more global sort, away from the leading edge. Again, its occurrence depends on the airfoil or blade design, and it affects the overall flow field substantially. Of general concern, indeed, is the nonlinear stability of many related small- or large-scale disturbed flows, such as (Fig. 6) flow over a hump on a flat surface, flow past a concave or convex corner, breakaway separating flow, say from an airfoil surface, the stability of growing Tollmien-Schlichting waves themselves and of unsteady boundary layers, trailing-edge motions, and three-dimensional flows.

The types of major instabilities available which can be triggered off are (1) Tollmien-Schlichting waves, (2) Görtler vortices, (3) Kelvin-Helmholtz waves, on a broader scale, and (4) Rayleigh waves, of short or long scale. Which of these take precedent, if any, and whether they are significant or secondary, depends to a large extent on the typical frequencies and amplitudes of the disturbances present, through surface roughness, free-stream turbulence, vibration of the airfoil or from a ribbon, change in the Reynolds number, and so on. All the types are inter-linked, however, certainly through the Tollmien-Schlichting type (1) which can change form to produce types (2) (by means of three-dimensionality, Ref. 11), (3) and (4) (see for example Refs. 12, 13, and below), in a separating flow.

Let us take globally separating laminar flow as a starting state, typical, or the most dramatic, of those mentioned in the first paragraph, and consider the sorts of nonlinear phenomena that can happen with unsteadiness present. The first notable feature then is that the structure of such separations (Refs. 14, 15) is exactly the same as the structure of Tollmien-Schlichting instabilities (Ref. 16), namely the triple-deck structure (Fig. 7). In the separation, an incoming attached boundary layer, which is unstable only to viscous-inviscid Tollmien-Schlichting instabilities, is smoothly turned into an outgoing, separated, free shear layer by means of the steady nonlinear triple-deck problem. That provides the starting flow, and we examine next the effects of a slight kick or vibration to the system.

The influence of the nonparallelism of the starting flow and the nonlinearity present are mixed up together in general. That is, we have to solve the unsteady nonlinear triple-deck problem

$$u = \frac{\partial \psi}{\partial Y}, \quad v = \frac{-\partial \psi}{\partial X}, \tag{3.1a}$$

$$\frac{\partial u}{\partial T} + u \frac{\partial u}{\partial X} - \frac{\partial \psi}{\partial X} \frac{\partial u}{\partial Y} = \frac{-\partial p}{\partial X}(X,T) + \frac{\partial^2 u}{\partial Y^2}, \tag{3.1b}$$

with the displacement condition and surface constraint

$$u \sim Y + A(X,T) + F(X,T) \text{ as } Y \to \infty \tag{3.1c}$$

$$\text{no-slip at } Y = 0 \tag{3.1d}$$

and the pressure-displacement law, from the subsonic/incompressible outer potential flow,

$$p(X,T) = \frac{1}{\pi} \fint_{-\infty}^{\infty} \frac{\partial A}{\partial \xi}(\xi,T) \frac{d\xi}{(X-\xi)}. \qquad (3.1e)$$

Here (X,Y), (u,v), p, T are scaled coordinates, velocities, pressure and time, with respect to Re^{-n} ($n = 3/8$, $5/8$, $1/8$, $3/8$, $1/4$, $1/4$, in turn), while $A(X,T)$ is a scaled decrement in the displacement and $F(X,T)$ denotes the surface deformation if any. The reduced problem (3.1a-e) is both nonparallel, being nontrivially dependent on X, and nonlinear, and it therefore poses a difficult numerical task generally: see Refs. 10, 16, 19. Linear Tollmien-Schlichting instabilities are present even for an unseparating basic boundary-layer motion, e.g. Blasius flow, $U=Y$, $V=P=A\equiv 0$, once the reduced frequency Ω of the disturbances exceeds $\Omega_c = 2.30$ (Refs. 16, 17); and the presence of separating motion instead as the base state can be expected to worsen matters here.

One way to gain some useful analytical insight, however, is to consider sending a relatively high-frequency wave through the motion, or to examine the downstream movement of a disturbance of given frequency (Refs. 10, 19). The two processes are equivalent because of the nonparallelism existing in the underlying boundary-layer flow. They correspond to an increased frequency parameter or fast time-dependence, $\Omega \equiv O(|\partial/\partial T|)$ then being large.

For fast time-dependence, multiple scales operate in the form

$$\left.\begin{array}{l} \dfrac{\partial}{\partial T} \rightarrow \Omega \dfrac{\partial}{\partial T_o} + \Omega^{1/2} \dfrac{\partial}{\partial T_1} + \dfrac{\partial}{\partial T_2} + \ldots, \\[2ex] \dfrac{\partial}{\partial X} \rightarrow \Omega^{1/2} \dfrac{\partial}{\partial X_o} + \dfrac{\partial}{\partial X_1} + \Omega^{-1/2} \dfrac{\partial}{\partial X_2} + \ldots, \end{array}\right\} \qquad (3.2)$$

so that the induced streamwise length scale is also shortened. If the initial kick to the system is not too large the local flow (3.1a-e) simplifies at leading order to a linear balance $\partial u/\partial T = -\partial p/\partial X + \partial^2 u/\partial Y^2$, in effect, within a Stokes layer, and the boundary conditions stay much as before except that the basic starting flow plays little part so far. Letting $Y \to \infty$ then we have

$$\partial A/\partial T = -\partial p/\partial X. \qquad (3.3)$$

This inviscid result, coupled with the inviscid law (3.1e), fixes the induced effective wavenumber and gives neutral stability,

$$\partial/\partial X_o = -\partial/\partial T_o. \qquad (3.4)$$

Hence the flow response, to leading order, is a neutral wave travelling downstream. Amplitude growth or decay is then decided at higher order by a combination of weak nonlinearity and viscous forces. If the pressure has the size

$$p = [\Omega^{1/2} p_o + p_1 + \ldots] E + \begin{Bmatrix} \text{complex} \\ \text{conjugate} \end{Bmatrix} + (E^2, \ldots), \qquad (3.5)$$

where $E \equiv \exp(i(X_o - T_o))$ from (3.4), we find that the slower-scale variation, in a downstream-travelling frame of reference, is controlled by a generalized cubic Schrödinger equation of the form

$$\left. \begin{aligned} &\left(\frac{\partial p_o}{\partial T_2} \; \frac{-2\partial p_o}{\partial X_2} \; \frac{-i \, \partial^2 p_o}{\partial X_1^2} \right) + \left(\frac{\partial p_1}{\partial T_1} \; \frac{-2\partial p_1}{\partial T_1} \; -i\overline{A} \; p_1 \right) \\ &-2\overline{A} \, \frac{\partial p_o}{\partial X_1} = \left[\left(\frac{1-i}{\sqrt{2}} \right) + \frac{d\overline{A}}{dX_1} \right] -2i\overline{p} \;\; p_o - \frac{5i}{4} \, p_o |p_o|^2. \end{aligned} \right\} \qquad (3.6)$$

Here nonparallelism is active through the starting steady-flow displacement and pressure, $-\bar{A}(X_1)$ and $\bar{p}(X_1)$ respectively, while nonlinearity figures in the amplitude-cubed term on the right-hand side.

At this stage the pressure response still depends on the disturbance amplitude that is introduced.

First, if the amplitude is not sufficiently small then nonlinearity is found to cause the distribution of $|p_o|$ to grow and spread exponentially fast: Ref. 19. Nonparallelism has little effect and can be absorbed into the correction pressure p_1. The main response, in p_o, involves a massive, exponentially large, distribution of the amplitude $|p_o|$, taking on an elliptical shape of exponentially large streamwise extent [see sketch in Fig. 8a] as the whole disturbance packet travels downstream. That is with completely <u>two-dimensional</u> motion assumed. The <u>three-dimensional</u> extension of the theory applied to oblique waves is found to change the sign of the double spatial derivative in (3.6), for wavefronts inclined at angles θ greater than

$$\theta = \tan^{-1}\left(\sqrt{2}\right) \qquad [= 54.7°] \tag{3.7}$$

to the free stream, and this change leads instead to a much faster-growing and apparently chaotic nonlinear behavior [Ref. 25, and sketched in Fig. 8b] in the flow solution, because of sideband instabilities. The apparent chaos and nonlinear growth are enhanced if the starting flow is itself three-dimensional also, and in addition fast growth occurs in resonant-triad interactions:

Ref. 25. There is good agreement here between (3.7) and the experimental findings of fast growth for oblique waves at angles θ beyond approximately 60° [Refs. 20, 21] to the stream. Even further downstream the Euler equations can eventually come into play with increased nonlinearity, but subject to an unstable viscous sub-layer, at the surface, which forces vorticity eruptions into the main unsteady boundary-layer flow. Again, this aspect agrees at least qualitatively with computational and experimental observations in transitional boundary layers: Refs. 23, 24.

On the other hand, if the disturbance amplitude is kept small initially, then nonparallel flow effects remain substantial. For (3.6) shows that the growth rate is positive or negative depending on whether the local slope of the displacement is above or below a certain critical value (Ref. 10), the criterion for instability being

$$\frac{d\overline{A}}{dX_1} < \frac{\sqrt{2}}{3}. \tag{3.8}$$

The associated stabilization or de-stabilization of the motion makes good physical sense in terms of corner flows, trailing-edge flows and flow over a hump, for instance. In particular, in the breakaway separating flow of present concern the displacement $-\overline{A} \propto X_1^{3/2}$ downstream: Ref. 15. Hence the flow is increasingly destabilized there. In fact, non-parallel-flow effects increase fast downstream and they affect the leading-order balance represented by (3.3) when the distance X_1 is large, positive and $O(\Omega^{1/3})$.

At that distance beyond separation the unsteady-flow behavior revolves around the stability of a simple quasi-parallel basic flow (Fig. 9), namely the uniform shear $\bar{u} = Y-d$ above the now-detached shear layer, for $Y > d$, and the negligible flow $\bar{u} = 0$ underneath, for $0 < Y < d$. This base state is the downstream form of the steady breakaway-separating motion (Ref. 15), with $d(X_1) \propto X_1^{3/2}$ standing for the distance of the free shear layer from the smooth surface. The associated linear and nonlinear stability properties are predominantly inviscid, or rather, in the nonlinear regime, will be taken to be so at the first approximation: see also later.

If we examine <u>linear</u> theory first, for a small travelling-wave disturbance of scaled wavenumber α and wavespeed c, then the perturbation's stream function above the shear layer has the form

$$\tilde{\psi} = \tilde{p} + \tilde{A}(Y-d-c) \tag{3.9a}$$

for $Y > d$, while underneath, in the slow-turning fluid occupying $0 < Y < d$,

$$\tilde{\psi} = \tilde{p}Y/c \tag{3.9b}$$

to preserve tangential flow at or near the surface. Tildas here denote perturbation quantities. The subsonic interaction law (3.1e) now becomes

$$\tilde{p} = \frac{\alpha^2}{|\alpha|} \tilde{A}. \tag{3.9c}$$

Here Real (α) is taken below to be positive, for downstream-moving disturbances, although it is of interest that some instability also results with Real (α) negative. Conservation of mass flux (the pressure is already continuous in (3.9a,b)) across Y=d therefore yields the eigen-relation $c^2 - \alpha c + \alpha d = 0$, from (3.9a-c): c.f. Ref. 9. Physically it seems more sensible to convert this relation to one determining α for a given prescribed frequency $\Omega(= \alpha c)$, which yields the cubic equation

$$\alpha^3 d - \alpha^2 \Omega + \Omega^2 = 0 \qquad (3.10)$$

for α. The equation has certain notable properties. First, it provides the continuation of the Tollmien-Schlichting mode described previously: the solution for α can match to the incoming Tollmien-Schlichting response in (3.4), upstream, since there d → 0 and the second and third terms in (3.10) can then dominate, giving $\alpha \to \Omega^{1/2}$. Second, inviscid Rayleigh-type waves are possible also, these occurring when interaction is suppressed ($\tilde{A} \to 0$ in (3.9a)) and the first two terms in (3.10) become dominant, yielding $\alpha \to \Omega/d$. These waves are neutral and correspond to shortened length scales, $\alpha \gg 1$, again for d small when Ω is fixed, or, more generally, for small values of the parameter d^2/Ω governing (3.10). Third, and in contrast, for large d (or d^2/Ω), i.e. further downstream, the first and third contributions in (3.10) can dominate and yield instability of a Kelvin-Helmholtz variety, since then $\alpha \sim (\Omega^2/d)^{1/3} \exp(\pm i\pi/3)$. Fourth, and perhaps the most significant feature, is the criterion for instability, i.e. complex roots for α, from (3.10): see Fig. 10. This is $d > d_{crit}$ where $d_{crit} = 2\sqrt{\Omega/27}$, corresponding to $X > X_{crit}$ where

$$X_{crit} = \left(\frac{\Omega}{3b^2}\right)^{1/3} \qquad [\text{and } b=0.44] \qquad (3.11)$$

in normalized terms, since $d = (2b/3)X^{3/2}$ [Ref. 15]. So on linear grounds there is a definite starting position, given by (3.11), at which abrupt disturbance growth [of an inviscid kind, much faster than in the flow upstream, at least for sufficiently small disturbances] is predicted downstream of the separation point.

It would be interesting to make a quantitative comparison between the prediction (3.11) and a controlled experiment measuring the breakdown position for a fixed-frequency vibration started ahead of the flow separation.

Concerning the <u>nonlinear</u> development next, further study is under way on the effect on the weakly nonlinear amplitude equation (3.6) when (3.4) is replaced by (3.10) in separating flow. For stronger nonlinearity, however, the main difference from the above is that the free shear layer or vortex sheet can move substantially (Fig. 11), its unknown scaled position being given by $Y = S(X,T)$, say. So, with the local flow variables u, ψ, p, A, X, Y, T now scaled with respect to Ω^m ($m = 1/2, 1, 1, 1/2, -1/2, 1/2, -1$ in turn), and at the same distance $O(\Omega^{1/3})$ beyond separation as before, the nonlinear momentum equation for the unsteady flow II between the free shear layer and the airfoil surface is the inviscid version of (3.1b), which forces the vorticity $\partial u/\partial Y$ to remain constant at any given fluid particle for all time. The scaled streamwise range here is $-\infty < X < \infty$, and the appropriate boundary conditions are $\psi = 0$ at $Y = 0$, for tangential flow at the surface, and the kinematic constraint $v = \partial S/\partial T + u\partial S/\partial X$ at $Y = S(X,T)-$. Above the shear layer, in zone I, the same thin-layer-Euler equation applies, subject to the same kinematic constraint at Y = S+ and to the outer shear condition (3.1c), with now F ≡ 0. In addition the pressure-displacement law (3.1e) holds. For given starting conditions, then, and suitable far-field requirements, an apparently closed, nonlinear, interacting-flow problem is posed for the two outer and inner flows, I, II, on either side of the unknown free shear layer.

As an example, suppose the vorticity in the inner zone II is initially zero and that in the outer zone I is unity. Then in zone II, for all times T at which the solution continues to exist,

$$u = U^-(X,T), \quad \psi = YU^-(X,T)$$

with the velocity U^- being an unknown function of X,T, and as a result the momentum balance and the kinematic condition become

$$\frac{\partial U^-}{\partial T} + U^- \frac{\partial U^-}{\partial X} = -\frac{\partial p}{\partial X} \tag{3.12a}$$

$$\frac{\partial S}{\partial T} + \frac{\partial}{\partial X}(U^- S) = 0, \tag{3.12b}$$

respectively. Meanwhile, zone I yields the solution

$$\psi = \frac{1}{2}(Y + A)^2 + p + Q(X,T),$$

where $\partial Q/\partial X = \partial A/\partial T$, and the kinematic condition for zone I then yields the relation

$$\frac{\partial}{\partial T}(S + A) + (S + A)\frac{\partial}{\partial X}(S + A) = \frac{-\partial p}{\partial X}, \qquad (3.12c)$$

while from (3.1e) we have

$$p(X,T) = \frac{1}{\pi} \fint_{-\infty}^{\infty} \frac{\partial A(\xi,T)}{\partial \xi} \frac{d\xi}{(X-\xi)}. \qquad (3.12d)$$

Hence the four unknown functions U^-, S, A, P evolve nonlinearly according to the four equations (3.12a-d), with suitable initial and farfield conditions. The linearized version, we note, with those four functions perturbed about zero, d, -d, zero in turn, reproduces the results (3.9a) - (3.11).

There are numerous aspects of the evolution problem (3.12a-d) still to be followed through. These include the possible existence of nonlinear travelling waves at large times, if the pressure amplitude can remain within bounds at finite times (see below); the likelihood of vorticity bursts emerging from the viscous wall layer (a classical unsteady boundary layer between zone II and the airfoil surface), once the slip velocity U^- becomes too extreme (again, see below); and the counterparts in supersonic separating flow, which is found to be de-stabilized, in channel flow, in wind-driven water flow, and elsewhere. The main aspect in the current, subsonic, separating flow, however, is that once significant deviation occurs, analogous to the criterion (3.11), then the flow solution seems most likely to break down within a finite time.

One such breakdown occurs for the more downstream flow, where (3.12c) is suppressed and A is replaced by $-S$ in (3.12d), leaving three equations, not unlike the shallow-water equations, for U^-, S, P. In that case, and for restricted initial conditions (Ref. 22), the flow solution focusses as T approaches a breakdown time T_o, in the form

$$[U^-, S, P] \sim [(T_o-T)^{n-1} U^*, (T_o-T)^{3n-2} S^*, (T_o-T)^{2n-2} P^*] \quad (3.13)$$

around a station $X=X_o$, with the starred variables dependent only on $X^* \equiv (X-X_o)(T_o-T)^{-n}$, so that $n > 0$ since $X-X_o$ is small. Here S^* is necessarily positive. The governing equations now reduce to the nonlinear similarity system

$$(2-3n)S^* + nX^* \frac{dS^*}{dX^*} + \frac{d}{dX^*}(U^* S^*) = 0 \quad (3.14a)$$

$$(1-n)U^* + nX^* \frac{dU^*}{dX^*} + U^* \frac{dU^*}{dX^*} = \frac{-dP^*}{dX^*} \quad (3.14b)$$

$$P^*(X^*) = \frac{-1}{\pi} \int_{-\infty}^{\infty} \frac{dS^*}{dq} \frac{dq}{(X^*-q)} \quad (3.14c)$$

for U^*, S^*, P^*. Many options are available for the similarity constant n, but the value we tend to favor, from some tentative analysis, is $n = 2/3$.

This value is also implied in the linearized unsteady version described previously, and for $n = 2/3$ a linearized version of (3.14a-c) where $[U^*, S^*, P^*]$ are slightly perturbed about $[0, S_o^*, 0]$ (S_o^* = constant) is found to allow solutions (Fig. 12) which are well-behaved at infinity. For $n \leq 1/2$, on the other hand, (3.14a-c) lead to a contradiction. With $n = 2/3$ provisionally, then, we have the induced speed $|U^-|$ and pressure p being very large at the breakdown, while the eddy shape S remains finite and produces a jump (Fig. 13), from $X^* = -\infty$ to $X^* = \infty$, or in effect a "shock".

The same three unsteady governing equations that led to (3.14a-c), that is, (3.12a,b) and

$$p(X,T) = \frac{-1}{\pi} \fint_{-\infty}^{\infty} \frac{\partial S(\xi,T)}{\partial \xi} \frac{d\xi}{(X-\xi)} , \qquad (3.15)$$

also apply in a modified sense to separating motion, with thin eddies, on a broader length scale comparable with the airfoil chord: Fig. 14. This corresponds to nonlinear unsteady effects acting on the separated-airfoil-flow model of Ref. 26, for example, with the modification that among other things the integration range should now take account of the moving separation position and the finiteness of the thin airfoil. Thus a kick to the separating flow, from free-stream turbulence for instance, has most effect in a slower-moving eddy and indices nonlinear motion there, given by (3.12a,b), whereas outside only linear disturbances are felt, as covered by the modified form of (3.15). The evolution of the flow (see Fig. 14), with interaction, then brings about the above shock-like collapse in the eddy boundary.

The suggested shock-like effect appearing abruptly in the free-shear-layer shape or eddy boundary, when the separating flow is nonlinearly disturbed, seems to be the main physical feature predicted theoretically so far, and it tends to agree qualitatively with experimental and computational findings (e.g. Refs. 1-5) on the collapse of separating flow. A quantitative comparison would be more helpful, however, and to that end among others the analysis should be pursued further.

SECTION 4 - FURTHER COMMENTS

Concerning Sections 2, 3, and the starts made there in describing the nonlinear evolution of significantly disturbed laminar separating flow, there is clearly much theoretical and numerical study still to be done. Some of this has already been mentioned in the text. Additional questions are, for example: how do the still shorter length-scale properties, implied by the finite-time collapses in both of Sections 2, 3, evolve as new and faster physical mechanisms come into the reckoning locally; is the nonlinear system (3.12) an over-simplification of the more general nonlinear problem; does nonzero eddy vorticity substantially alter the breakdown properties there (preliminary analysis suggests not, in the case of a nonzero uniform vorticity); is nonsymmetric breakdown more violent than the symmetric airfoil case referred to just after (3.15); and what of the effects of three-dimensionality, often so significant in boundary-layer transition? In the meantime, however, it seems that some degree of encouragement can be drawn from the studies in Sections 2, 3, as regards obtaining clear theoretical accounts of leading-edge stall and mid-chord or trailing-edge stall. The former is somewhat boundary-layer-controlled initially (or viscous-inviscid in Ref. 7), until interaction comes in locally, whereas the latter is mainly inviscid, although vorticity eruptions from a viscous sub-layer play a part. In both cases, nevertheless, the flow theory leads to the formation of a <u>collapse</u> in the separating motion or a shock-like effect, an essentially nonlinear phenomenon, and one which is in line with experimental results.

The above features suggest that appropriate computational schemes for unsteady flow, based on the interacting boundary-layer equations or a composite set for example, should be able to capture many of the properties of separating-flow collapse observed experimentally (Refs. 1 through 4). There is also the suggestion in Section 3 that experimental measurements of the breakdown position, beyond laminar separation, when a fixed-frequency vibration is imposed upstream, would be desirable (if they have not been done already), this to compare with the analytical prediction.

A further matter is that viscous forces still have a substantial role in the breakdown processes, despite the tendency of nonlinear inviscid dynamics to appear dominant at first. For viscous forces cause the eruptions of vorticity from the surface sub-layer which are an experimentally recognized part of the local transition in boundary layers, separating or otherwise. Accordingly, near the end of Section 3, there is something of a limitation, according to rational theory if not in practical terms, on how close to the breakdown time the non-linear inviscid evolution equations such as (3.12a-d) continue to apply, before viscous effects re-assert themselves through the sub-layer vorticity burst. This is rather an involved matter. But, against that, there is a possibly hopeful sign in that (Ref. 18) the initial burst of the sublayer is itself controlled, in a rational sense, by the same thin-layer Euler equations and pressure-displacement interaction law as hold in the latter parts of Section 3. So, although the boundary conditions are different here, a breakdown and perhaps a shock-like effect are distinct possibilities for the bursting sublayer also, in much the same vein as or similar to those in Section 3.

In the above flows nonparallel effects can exert a strong influence, at least in getting disturbances started, although once nonlinearity comes into operation shortened length scales tend to be induced and nonparallelism then takes second place. Another type of linear and nonlinear instability where nonparallelism matters is currently being studied, by J. Bennett, R. V. Brotherton-Ratcliffe, O. R. Burggraf and the author. This is where a fixed-frequency wave is driven through a basic starting flow which, in spatial terms, is nonparallel but is stable upstream and then accelerates downstream. Two examples are trailing-edge flow and flow through corners in pipes. The former is initially stable to Tollmien-Schlichting waves of sufficiently low frequency and one might expect the accelerating wake motion to have a stabilizing effect on these, although destabilizing with respect to Rayleigh modes. Computations and subsequent analysis show that instead the wake flow becomes destabilized, because any maintained time-dependence eventually overwhelms the inertial forces, as these decrease downstream despite the acceleration present. In the context of cornered pipe flows a similar process occurs. Ahead of a slight corner, say in fully-developed Hagen-Poiseuille flow with axisymmetry present, any boundary-layer motion is like plane Couette flow and is linearly stable; so there the fixed-frequency disturbance decays spatially with distance downstream. Beyond the corner the flow accelerates if the corner is concave, constricting the flow, and intuition suggests a stabilizing effect. The opposite is true. The cornered flow admits viscous-inviscid instabilities of the Tollmien-Schlichting kind due to the curved attached boundary-layer profiles produced there. Moreover, the disturbance blows up in a singularity within a finite distance of the corner, both in the linear and nonlinear regimes. A similar de-stabilization and blow-up may occur also in front-stagnation-point motion.

The interest of Dr. M. J. Werle (UTRC) and Dr. R. E. Whitehead (Office of Naval Research) in this research is gratefully acknowledged.

REFERENCES

1. Tani, I., 1964, Progr. Aeron. Sci., $\underline{5}$, 70.

2. Gaster, M., 1966, AGARD Conf. Proc. No. 4, p. 819, and Aero. Res. Counc. Rept. & Memo., No. 3595, 1969.

3. Mueller, T. J. & Batill, S. M., 1980, AIAA Paper No. 80-1440; and Mueller, T. J., 1984, AIAA Paper No. 84-1617.

4. Van Dyke, M., 1982, "An Album of Fluid Motion", Stanford: Parabolic Press.

5. Davis, R. L. & Carter, J. E., 1984, AIAA Paper No. 84-1613 (presented, Snowmass, Col., June 1984); and NASA Conf. Rept., No. 3791.

6. Stewartson, K., Smith, F. T. & Kaups, K., 1982, Stud. in Appl. Math., $\underline{67}$, 45.

7. Smith, F. T., 1982, Aeron. Quart., November/December, 331.

8. Smith, F. T. & Elliott, J. W., 1985, Proc. Roy. Soc. A (in press).

9. Drazin, P. G. & Reid, W. H., 1981, "Hydrod. Stability", Cambr. Univ. Press.

10. Smith, F. T. & Burggraf, O. R., 1985, Proc. Roy. Soc. A, $\underline{399}$, 25.

11. Hall, P. & Smith, F. T., 1984, Stud. in Appl. Math., $\underline{70}$, 91.

12. Bodonyi, R. J., 1985, Proc. Symp. on Stability of Spatially-Varying and Time-Dependent Flows, NASA Langley Res. Ctr., August 19-20, 1985.

13. Smith, F. T., & Bodonyi, R. J., 1985, Aeron. Journl., Summer Issue.

14. Sychev, V. V., 1972, Izv. Akad. Nauk. SSSR, Mekh. Zh. Gaza, $\underline{3}$,47 [transl. Fluid Mech., 407-419, 1974, Plenum Pub.].

15. Smith, F. T., 1977, Proc. Roy. Soc. A $\underline{356}$, 433.

16. Smith, F. T., 1979, Proc. Roy. Soc. A $\underline{366}$, 91 and A $\underline{368}$, 573.

17. Reid, W. H., 1965, in "Basic Develop. in Fluid Dyn." (ed. M. Holt), Vol. 1, p. 249, Academic Press.

18. Elliott, J. W., Cowley, S. J. & Smith, F. T., 1983, Geo. Astro. Fluid Dyn., $\underline{25}$, 77.

REFERENCES (Cont'd)

19. Smith, F. T., 1985, United Tech. Res. Ctr., E. Hartford, CT, Report 85-36.

20. Saric, W. S., Kozlov, V. V. & Levchenko, V. Ya., 1984, AIAA Paper No. 84-0007 (presented January 1984, Reno, Nevada).

21. Herbert, T., 1984, AIAA Paper No. 84-0009 (presented January 1984, Reno, Nevada).

22. Ryzhov, O. S. & Smith, F. T., 1984, Mathematika, December Issue, $\underline{31}$, No. 62.

23. Walker, J. D. A. & Abbott, D. E., 1977, in "Turb. in Int. Flows," (ed. S. N. B. Murthy), 131, Hemisph. Pub. Corp., Wash..

24. Walker, J. D. A. & Scharnhorst, R. K., 1977, in "Recent Advs. in Eng. Sci." (ed. G. C. Sih), 541, Univ. Press, Bethlehem, PN..

25. Smith, F. T., 1985, United Tech. Res. Ctr., E. Hartford, CT, Report 85, in preparation.

26. Cheng, H. K. & Smith, F. T., 1982, Z.A.M.P., $\underline{33}$, 151.

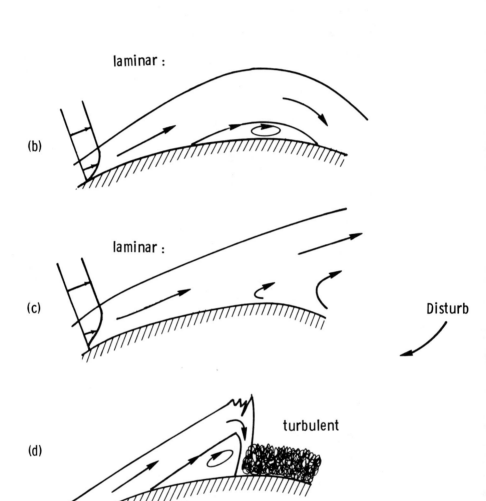

Fig. 1 Concerning leading-edge separation and collapse: the representative effect in practice of a significant disturbance, to small- or larger-scale laminar separating flow (a)-(c), producing turbulent reattachment in (d).

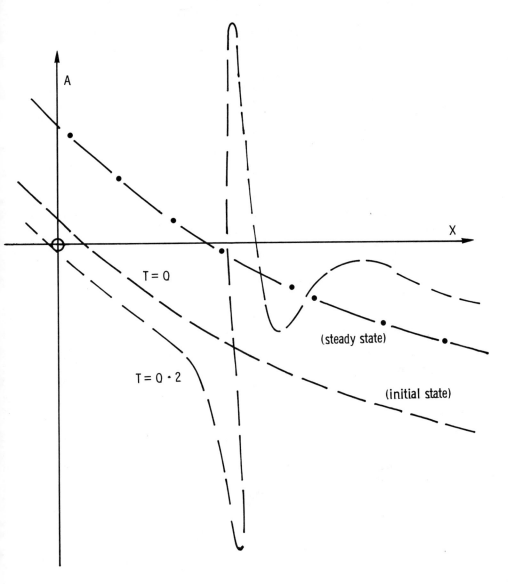

Fig. 2 Sketch [see Ref. 8] of the evolution of the skin friction $A(X,T)$, according to (2.1), from an initial state, and the violent behavior encountered when reversed flow is present.

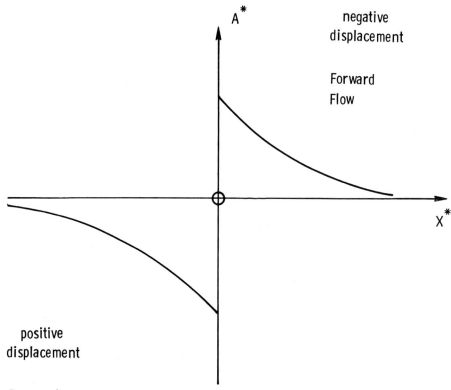

Fig. 3   The breakdown solution, diagrammatically, for A^* versus X^* from (2.2a,b), Ref. 8, showing the shock-like jump (at the origin of X^*) from reversed flow upstream to forward flow downstream.

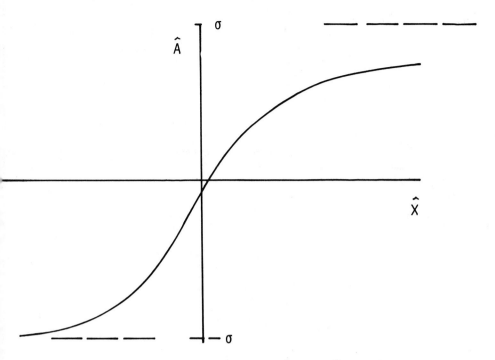

Fig. 4 Diagram of the inner-shock solution $\hat{A}$ versus $\hat{X}$, from (2.4a,b) and Ref. 8, showing the transition from reversed to forward motion.

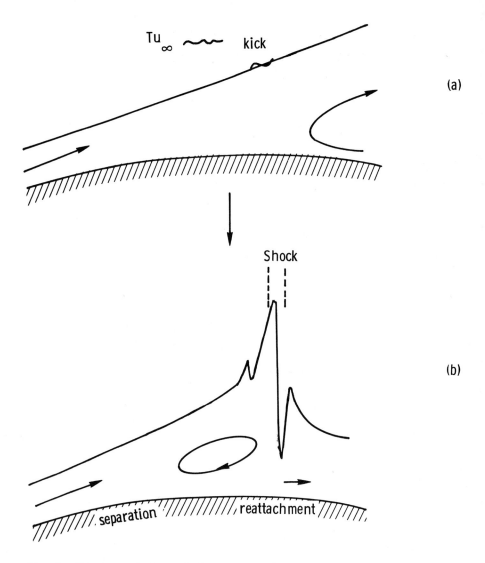

Fig. 5 The theoretical prediction from Section 2, of a laminar leading-edge separation (a) being significantly disturbed and producing the "shock" effect and abrupt reattachment in (b).

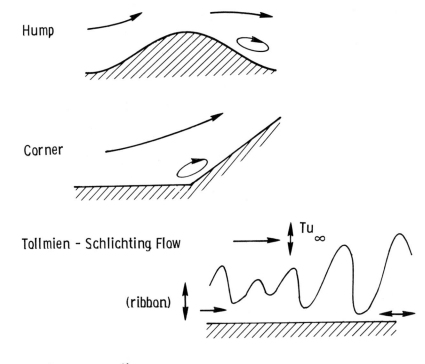
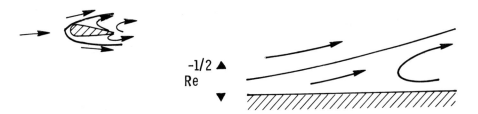

Fig. 6 Basic flows of concern (Section 3) with regard to their nonlinear instability properties: flow past a hump or a corner, nonlinear Tollmien-Schlichting flow, and breakaway separating motion.

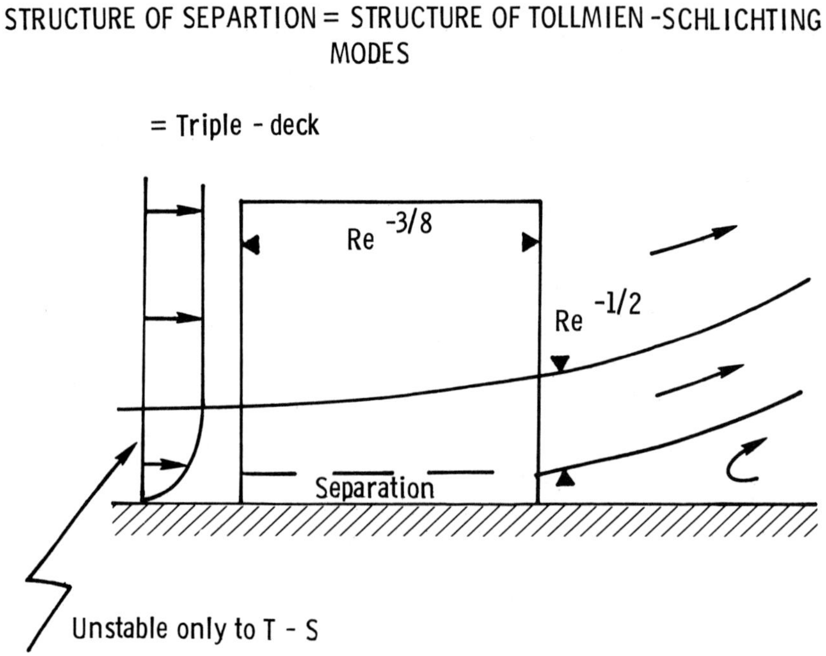

STRUCTURE OF SEPARTION = STRUCTURE OF TOLLMIEN -SCHLICHTING MODES

= Triple - deck

$Re^{-3/8}$

$Re^{-1/2}$

Separation

Unstable only to T - S

Fig. 7 The triple-deck structure, describing both breakaway separation and Tollmien-Schlichting instabilities.

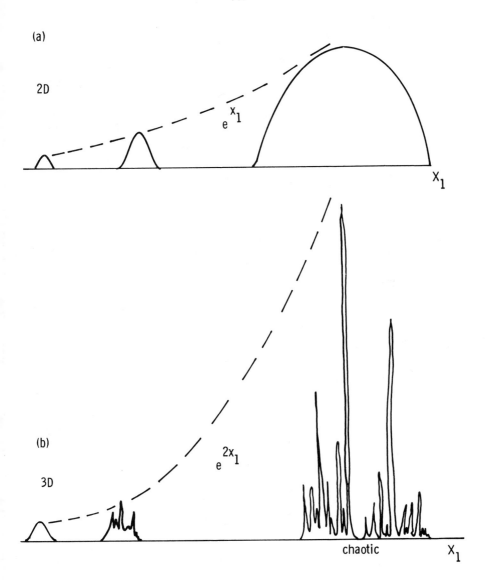

Fig. 8 Sketches of the nonlinear evolution of: (a) two-dimensional disturbances, governed by (3.6), for the pressure amplitude $|p_0|$ (Ref. 19); (b) three-dimensional oblique waves (Ref. 25) at angles from the free-stream direction exceeding θ in (3.7). The growth rate in the (chaotic) case (b) is the greater.

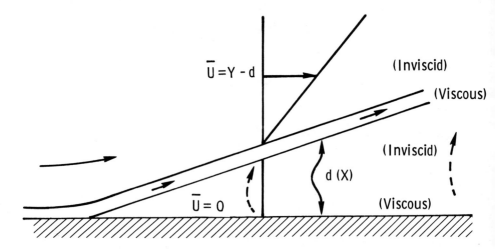

Fig. 9 The basic detached flow just beyond separation, at the downstream end of the triple-deck.

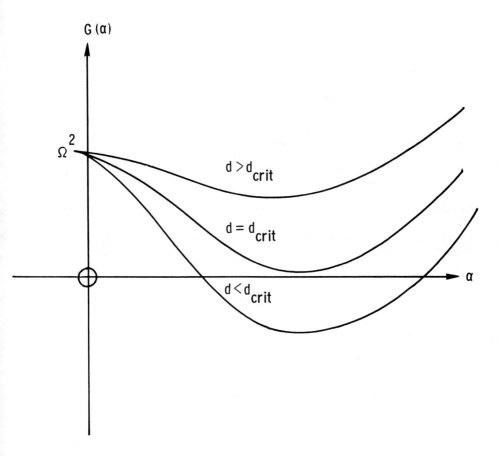

Fig. 10 The left-hand side of Eq. (3.10), $G(\alpha)$, versus α, for various values of the shear-layer height d, including the critical value d_{crit} which leads to the prediction (3.11). For $d > d_{crit}$ the two roots of $G(\alpha) = 0$ with Real(α) > 0 are both complex, leading to linear instability.

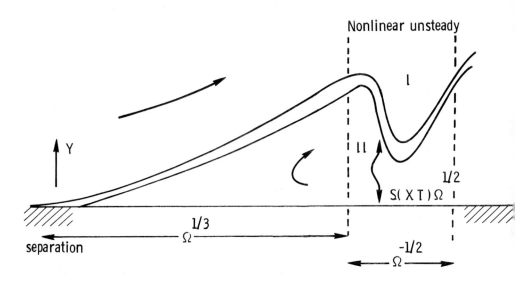

Fig. 11 The structure of the nonlinear unsteady separating flow, leading in a special case to (3.12a-d).

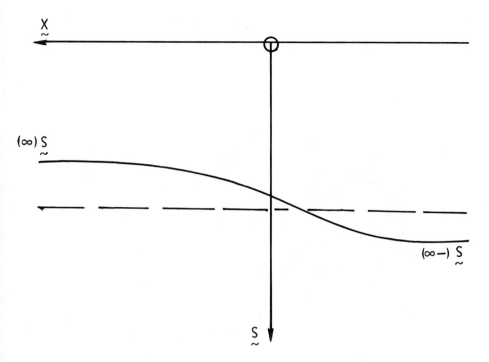

Fig. 12 Linearized solution, sketched, of the nonlinear breakdown equations (3.14a-c) when n = 2/3, with a jump produced between $X^* = -\infty$ and $X^* = \infty$.

Effect

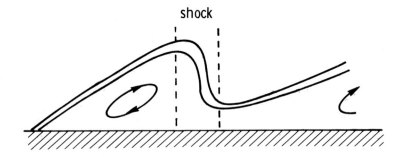

Fig. 13 The theoretical prediction, from Section 3, of a nonlinear shock-like breakdown of the separating flow.

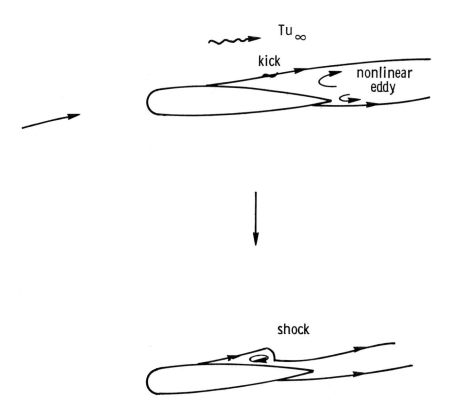

Fig. 14 The theoretical nonlinear effects of free-stream turbulence or other significant disturbances to separated airfoil flow, forcing the shock-like collapse of the eddy.

On Short-Scale Inviscid Instabilities in the Flow Past Surface-Mounted Obstacles

R.J. BODONYI AND F.T. SMITH

1. Introduction

The steady and unsteady effects of small surface-mounted obstacles on the boundary-layer flow over a surface have been of experimental concern for many years. These effects include most notably separation and instability, often leading to transition to turbulence. Two main reasons possible for such transition have been suggested: either the sur distortion produces in effect a locally separated shear flow which is susceptable to inviscid instabilities associated with the inflectional velocity profile or vortex sheet present; or there is a sensitive interaction between the surface distortion and the basic flow, possibly with unsteadiness/turbulence in the free stream, which can readily accentuate the viscous-inviscid growth of the Tollmien-Schlichting instabilities which are usually present in boundary layers in any case.

On the theoretical side the major steady-flow phenomenon observed, separation, is now well understood in laminar flow. It is generally of an interactive viscous-inviscid type in which the flows inside and outside the boundary layer affect each other significantly within a relatively short length scale. The question of the stability of separating flow or other locally distorted steady or unsteady motions is always present, however, and this has started to receive increased attention recently, in part because of modern developments in boundary-layer methods. Some studies of the influence of surface distortions or of laminar separation on the behavior of Tollmien-Schlichting instabilit have been made: e.g. [3], [9]. In this paper we shall be concerned with the alternative (which is far from new: see e.g. [16]) of essentially inviscid, i.e. Rayleigh, instability occuring due to inflectional velocity profiles being produced locally. As in the studies just mentioned, the basic motion is necessarily a nonparallel one. However, since the typical length scales of Rayleigh instability tend to be much shorter initially than those of Tollmien-Schlichting instability the nonparallelism is then a secondary feature to a large extent. Whether in a global sense the Rayleigh instability is more significant than distorted Tollmien-Schlichting behavior in nature is not entirely clear, although the former does readily represent a bursting phenomenon due to the fairly sudde production of faster spatial and temporal effects and also the former instability can develop as a secondary instability of the distorted Tollmien-Schlichting behavior (see e.g.[15]).

Two-dimensional unsteady flow properties are examined in this study, with velocity components $U^*(u,v)$ in the Cartesian frame $(x^*,y^*) = L^*(x,y)$, where U^* is a typical main stream speed and L^* is a characteristic streamwise length scale of the undistorted laminar boundary layer adjoining the solid surface $y = 0$. The pressure is written $\rho^* U^{*2} p$ and the Reynolds number $Re = U^* L^*/\nu^*$ is large, with ρ^*, ν^* denoting the density and kinematic viscosity, respectively, of the incompressible fluid. The nondimensional stream function is ψ and time is written as $L^* t/U^*$.

2. Scalings and Governing Equations

Our interest is in the stability of nonparallel flows and especially those with local regions of flow reversal, in view of the potential applications to trip-wire transition and wall-roughness effects. It is appropriate therefore to take steady nonlinear viscous-inviscid interactive solutions, of the triple-deck and similar kinds, for the basic steady motion. This is because, as is now well-known, flow reversal for small or large-scale separations in such motions is not a catastrophic event: the solution at the separation point is regular due to the presence of interaction, unlike that in steady classical, i.e. noninteractive, boundary layers for instance. Hence a steady nonparallel basic flow smoothly exhibiting a finite or semi-finite range of reversed motion can be described fully, with the classical boundary-layer equations holding subject to an unprescribed pressure gradient. With inflection points then appearing in the basic velocity profiles, with or without flow reversal also, short-wave linear instabilities of inviscid type may become possible. The governing equation for these inviscid instabilities is the classical Rayleigh equation provided that the wavelengths involved are much shorter than the typical length scale of the basic flow. With this proviso, the basic flow can be treated in a quasi-parallel fashion as far as these instabilities are concerned, to leading-order. The downstream nonparallel development of a small disturbance then follows formally from integration in the streamwise direction.

A central example of the basic steady flows mentioned above is that past a short smooth obstacle mounted on a flat surface (Fig. 1), a flow which is also relevant to the transition processes mentioned earlier. If the length and height scales of the obstacle, centered at the position $x = x_0 > 0$, are $O(Re^{-3/8})$ and $O(Re^{-5/8})$ respectively, then the local steady motion [10] is controlled by the triple-deck structure, which reduces the flow problem to solving the nonlinear boundary-layer equations

$$u_X + v_Y = 0, \tag{2.1a}$$

$$uu_X + vu_Y = -p_X + u_{YY}, \tag{2.1b}$$

applying in the lower deck. The boundary conditions on the unknown functions (u,v) (X,Y), $p(X)$, $A(X)$ are

$$u = v = 0 \quad \text{on } Y = 0, \tag{2.1c}$$

$$u - Y \to A(X) + F(X) \quad \text{as } Y \to \infty, \tag{2.1d}$$

$$(u,v,p,A) \to (Y,0,0,0) \quad \text{as } |X| \to \infty, \tag{2.1e}$$

$$p(X) = 1/\pi \int_{-\infty}^{\infty} A_\xi/(X - \xi) d\xi. \tag{2.1f}$$

Here (2.1c) is the no-slip condition on the obstacle surface, (2.1d) defines the scaled displacement function $-A(X)$, and (2.1e) fixes the uniform shear flow holding far upstream and beyond the bounded obstacle. The principle value/Cauchy-Hilbert integral in (2.1f) is the interaction law between the pressure p and the displacement, stemming from the potential flow properties holding just outside the boundary layer.

The scalings leading from the Navier-Stokes equations to (2.1) are

$$(u,v,p) = (\lambda^{1/4} Re^{-1/8} u, \lambda^{3/4} Re^{-1/8} v, \lambda^{1/2} Re^{-1/4} p), \tag{2.2a}$$

$$(x,y) = (x_0 + \lambda^{-5/4} Re^{-3/8} X, \lambda^{-3/4} Re^{-5/8} Y), \tag{2.2b}$$

where $\lambda(x_0)$, of order unity, is the scaled skin friction of the oncoming classical boundary layer as $x \to x_0-$. In addition, the Prandtl shift $Y \to Y - F(X)$ has been applied to yield (2.1c,d), where $F(X)$ is the given scaled shape of the obstacle. In dimensional form the equation of the obstacle is given by

$$y^* = L^* Re^{-5/8} \lambda^{-3/4} F((X^* - L^* x_0) Re^{3/8} \lambda^{5/4}/L^*). \tag{2.2c}$$

We should emphasis here, however, that the fundamental steady flow problem (2.1) applies in a limiting sense to a much wider range of obstacle sizes than (2.2a-c) tends to suggest. The full range is discussed in [12]. For our purposes it suffices to note that for an extensive range of shorter obstacles the triple-deck interaction law (2.1f) is simply replaced by the constraint

$$A(X) = 0, \tag{2.3}$$

in effect. In this range the streamwise length scale of the obstacle lies between $O(Re^{-3/4})$ and $O(Re^{-3/8})$, while the y-scale is between $O(Re^{-3/4})$ and $O(Re^{-5/8})$ and the typical obstacle slope $O(y/(x-x_0))$ is larger than the $O(Re^{-1/4})$ value in (2.2b) but still small. Flow solutions of (2.1a-e), with the triple-deck law (2.1f) or the condensed flow (2.3), are described later.

We now wish to consider the possibility of short-scale inviscid instability arising when the basic flow has the form in (2.2a,b). We observe that such instability usually occurs on an x-length scale comparable with a main y-scale of the motion. The latter scale has three possible values corresponding to the three decks of the triple-deck solution, but of these the lower-deck one in (2.2b) seems the most important since there the nonlinear basic-flow solution can admit inflectional velocity profiles of shortest scale in y. Inflection points can also appear in the upper-deck profiles, it is interesting to note, even when only relatively weak obstacles are present (i.e. $|F|$ small in (2.1d)), but there the corresponding instability properties are fully nonparallel having spatial or temporal growth rates much less than those associated with the lower-deck profiles. So the most dangerous x- (and y-)scale for possible inviscid instability is $O(Re^{-5/8})$ (Fig. 1), suggesting a small unsteady perturbation of the underlying form

$$\bar{u} = Re^{-1/8} u(X,Y) + \Delta\tilde{u}(Y) e^{i(\alpha X - \omega T)} G(X) + O(\Delta^2). \qquad (2.4)$$

Here we take $\lambda = 1$ without loss of generality. The typical amplitude Δ of the disturbance in (2.4) is small ($|\Delta G| << Re^{-1/8}$), with $G(X)$ denoting the relatively slowly varying nonparallelism present compared with the shorter scale $x - x_o = Re^{-5/8}\bar{X}$. Further, the wavenumber α and frequency $\omega = \alpha c$ are assumed to be $O(1)$, the time scale involved being $t = Re^{-1/2} T$. The expansions for $\bar{v}, \bar{p}$ are similar to (2.4) but v, p are negligible to leading order. The disturbance properties are then controlled by Rayleigh's equation for the disturbance stream function $\tilde{\psi}$ (defined by $\tilde{u} = \tilde{\psi}_Y$, $\tilde{\psi} = 0$ at $Y = 0$),

$$(u - c)(\tilde{\psi}_{YY} - \alpha^2 \tilde{\psi}) = u_{YY} \tilde{\psi} , \qquad (2.5a)$$

subject to the constraints

$$\tilde{\psi} = 0 \quad \text{at } Y = 0, \text{ and as } Y \to \infty . \qquad (2.5b,c)$$

Here (2.5b) is the tangential flow condition characteristic of an inviscid analysis, while (2.5c) is effectively a condition of zero displacement, appropriate for the current relatively short scale in $x - x_o$. The condition (2.5c) is typical for all length scales shorter than the triple-deck scale. The condition also agrees with the necessary avoidance of exponential growth from the asymptotes $\tilde{\psi} \propto \exp(\pm Y)$ of (2.5a) for large Y, given (2.1d).

The same governing equation and boundary conditions (2.5a-c) also describe the short-scale instability of the basic flows for which

(2.3) holds, although then the expansion (2.4) must be modified to account for the altered x- and y-scales applying. Nevertheless, the disturbance's x-scale remains much less than the basic flow's for the whole range of obstacle lengths between $O(Re^{-3/4})$ and $O(Re^{-3/8})$ and so (2.5a) is obtained to leading order once again.

3. A Model Problem

Before solutions for certain basic steady flows u,v,p and then of (2.5a-c) for the inviscid disturbance properties are dessribed in Section 4 below, a model problem is discussed.

It is noted first here that even for comparatively small obstacles with $|F| \ll 1$ inflection points can arise in the basic velocity profiles u determined from (2.1a-e), with (2.1f) or (2.3). For when $|F|$ is small, say typically $O(h)$ where $h \ll 1$, then the linearized analytic solutions of [10] apply. These exhibit certain intervals of X within which the increment in basic skin friction, $u_Y(X,0)-1$, and the combined displacement, $(A+F)$, have opposite signs, thus yielding inflection points. The inflection points corresponding to a maximum vorticity u_Y are therefore candidates [5] for Rayleigh instability. To gain some firm analytical feeling for this physically rather significant question of whether the instability can occur for any value of h, no matter how small, we consider a model inflectional velocity profile reasonably typical of those obtained in the flow past an obstacle. The model shows up some interesting features of the inviscid stability problem.

If the Rayleigh equation is written in the form

$$\tilde{\psi}_{YY} = (\alpha^2 + g(Y))\tilde{\psi}, \quad \text{where } g(Y) = u_{YY}/(u-c) \qquad (3.1a,b)$$

and α is assumed real and non-negative, then the function $g(Y)$ is generally of order h, for h of $O(1)$ or less. Also, $g(Y)$ tends to zero exponentially fast as $Y \to \infty$, from analysis of the properties of u in (2.1a,b,d), and $g(Y)$ is smooth for all $Y > 0$ in the neutral case where α,c are both real, since then $u = c$ at the single inflection point where u_{YY} is zero. A representative example for $g(Y)$ is therefore

$$g(Y) = -he^{-2Y}. \qquad (3.2)$$

Thus we take effectively the inverse problem of finding u, given g. The form of the local velocity profile u implied by (3.1b) with (3.2) is found to be

$$u = c - \tfrac{1}{2}\pi Y_0(z) + bJ_0(z), \qquad (3.3a)$$

where J_0, Y_0 are the standard Bessel functions of zero order and

$$z = h^{1/2}e^{-Y}, \qquad (3.3b)$$

$$c = \frac{(A + F + \tfrac{1}{2}\ln(\tfrac{1}{4}h) + \gamma)J_0(h^{1/2}) - \tfrac{1}{2}\pi Y_0(h^{1/2})}{J_0(h^{1/2}) - 1} \tag{3.3c}$$

$$b = (\tfrac{1}{2}\pi Y_0(h^{1/2}) - c)/J_0(h^{1/2}). \tag{3.3d}$$

The coefficients of Y_0, J_0 in (3.3a) are such that, as $Y \to \infty$, (2.1d) holds, i.e. $u \sim -\ln z + O(1)$ as $z \to 0$, from (3.3b), with the $O(1)$ term here satisfying the displacement effect $(A+F)$; and the no-slip condition is satisfied on $Y = 0$. The velocity profile has an inflection point at $u = c$, as required. Further, the profile can exhibit flow reversal $u < 0$, depending on the value of $(A+F)$. The solution for $\tilde{\psi}$ follows in a similar way. With (3.2) holding, the substitution (3.3b) converts (3.1a) to the Bessel equation

$$z^2 \tilde{\psi}_{zz} + z \tilde{\psi}_z + (z^2 - \alpha^2)\tilde{\psi} = 0, \tag{3.4a}$$

subject to the conditions $\tilde{\psi} = 0$ at $z = 0$, and $z = h^{1/2}$, from (2.5b,c). Hence, with a normalization applied, $\tilde{\psi} = J_\alpha(z)$, to satisfy the condition at $z = 0$. The condition at $z = h^{1/2}$ then requires $J_\alpha(h^{1/2}) = 0$, so that

$$h = z_\alpha^{(m)2} \tag{3.4b}$$

determines α values implicitly in terms of h, where the $z_\alpha^{(m)}$ for $m = 1, 2, \ldots$ are the positive zeros of the Bessel function $J_\alpha(z)$. Since α is real, however, the smallest root possible is the first root $z_0^{(1)}$ of J_0. Therefore, the condition

$$h \geq h_c = z_0^{(1)2} (\doteq 5.7831) \tag{3.4c}$$

must hold, if there is to be a neutral solution.

Assuming that Rayleigh instability can appear only for a range of positive values of α, requiring at least one neutral case to exist then, we may surmise from (3.4c) that there is a cut-off value $h = h_c$ below which the profile (3.3a) is inviscidly stable. This demonstrates that in the present context the existence of an inflection point in the local velocity profile is necessary but not sufficient for Rayleigh instability to occur; c.f.[5,6]. Also, in particular, the instability need not occur for small values of h such as those yielding a linearized solution for the basic flow in Section 2, despite the presence of inflection points then.

Varying the parameter h corresponds in a certain sense to travelling over the obstacle, as the velocity profiles vary. At a range of positions where in effect (3.4c) is satisfied we would expect local Rayleigh instability to be possible, but sufficiently far upstream and downstream the basic steady flow returns to its original uniform-shear state with $h \to 0$ then, and thus the flow is inviscidly stable there. Conversely, varying h represents a variation in the obstacle height, at fixed X, with (3.4c) then determining the minimum height which allows Rayleigh instability, for the current model at least. It is interesting that at the marginal stage, where h just exceeds h_c, the marginally unstable wavenumbers α are small, from (3.4b,c), corresponding to long-wavelength instability of the Rayleigh type.

Other models could be constructed in a similar fashion, but the one above is sufficient for present purposes. It suggests tentatively the interesting property that Rayleigh instability, via (2.5a-c), will occur only when the obstacle-height parameter h is sufficiently large, beyond an O(1) cut-off value. Since the basic flow problem is then nonlinear however, numerical solutions for u(X,Y) are strictly required, followed by a numerical treatment of the Rayleigh stability problem in general. This is taken up next.

4. Numerical Results

Guided by the model of the previous section, which suggests a finite cut-off value of h at which Rayleigh instability first appears, we have obtained numerical solutions of (2.1a-e) with (2.1f) or (2.3) for finite values of h by a finite-difference box scheme described in detail by Smith & Bodonyi [13].

Briefly, the governing equations (2.1a,b) are replaced, to a second-order level of accuracy, by a first-order difference representation for $\tau = u_Y$, $u = \psi_Y$, p, and A with uniform steps in X,Y. The calculation region extends from $X = X_1(<0)$ to $X = X_2(>0)$ and from $Y = 0$ to $Y = Y_2$, with starting conditions (2.1e) specified in effect at $X = X_1$. The remaining equation necessary to complete the problem comes from the finite-difference form of the interaction law (2.1f) or (2.3).

The nonlinear difference equations at a streamwise location X are solved to within a specified tolerance q_1 by Newton iteration involving Gaussian elimination. A single streamwise sweep is sufficient to fully determine the solution for the case using (2.3) if separation does not occur, but forward-marching sweeps must be repeated until a tolerance q_2 between succesive values obtained for p(X) is satisfied when (2.1f) is used or flow reversal is present. The diagonally dominant nature of the finite-difference form of the interaction law makes this multi-sweeping process fast and stable. Whenever flow reversal occurs, i.e. u < 0, windward differencing was used to represent uu_X in finite-difference form. In this work the tolerances q_1, q_2 were set at 10^{-7}, 10^{-5}, respectively, while the uniform grid sizes in X and Y were generally taken to be 0.2 and 0.1, respectively.

With the steady basic-flow solution known, the inviscid instability problem (2.5a-c) was treated in the following way. For each X, since the u values are known only at the grid points, an approach assuming u to be linear in Y between each grid point seems quite appropriate, i.e. approximating u by a series of continuous straight lines. Thus the stability problem could be reduced to numerically solving a complex algebraic system. For further details the interested reader is referred to [13].

Results from the steady flow and the corresponding stability calculations are given in Figs. 2 and 3. We chose the example of

$$F(X) = \begin{cases} h(1 - X^2)^2 & \text{for } |X| < 1 \\ 0 & \text{otherwise} \end{cases} \quad (4.1)$$

for the obstacle shape, for various values of h. The basic flow results agree well with those of [14] for h = 3 in the triple-deck case of Fig. 2. For h = 4 the separation zone is quite elongated, while for h = 5 reliable results were difficult to obtain because of the extent of the flow reversal occuring. Similar effects arise for increasing h in the condensed case of Fig. 3. Increasing the value of the height parameter also increases the range of unstable profiles present in Figs. 2 and 3, as expected physically. In this respect the inviscid instability results in fact accord quite fully with the predictions from the model problem of the previous section. The results also show the existence of a finite cut-off value h = h_c below which the basic flow is inviscidly stable according to (2.5a-c). For values of h just above h_c the Rayleigh instability first appears as a relatively long-wave phenomenon, i.e. for α small, in line again with Section 3. Larger values of h then destabilize modes with higher values of α and hence shorter wavelengths.

5. Concluding Remarks

The basic steady boundary-layer flow past obstacles, corners, trailing edges and other local distortions have themselves long been of practical concern and have attracted increased theoretical attention in recent years due mainly to the ability of viscous-inviscid interaction theory to describe them and in particular to incorporate regions of flow reversal. This has enabled considerable advances to be made in our understanding of small- or large-scale separated flow, a common occurance at high Reynolds numbers. But, more recently, there seems to be enhanced theoretical interest in unsteady effects with an increasing awareness of the possibility, as well as the need, for gaining insight by means of interactive theory into the boundary-layer instability, and transition to turbulence, which often happens in nature when local distortions like those above are present. One possible and physically clear mechanism of instability is addressed in this study, namely the development of inviscid Rayleigh instability of short wavelength when the local streamwise velocity profiles are sufficiently inflectional. This happens in particular near and inside regions of reversed flow, according to both experiments (see e.g., [11], Figs.38 & 39 of [1] and to the present analysis and calculations for flow over obstacles.

Yet the production of an inflectional velocity profile is not sufficient to generate Rayleigh instability. Otherwise, the slightest surface distortion could give inviscid instability, since the undistorted boundary-layer flow has identically zero curvature to leading order in the viscous wall layer ($u = Y$ in the lower deck), so that the slightly distorted case can readily provoke inflection points. Instead, the model in Section 3 above and the numerical results in Section 4 show that there is a finite cut-off value for the obstacle height. This cut-off value, we should stress, is in the nonlinear range as far as the basic flow is concerned, i.e. the orginal motion must be distorted nonlinearly, for Rayleigh instability to occur. In a sense this is a counter-example to Tollmien's [6] suggestion that in boundary layers, and in channel flows, the occurance of an inflection point in the velocity profile is sufficient for Rayleigh instability. Further, it is interesting that at the cut-off value at which Rayleigh instability first appears the instability is found to have a relatively long wave form, with wavenumber α small and wavespeed c satisfying $\int_0^\infty (u-c)^{-2} dY = 0$ from (2.5a-c). That particular point is made well in a related study ([2], see also [7]) which, with [4], notesthat the instability can occur within the unsteady interactive boundary-layer equations alone, thus rendering any time-marching calculations of the latter equations ill-posed or difficult then. A composite numerical treatment such as Smith's et. al. [8] should capture all the shorter-scale instabilities appearing, however. What is perhaps more significant is the question of nonlinear growth of these short-scale disturbances, with nonlinear critical layers for example coming into play in the case of Rayleigh instabilities.

The present work may also give reasonable estimates for the transition properties of trip-wires, since the inviscid disturbances studied are of short length scale and high growth rate compared with most viscous-inviscid Tollmien-Schlichting disturbances. For a triple-deck hump, for instance, in dimensional terms, the inviscid instabilities here have typical length scales as short as $L^* Re^{-5/8} \lambda^{-3/4}$, compared with the longer typical length scale $L^* Re^{-3/8} \lambda^{-5/4}$ of an undisturbed Tollmien-Schlichting wave. The latter can be affected significantly, however, by the nonparallelism of the flow past an obstacle. The present approach also re-emphasizes the major effects (separation, instability) of even tiny obstacles or wall roughness on the boundary-layer character, if the surface distortion nonlinearly distorts the viscous wall-layer flow sufficiently to produce Rayleigh instability. The cut-off case for Rayleigh instability is in fact an intriguing marginal case throughout. Similar considerations apply to other basic flows such as unsteady boundary layers, internal flows, and three-dimensional flows, and to the first stages of boundary-layer instability and transition (see e.g. [3]) where again the viscous wall-layer flow develops nonlinearly. In this last context it seems an open issue still whether the Rayleigh instability is a secondary or a primary influence but it is noteworthy that the nonlinear developments ([3]) of viscous-inviscid instability and Rayleigh instability both tend to point to the Euler stage of shortened wavelengths comparable with the boundary-layer thickness. It is noteworthy also, on linear theory, that the Rayleigh inviscid instability produces in some sense a bursting process as it is associated with much shortened length scales. In the triple-deck flow for example, whether over a hump or for a nonlinear Tollmien-Schlichting wave, the relative frequency is enhanced by a factor $O(Re^{1/4})$, which compares not unfavorably at typical large Reynolds numbers with the approximate factor of 10 noted in [15] in connection with bursting in Tollmien-Schlichting instabilities.

REFERENCES

1. Van Dyke, M. 1982 An *Album of Fluid Motion*, Parabolic Press, Stanford.
2. Tutty, O.R. & Cowley, S.J. 1984 submitted to *J. Fluid Mech.*
3. Smith, F.T. & Elliot, J.W. 1984 submitted to *Proc. Roy. Soc. A.*
4. Ryzhov, O.S. & Smith, F.T. 1984 *Mathematika*.
5. Drazin, P.G. & Reid, W.H. 1981 *Hydrodynamic Stability*, Cambridge University Press, Cambridge.
6. Tollmien, W. 1936 Tech. Memo. Nat. Adv. Comm. Aero., Wash., No. 792.
7. Drazin, P.G. & Howard, L.N. 1962 *J. Fluid Mech.* 13, 257.
8. Smith, F.T., Papageorgiou, D. & Elliott, J.W. 1984 *J. Fluid Mech.* 146, 313.
9. Hall, P. & Smith, F.T. 1982 *Stud. in Appl. Math.* 66, 241.
10. Smith, F.T. 1973 *J. Fluid Mech.* 57, 803.

11. Hall, D.J. 1968 Ph.D Thesis, University of Liverpool.

12. Smith, F.T., Brighton, P.W.M., Jackson, P.S. & Hunt, J.C.R. 1981 J. Fluid Mech. 113, 123.

13. Smith, F.T. & Bodonyi, R.J. 1985 Aeronaut. J., to appear.

14. Sykes, R.I. 1978 Proc. Roy. Soc. A361, 225.

15. Maslowe, S.A. 1981 ch. 7 "Hydrod. Instabilities and the Transition to Turbulence" (ed. H.L. Swinney & J.P. Gollub), Springer-Verlag.

16. Greenspan, H.P. & Benney, D.J. 1963 J. Fluid Mech. 15, 133.

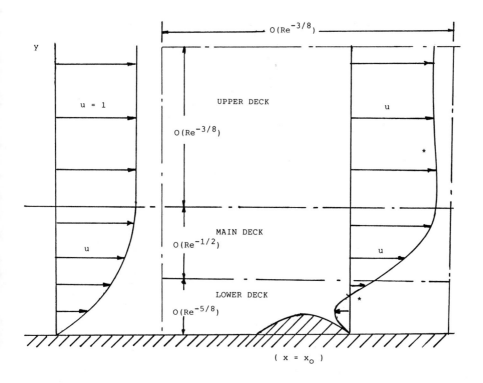

Fig. 1 Sketch of the basic flow structure in the triple-deck case (2.1f), comprising the upper, main and lower decks, for an oncoming boundary layer of thickness $O(Re^{-1/2})$. Inflection points (*) can appear in the lower or upper deck flows, depending on the obstacle size.

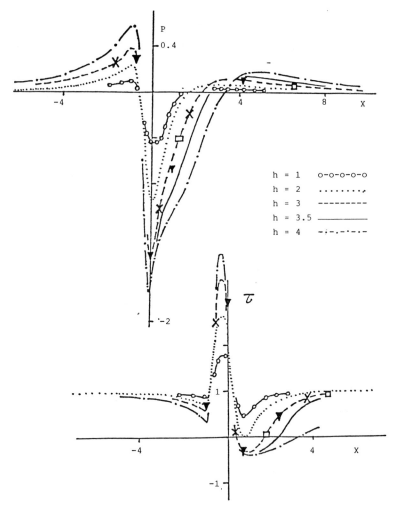

Fig. 2a The basic flow solutions for the triple-deck case for the pressure P and the scaled skin friction $\tau = U_Y(X,0)$, versus X, for the hump defined by (4.1) The computational grid had $\Delta = 0.2$, $\delta = 0.1$, $X_1 = -10$, $X_2 = 10$, and $Y_2 = 8$. Also shown, for h = 3, are ref. 14's results (▼) and our results obtained when the X_1, X_2, Y_2 values were doubled (✗) and when the step sizes Δ, δ were halved (□) for comparison purposes.

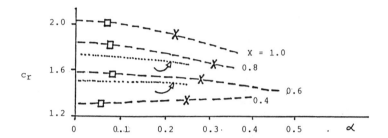

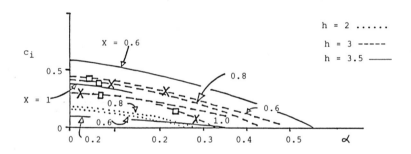

Fig. 2b The inviscid instability properties, showing the real and imaginary parts (c_r, c_i) of c, versus α at the values of X indicated. The values ✗ , ☐ correspond to the checks applied in Fig. 2a.

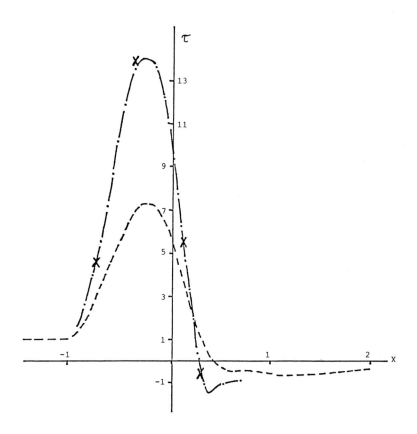

Fig. 3a. The scaled skin friction τ versus X for the condensed case in the basic flow past the hump (4.5) for h = 3 (---) and h = 5 (—.—.—). Note the lack of upstream influence in X< -1. The crosses show results obtained by halving the usual grid sizes Δ = 0.1, δ = 0.05, the usual value of Y_2 = 10 being kept fixed.

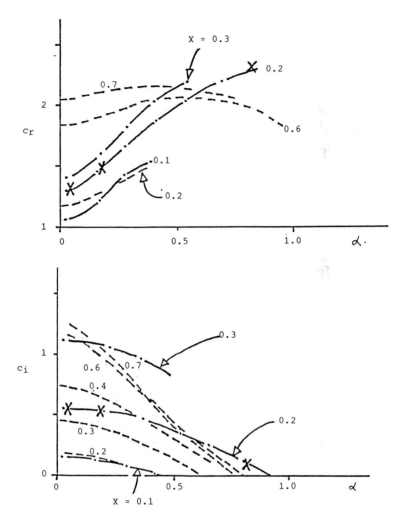

Fig. 3b. The inviscid stability properties for the condensed problem, giving the wavespeed c_r and the instability component c_i, versus α, for $h = 3$ (----), $h = 5$ (—·—·—), at the X values indicated. The crosses show grid size effects, as in Fig. 3a.

Review of Linear Compressible Stability Theory

Leslie M. Mack

Summary

A review is given of some aspects of the linear compressible stability theory. Major attention is given to the inviscid theory and the additional solutions that arise when there is a region of supersonic flow relative to the phase velocity. For highly cooled flat-plate boundary layers at Mach number 5.8, the unstable region includes supersonic outgoing waves. A previously unknown neutral incoming wave has also been found. An example of viscous multiple solutions is given, along with calculations of higher viscous discrete modes and the compressible counterpart of the Squire mode.

1. Introduction

Compressible stability theory has attracted much less attention than the incompressible theory in the 39 years since it was founded by Lees and Lin [1]. However, there is now a renewed interest in supersonic and hypersonic flight, and it is to be hoped that this will encourage further theoretical and experimental work. The latter is important because not enough is known about mechanisms of transition in this speed range to serve as a guide for theories looking beyond the linear regime. As a background for further work, it may be useful to review the linear theory with emphasis on those aspects that differ from the incompressible theory, and on the multiplicity of solutions that exist. Most of the results to be presented are from the inviscid theory; these are supplemented by a few recent viscous calculations. One reason the compressible theory is less well known than the incompressible theory may be because the basic references are more difficult to obtain. Some of the material from the older comprehensive account by Mack [2] is now available in a more recent report [3]. Both theory and experiment are treated in a 1980 book (in Russian) by Gaponov and Maslov [4], which is a good source for Soviet theoretical and experimental work on this subject.

A major contribution of Lees and Lin [1] was the development of an inviscid stability theory, and this theory later proved to be invaluable in unravelling the intricacies of the compressible theory. Lees and Lin considered only boundary layers, but free shear flows have also been treated using the inviscid theory (see, e.g., Lees and Gold [5]). The direct numerical solution of the linearized viscous parallel-flow stability equations was initiated by Brown [6], and the detailed working out of inviscid and viscous numerical results for flat-plate boundary layers was accomplished by Mack [2,7]. Extensive numerical results for the compressible mixing layer were obtained from the inviscid spatial stability theory by Gropengiesser [8]. A number of significant differences from incompressible theory were uncovered in the numerical work that had not been noted in the preceding analytical work.

Since the numerical work of Mack and Gropengiesser, the spatial stability theory has been used to study the problem of transition in a supersonic wind tunnel [9,10], and linear nonparallel theories were worked out by Nayfeh and El-Hady [11] and Gaponov [12]. The mixing layer was taken up again by Drazin and Davey [13], but on the basis of the constant-temperature hyperbolic-tangent profile rather than the exact solutions of Gropengiesser. The interest in Laminar Flow Control on transonic aircraft led to the application of the compressible stability theory to three-dimensional boundary layers on infinite swept wings by Mack [14] and Lekoudis [15]. A compressible computer code which allows N-factor calculations to be readily carried out was developed by Malik [16] and is in routine use in design studies. Crossflow instability in a compressible boundary layer at transonic Mach numbers is little different than at low speeds, but in general the compressible theory must be used at other than quite low subsonic Mach numbers.

There has been some work on nonlinear inviscid theories for wakes, but these are of the single-mode type. Nonlinear theories with wave interactions or resonances, or linear theories of secondary instability such as the incompressible theory of Herbert [17,18], have not yet been developed for compressible flows. In the incompressible secondary instability theory, a hitherto neglected mode, termed the Squire mode by Herbert, turned out to be responsible for the dominant mode of subharmonic instability. One feature of the compressible theory is the existence of a number of additional damped solutions. Because of the possibility that some of these additional solutions may be important in amplitude-dependent interactions, it is one of the purposes of this review to examine all of the known inviscid and viscous solutions.

2. Formulation of Theory

The compressible stability theory is based on the Navier-Stokes equations for a compressible, heat-conducting perfect gas. Only a summary will be given here of the derivation of the governing equations as a detailed derivation has been given elsewhere [3]. Each flow quantity is written as a mean-flow term plus a small unsteady perturbation, and the equations are linearized with respect to the perturbations. When the mean-flow equations are subtracted out, the stability equations in Cartesian coordinates for the fluctuations u, v, w, p, ρ and θ are obtained, where u,v,w are the velocity components in the x (streamwise), y (normal) and z (spanwise) directions, respectively, p is the pressure, ρ is the density and θ is the temperature. The mean flow is represented by the velocity components U, V, W and the temperature T. All quantities are made dimensionless with respect to a length scale L^* (an asterisk refers to a dimensional quantity), the freestream velocity U_1^*, and, for the thermodynamic variables, the corresponding freestream values. The dimensionless equations contain the viscosity coefficient and its first and second derivatives;

the second viscosity coefficient; the thermal conductivity coefficient and its first and second derivatives; the ratio of specific heats, γ; the Prandtl number, which is taken to be a function of temperature; the freestream Mach number, M_1; and the Reynolds number $R = U_1^* L^* / \nu_1^*$, where ν_1^* is the freestream kinematic viscosity. With the usual choice of $L^* = (\nu_1^* x^* / U_1^*)^{1/2}$, the Reynolds number R is $(U_1^* x^* / \nu_1^*)^{1/2}$.

When the equations are further simplified by considering the mean flow to be parallel, the fluctuations can be written as normal modes. For instance,

$$u(x,y,z,t) = \hat{u}(y) \exp[i(\alpha x + \beta z - \omega t)], \tag{1}$$

with similar expressions for v, w, p, ρ and θ, where α and β are the wavenumber components in the x and z directions, and ω is the frequency. In general α, β and ω are all complex and the physical quantity is the real part of eq. (1). The linearized, parallel-flow equations reduce to a system of ordinary differential equations for the eigenfunctions $\hat{u}$, $\hat{v}$, $\hat{w}$, $\hat{p}$, $\hat{\rho}$ and $\hat{\theta}$. With $D = d/dy$ and the dependent variables

$$\begin{aligned} Z_1 &= \alpha \hat{u} + \beta \hat{w}, & Z_2 &= DZ_1, & Z_3 &= \hat{v}, & Z_4 &= \hat{p}/\gamma M_1^2, \\ Z_5 &= \hat{\theta}, & Z_6 &= DZ_5, & Z_7 &= \alpha \hat{w} - \beta \hat{u}, & Z_8 &= DZ_7, \end{aligned} \tag{2}$$

the equations can be written

$$DZ_i(y) = \sum_{j=1}^{8} a_{ij}(y) Z_j(y), \quad (i=1,8). \tag{3}$$

The coefficient matrix a_{ij} is given in [3]. Consequently, compressible stability is governed by an eighth-order system of equations. For a two-dimensional (2D) wave in a 2D boundary layer, the equations reduce to a sixth-order system. With the boundary conditions

$$\begin{aligned} & Z_1(0) = 0, \quad Z_3(0) = 0, \quad Z_5(0) = 0, \quad Z_7(0) = 0 \\ & Z_1(y), \quad Z_3(y), \quad Z_5(y), \quad Z_7(y) \text{ bounded as } y \to \infty, \end{aligned} \tag{4}$$

we have an eigenvalue problem. The solution of the eigenvalue problem establishes the complex dispersion relation

$$\omega = \Omega(\alpha, \beta) \tag{5}$$

for the given boundary-layer profile (U,W) at Reynolds number R.

The dependent variables of eq. (2), except for Z_5 and Z_6, reduce the sixth-order incompressible stability equations to fourth order for the determination of eigenvalues. A corresponding reduction in order does not take place here because of the term a_{68} which couples the Z_7 and Z_8 equations to the first six equations. This coupling is absent in the incompressible theory. A feature of the incompressible theory is that there is an independent eigensolution of the Z_7 and Z_8 equations, called the Squire mode by Herbert [17], for which Z_1, Z_3 and Z_4 are all zero. With the interpretation of α and β as x and z partial derivatives, respectively, in the normal mode representation, Z_7 is the vertical vorticity fluctuation. Because v is zero in the Squire mode, so is the Reynolds stress, and this mode must be always stable as was proved by Squire [19]. It was a remarkable finding of Herbert [17] that this highly damped mode becomes a dominant mode of subharmonic instability for finite, but small, amplitudes of the primary unstable Tollmien-Schlichting waves. In the final section of this paper, we give some calculations concerning the compressible counterpart of the Squire mode.

Nonparallel Theory

Boundary layers and free shear flows are not parallel flows. The effect of the (usually) slowly varying spatial growth has been taken into account by linear nonparallel theories developed by Nayfeh and El Hady [11] and Gaponov [12] for the boundary layer, and by Tam and Morris [20] for free shear flow. The paper by Nayfeh and El Hady [11] gave inexplicably large nonparallel effects at $M_1 = 4.5$. A later unpublished calculation by El Hady [21] gave instead a rather small nonparallel effect, and this more reasonable result agrees, at least in magnitude, with Gaponov [12]. Gaponov also calculated only a small nonparallel effect at $M_1 = 2.2$. Although further work is needed to clarify a rather confused numerical situation, it seems that the complications of the nonparallel theories may justifiably be avoided in developing the basic ideas of compressible stability theory. This is not to say that the nonparallel theory is not needed for specific calculations. In particular, any calculation of wave amplitude over many wave lengths is subject to significant nonparallel effects, and any detailed comparison of amplitude growth with experiment is best done with the nonparallel theory.

Quasi-parallel Theory

The form of the theory that is actually used is properly termed quasi-parallel rather than parallel. The normal modes of eq. (1) are replaced, for a 2D boundary layer or a 3D boundary layer independent of z, by the constant-frequency wave-train solution

$$u(x,y,z,t) = A_0 \hat{u}(y;x) \exp[i(\int^x \alpha(x)dx + \beta z - \omega t)], \tag{6}$$

where the eigenvalue $\alpha(\beta,\omega)$ and the corresponding eigenfunction $\hat{u}(y)$ are slowly varying functions of x. For the assumed boundary layers, the specification of constant dimensional β satisfies the irrotationality condition of kinematic wave theory and eq. (6) represents a physically realizable wave. In the nonparallel theory, the constant A_0 is replaced by a slowly varying function $A(x)$, and the product $A\hat{u}$ is independent of any eigenfunction normalization.

3. Inviscid Theory

In contrast to a Blasius boundary layer, a compressible flat-plate boundary layer is unstable to purely inviscid waves. Indeed, above a Mach number of about three inviscid instability becomes dominant and viscosity is stabilizing at all Reynolds numbers. It is this fact that allows the inviscid theory to be used to study the most characteristic features of the compressible theory.

Relative Mach Number

As the distinction between subsonic and supersonic flow is fundamental in compressible stability theory, it is pertinent to write the equations in terms of the relative Mach number defined by

$$\overline{M} = \frac{\alpha U + \beta W - \omega}{(\alpha^2+\beta^2)^{1/2}} \frac{M_1}{T^{1/2}} = (\tilde{U} - \frac{\omega}{\tilde{\alpha}}) \frac{M_1}{T^{1/2}} , \qquad (7)$$

where we have made use of two of the tilde variables defined by

$$\tilde{\alpha} = (\alpha^2+\beta^2)^{1/2}, \quad \tilde{\alpha}\tilde{U} = \alpha U + \beta W, \quad \tilde{\alpha}\tilde{W} = \alpha W - \beta U. \qquad (8)$$

When α and β are real, as in the temporal theory, $\tilde{\alpha}$ is the same as k, the magnitude of the wavenumber vector, and $\tilde{U}$ and $\tilde{W}$ are the mean velocities in the direction of $\vec{k}$ and normal to it in the plane of flow. In general, however, except for a neutral wave all of these quantities, as well as $\overline{M}$, are complex.

Inviscid Equations

In the limit of infinite Reynolds number, the eighth-order viscous equations reduce to second order. These equations may be written in various forms. The equation for the velocity fluctuation normal to the wall, which is the compressible counterpart of the Rayleigh equation is

$$D\left[\frac{(\tilde{\alpha}\tilde{U}-\omega)D\hat{v} - \tilde{\alpha}D\tilde{U}\hat{v}}{1-\overline{M}^2}\right] - \tilde{\alpha}^2 (\tilde{\alpha}\tilde{U}-\omega)\hat{v} = 0 . \qquad (9)$$

The boundary conditions are

$$\hat{v}(0) = 0, \quad \hat{v}(y) \text{ bounded as } y \to \infty . \qquad (10)$$

Equation (9) and its boundary conditions constitute an eigenvalue problem whose solution is the inviscid dispersion relation $\omega=\Omega(\alpha,\beta)$. This relation differs from the viscous dispersion relation in the absence of the Reynolds number as a parameter.

With $\hat{v}$ and $D\hat{v}$ known, the other flow variables are obtained from explicit expressions. A useful equation for the pressure fluctuation is:

$$D^2\hat{p} - D(\ln\overline{M}^2)D\hat{p} - \tilde{\alpha}^2(1-\overline{M}^2)\hat{p} = 0 . \qquad (11)$$

It is evident from this equation that the analytic nature of the solution depends on whether the real part of $\overline{M}$ is subsonic or supersonic. There are two cases to distinguish: real $(\overline{M}) > 1$ and real $(\overline{M}) < -1$. In the first, the relative supersonic region is the freestream and the outer portion of the boundary layer. In the second, the supersonic region extends from the wall to the sonic line located well within the boundary layer. It is the second case that is the most interesting.

It can be noted that the tilde equations are identical in form to the equations for a 2D wave with wavenumber $\tilde{\alpha}$ and frequency ω in a mean flow at Mach number M_1 and with boundary-layer profiles $\tilde{U}$ and T. For a neutral wave, or temporal non-neutral wave, the governing velocity profile $\tilde{U}$ is real and in the direction of the wavenumber vector. In general $\tilde{U}$ is complex, but it is convenient even in this case to refer to it as the directional profile. The result that inviscid instability is governed by the directional profile for both 2D and 3D boundary layers is thus valid for compressible as well as incompressible boundary layers.

Classification of Neutral Waves

The various inviscid instability waves may be classified in different ways. Lees and Lin [1] classified them on the basis of the <u>freestream</u> relative Mach number. They distinguished three types of neutral waves:

subsonic $\quad 1-1/\tilde{M}_1 < \tilde{c} < 1+1/\tilde{M}_1$,

sonic $\quad \tilde{c} = 1-1/\tilde{M}_1$ or $1+1/\tilde{M}_1$, $\qquad(12)$

supersonic $\quad \tilde{c} < 1-1/\tilde{M}_1 \quad$ (no eigenvalues for $\tilde{c} > 1+1/\tilde{M}_1$),

where

$$\tilde{M}_1 = M_1 \cos\psi, \quad \tilde{c} = c^*/(U_1^* \cos\psi), \quad c^* = \omega^*/k^*. \qquad(13)$$

As all quantities are real for a neutral wave, c^* is the phase velocity. The dimensionless $\tilde{c}$ has been referred to $U_1^* \cos\psi$, where ψ is the waveangle, instead of to U_1^* as are all other quantities, in order to make eq. (12) have the same appearance for oblique waves as for 2D waves. In this way all of the results derived for 2D waves can be applied to 3D waves and 3D boundary layers by placing a tilde over the appropriate variables. Lees and Lin limited themselves to 2D waves with c<1.

Another classification of neutral waves can be made with respect to the <u>local</u> speed of sound. When $\overline{M}^2 < 1$, the relative flow is everywhere subsonic and the theory is similar to the incompressible theory. When $\overline{M}^2 > 1$ over some portion of the boundary layer, the theory has important differences from the incompressible theory. Yet a third classification is with respect to the freestream velocity. In the incompressible theory, the phase velocity must be less than (or at most equal to) the freestream velocity, but in the compressible theory, with $\overline{M}^2 > 1$, this is no longer true. Neutral waves are possible with $\tilde{c}$ between 1 and $1+1/\tilde{M}_1$.

The most straightforward way to derive some necessary conditions is by means of the Reynolds stress $\tilde{\tau} = -\rho \langle \tilde{u}\tilde{v} \rangle$. It was shown by Lees and Lin [1] that the Reynolds stress is everywhere constant for a neutral wave except possibly at the critical layer y_c, where

$$\tilde{\tau}(y_c+\varepsilon) - \tilde{\tau}(y_c-\varepsilon) = \frac{\pi}{\tilde{\alpha}} \frac{[D(\rho D\tilde{U})]_c}{(D\tilde{U})_c} \langle v_c^2 \rangle. \tag{14}$$

The Reynolds stress is zero at $y=0$ by the wall boundary condition, and for a neutral subsonic wave ($\tilde{c} > 1 - 1/\tilde{M}_1$) it must also be zero as $y \to \infty$. Consequently, we may immediately deduce from eq. (14) that a necessary condition for a neutral subsonic wave in a boundary layer with one critical point is that $D(\rho D\tilde{U})$ must be zero at the critical point. The point y_s where this condition holds is called the generalized inflection point. Thus the phase velocity $\tilde{c}_s$ is equal to $U_s + W_s \tan\psi$. If $\overline{M}^2 < 1$ everywhere, Lees and Lin showed that a generalized inflection point with $y_s > y_0$ is also a sufficient condition for instability, and that there is a unique wavenumber α_s corresponding to c_s. The numerical results indicate that this sufficient condition holds even when $\overline{M}^2 > 1$, but the uniqueness of $\tilde{\alpha}_s$ does not.

Neutral Sonic and Supersonic Waves

Most attention is paid to subsonic waves (in the Lees-Lin sense), but it is of interest to look briefly at sonic and supersonic waves. There are two sonic waves: the upstream running wave with phase velocity $\tilde{c}_u = 1 - 1/\tilde{M}_1$, and the downstream running wave with phase velocity $\tilde{c}_d = 1 + 1/\tilde{M}_1$. Both of these waves are neutral. When $\overline{M}_1^2 < 1$, only the first wave exists, and with zero wavenumber as found by Lees and Lin. When there is a region of relative supersonic flow, there is an infinite sequence of additional wavenumbers $\tilde{\alpha}_{un} > 0$ for $\tilde{c}_u$, and there is also an infinite sequence of wavenumbers, $\tilde{\alpha}_{dn}$, for the second sonic wave with phase velocity $\tilde{c}_d$. The first wavenumber in the latter sequence, $\tilde{\alpha}_{d1}$, is zero. Both upstream and downstream running waves with zero wavenumber have a constant pressure fluctuation from the wall to $y \to \infty$. However, the additional waves with non-zero wavenumbers have a constant pressure fluctuation only in the freestream, but not in the boundary layer.

Neutral supersonic waves have $\tilde{c} < 1 - 1/\tilde{M}_1$. Not much is known about these waves. A brief discussion of necessary conditions has been given by Mack [2] based mainly on the energy equation and the freestream solutions of eq. (11). There are two families of waves: outgoing waves and incoming waves. For both families, the Reynolds stress is constant from the critical point to $y \to \infty$. The energy transfer to infinity is $\tilde{\tau}(1-\tilde{c})$, and $\tilde{\tau} = \pm \tilde{\alpha}^2 (\tilde{M}_1^2 - 1)^{1/2}$. The upper sign corresponds to outgoing waves which transfer energy to infinity; the lower sign to incoming waves which transfer energy from infinity. It is apparent from eq. (14) that a necessary condition for an outgoing wave is that $D(\rho D\tilde{U})_c > 0$; and for an incoming wave, $D(\rho D\tilde{U})_c < 0$.

The critical layer cannot be at the generalized inflection point because $\tilde{\tau}$ cannot be zero. Outgoing neutral waves have been known for some time in highly cooled boundary layers (Mack [2]), but incoming neutral waves were obtained for the first time in calculations made for this paper and will be presented later when the effect of cooling is discussed. It was already pointed out by Lees and Lin [1] that a combination of incoming and outgoing neutral waves satisfies the boundary condition at y=0 for any $\tilde{\alpha}$. Extensive numerical solutions of this non-eigenvalue problem, which represent the reflection of travelling Mach waves from a boundary layer, have been given by Mack [9].

The distribution of D(ρDU) with y for a family of flat-plate boundary layers is shown in Fig. 1 at four supersonic Mach numbers. These boundary layers are for wind-tunnel conditions. The wall is insulated, and the freestream temperature conditions are that the stagnation temperature is 311°K or the static temperature is 50°K, whichever gives the higher static temperature. The outward movement of the generalized inflection point with increasing M_1 is apparent in Fig. 1. It can also be seen that a wave with $c=U_s$ is always subsonic. The necessary condition for a supersonic outgoing neutral wave is satisfied for all possible supersonic waves, but the condition for a supersonic incoming neutral wave cannot be satisfied for any c.

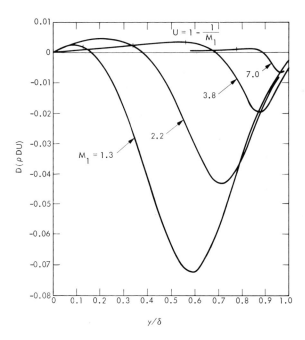

Fig. 1. Distribution of D(ρDU) through insulated-wall, flat-plate boundary layers for four Mach numbers.

Numerical Integration

For numerical integration it has been found convenient to use the following pair of first-order equations:

$$D\hat{v} = \frac{\tilde{\alpha}D\tilde{U}}{\tilde{\alpha}U-\omega} \hat{v} + i \frac{\tilde{\alpha}^2}{\tilde{\alpha}\tilde{U}-\omega} T(1-\bar{M}^2) \frac{\hat{p}}{\gamma M_1^2}$$

$$D(\frac{\hat{p}}{\gamma M_1^2}) = -\frac{i}{T} (\tilde{\alpha}\tilde{U}-\omega)\hat{v}$$

(15)

These equations are singular at the critical point where $\tilde{\alpha}\tilde{U}=\omega$. The critical point is on the real y axis for a neutral wave. For an amplified wave in a boundary layer with $DU_c > 0$, which is the usual case, the singularity is above the real axis; for a damped wave, it is below it. The application of Lin's argument that an inviscid solution must be the infinite Reynolds number limit of a viscous solution leads immediately to the conclusion that for $DU_c > 0$, the integration contour must pass below the singularity. Numerical solutions of the eigenvalue problem for neutral and damped waves can be readily obtained by integrating along an indented contour in the complex y plane (Mack [22]). Eigenvalues of unstable waves can be calculated by integrating along the real axis, and the eigensolution is valid everywhere on the real axis.

Inflectional Neutral Waves

The wavenumbers α_{sn} corresponding to the phase velocity c_s of 2D neutral subsonic waves are given in Fig. 2 as a function of M_1. For this value of c, a region of relative supersonic flow first appears in the boundary layer at about $M_1=2.2$. Below this Mach number there is a unique wavenumber; above it, there is an infinite sequence of wavenumbers (Mack [7]). The additional solutions, called higher modes, are primarily acoustic in origin and represent sound waves reflecting from the wall and relative sonic line. The mode 1 waves below $M_1=4.5$ are similar to low-speed instability waves and are waves of vorticity. Figure 2 suggests that the compressible counterparts of low-speed vorticity waves do not remain mode 1 waves at all Mach numbers. Instead, the transfer of the rising portion of the n=1 curve to successively higher modes with increasing M_1 implies that the mode identity is also transferred. If we adopt the terminology of Lees and call this mode the vorticity mode, and the modes where α_{sn} decreases with increasing M_1 the acoustic modes, we see that at $M_1=6.0$, the vorticity mode is mode 2; at $M_1=8.0$, it is mode 3; and at $M_1=10.0$, it is mode 4. The identity of these waves is confirmed by an examination of the eigenfunctions. In all cases, the number of sign changes of $\hat{p}$ in the relative supersonic region is one less than the mode number.

The distinction between vorticity and acoustic waves, and the transfer of the mode identity from one mode to another, is made quite graphic if we examine the behavior

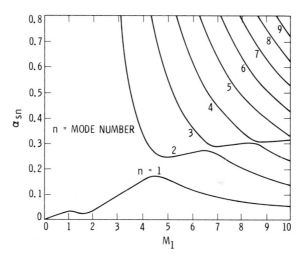

Fig. 2. Multiple inflectional 2D neutral wavenumbers vs M_1; insulated-wall, flat-plate boundary layers.

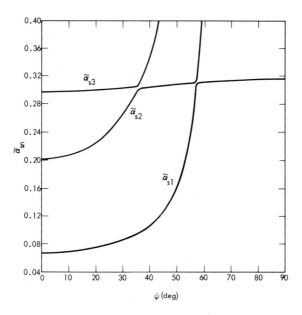

Fig. 3. Multiple inflectional neutral wavenumbers vs wave angle at $M_1=8.0$; insulated-wall, flat-plate boundary layer.

of $\tilde{\alpha}_{sn}$ as a function of wave angle at a fixed Mach number. A result at $M_1=8.0$ from [2] is shown in Fig. 3 for the first three modes. At $\psi=0°$, the vorticity mode is mode 3. Near $\psi=36°$ the mode 2 and mode 3 curves almost intersect. For larger ψ, the mode 3 curve appears to be a continuation of the mode 2 curve. The supposition that the vorticity mode has been transferred to mode 2 and remains there until the near intersection with the mode 1 curve near $\psi=57°$ is confirmed by the eigenfunctions. Above $\psi=57°$, the vorticity mode is mode 1.

Non-inflectional Neutral Waves

A major consequence of a region of relative supersonic flow in a boundary layer is the existence of a class of neutral waves with phase velocities from $\tilde{c}=1$ to $\tilde{c}_d = 1+1/\tilde{M}_1$. For _each_ of these phase velocities there is an infinite sequence of wavenumbers. All of the $\tilde{c} \geq 1$ waves are subsonic waves with respect to the freestream velocity. There is no discontinuity in the Reynolds stress even when $\tilde{c}=1$. Thus the necessary condition for a neutral wave is satisfied without there having to be a generalized inflection point in the boundary layer. The importance of the neutral wave with $\tilde{c}=1$ is that the neighboring wave with $\tilde{c}<1$ is always unstable. Therefore, whenever $\overline{M}^2>1$ a boundary layer is unstable to inviscid waves regardless of any other feature of the velocity and temperature profiles.

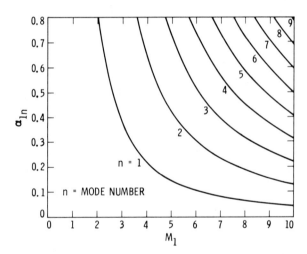

Fig. 4. Multiple noninflectional 2D neutral wavenumbers vs M_1; insulated-wall, flat-plate boundary layers; c = 1.

Because of the importance of the $\tilde{c}=1$ waves, the neutral wavenumbers α_{1n} for 2D waves with $c=1$ are shown in Fig. 4. The appearance of this figure confirms the distinction made between vorticity and acoustic modes. The vorticity mode requires a critical layer; the acoustic modes do not. Thus with $c \geq 1$ there is no critical layer and no vorticity mode. Only the acoustic modes exist, as seen in Fig. 4. The lowest mode has no sign change in $\hat{p}$ across the boundary layer. In this respect it is the same as mode 1 in Fig. 2 and so has also been designated mode 1. With this convention, all of the mode numbers are equal to the number of sign changes in $\hat{p}$. Because these modes can exist when there is no generalized inflection point, they are called non-inflectional neutral waves.

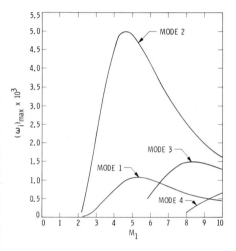

Fig. 5. Maximum temporal amplification rate of 2D waves vs M_1 for first four modes; insulated-wall, flat-plate boundary layers.

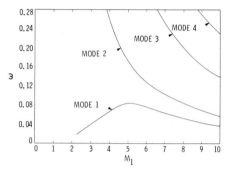

Fig. 6. Frequency of most unstable temporal 2D waves vs M_1 for first four modes; insulated-wall, flat-plate boundary layers.

Most-Unstable Waves

The primary interest in stability theory is in unstable rather than neutral waves. Figure 5 shows the maximum temporal amplification rate, $(\omega_i)_{max}$, as a function of M_1 for 2D waves. The figure for the maximum spatial amplification rate has the same general appearance. The most unstable mode is the second mode. The second and higher modes are most unstable as 2D waves, because they depend on the thickness of the relative supersonic region, but the first mode is most unstable as an oblique wave at all supersonic Mach numbers. For $M_1 > 5$, the most unstable oblique wave has an amplification rate about double that of the most unstable 2D wave, but for lower Mach numbers an oblique wave can have an amplification rate several times larger than a 2D wave. For $M_1 < 2.4$, an oblique first-mode wave is even more unstable than a 2D second-mode wave. However, in this Mach number range viscous instability is important, and the inviscid theory ceases to be a reliable quide to what happens at finite Reynolds numbers. For Reynolds numbers from R=1000 to 2000, the viscous theory gives the result that $(\omega_i)_{max}$ never drops below 1.3×10^{-3} at any Mach number.

The frequencies of the most unstable waves of Fig. 5 are shown in Fig. 6. This figure may be considered the equivalent for unstable waves of Figs. 2 and 4 for neutral waves except that here the phase velocity is not fixed at each Mach number. The figure is regular in appearance, and there is nothing comparable to the mode-identity changes of Figs. 2 and 3. The pressure-fluctuation eigenfunction is complex, and instead of the characteristic 180° phase change across the boundary layer of second-mode neutral waves, the phase change of the most unstable second-mode waves is about 100°.

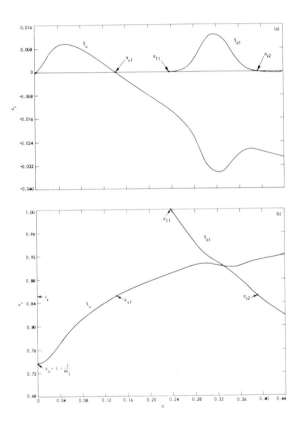

Fig. 7. Eigenvalue diagram of 2D temporal waves at $M_1=3.8$; (a) c_i vs α; (b) c_r vs α; insulated-wall, flat-plate boundary layer.

Eigenvalue Diagrams - Insulated Wall

Figures 2 and 4 give the multiplicity of neutral wavenumbers for a fixed value of the phase velocity. We turn now to a consideration of the multiplicity of eigenvalues for a fixed wavenumber. For this discussion it will be necessary to talk of families of eigensolutions. Diagrams for the complex temporal eigenvalue $c_r + ic_i$ as a function of α for 2D waves at $M_1 = 3.8$, 5.8 and 7.0 are given in Fig. 7, 8 and 9, respectively. In each figure, c_i is given in (a), and c_r in (b). The first family of solutions, called S_u, starts from the upstream-running Mach wave; the other families, called S_{dn}, start from the successive wavenumbers α_{dn} of the downstream running Mach wave. For these families, the neutral eigenvalues with $c>1$ are not shown, so the curves start from α_{1n}, $c=1$. In all cases, the S_u family contains the first-mode amplified solutions which extend from $\alpha=0$ to α_{s1}. At $M_1 = 3.8$ and 5.8, this family also contains what may be called second-mode damped solutions. These solutions have a local maximum in the damping rate at approximately the same wavenumber at which the second-mode amplified solutions are most unstable. The latter solutions belong to the S_{d1} family at $M_1 = 3.8$ and 5.8, but at $M_1 = 7.0$ the first and second-mode unstable regions have merged and they belong to the S_u family.

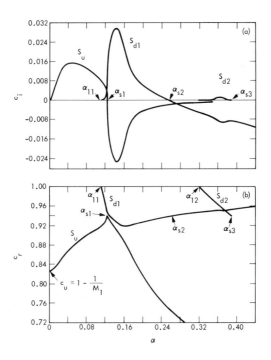

Fig. 8. Eigenvalue diagram of 2D temporal waves at $M_1 = 5.8$; (a) c_i vs α; (b) c_r vs α; insulated-wall, flat-plate boundary layer.

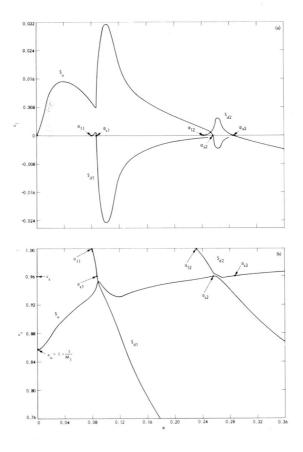

Fig. 9. Eigenvalue diagram of 2D temporal waves at $M_1=7.0$; (a) c_i vs α; (b) c_r vs α; insulated-wall, flat-plate boundary layer.

This merger process continues at higher Mach numbers. First, there is an overlap of unstable regions when $\alpha_{1n} < \alpha_{sn}$ and then a merger. At $M_1 > 7.0$, the third-mode unstable region, which at $M_1=7.0$ belongs to the S_{d2} family, merges with the second-mode unstable region and belongs to the S_u family. It is to be noted that the neutral wavenumber which marks the end of the merged unstable region is always the vorticity mode at that Mach number. After the merger, the neutral wavenumbers α_{1n} and α_{sn}, where n<nv and nv is the vorticity mode, still have neighboring unstable solutions, but these now form only the insignificant unstable region that belong to the S_{dn} families. The latter families contain mostly damped solutions which are supersonic waves when $c < c_u$.

The pattern of the phase-velocity curves also follows a definite sequence of changes as the Mach number increases. First, c_s is less than the c_r at which the S_u and S_{dn} curves intersect in a c_r vs α diagram, and the pattern is as in Fig. 7b. Then, when an overlap of unstable regions occurs, the pattern changes to the one seen for the S_u family in Fig. 8b. Finally, after merger, c_s is greater than the intersection c_r, and the pattern reverts to its original form. It may also be noted that there are multiple wavenumbers α_{un} for the sonic phase velocity c_u. The damped supersonic solutions with $\alpha > \alpha_{un}$ are incoming waves.

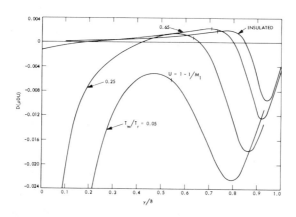

Fig. 10. Effect of cooling on the distribution of $D(\rho DU)$ through $M_1=5.8$ flat-plate boundary layer; $T_1^*=125°$ K for $T_w/T_r=0.05$; $T_1^*=50°$ K otherwise.

Effect of Cooling

The effect of cooling the wall on stability has been of interest ever since the celebrated prediction by Lees [23] that a boundary layer can be completely stabilized in this manner. However, this prediction was made without knowledge of the higher modes. Figure 10 shows the effect of cooling on the distribution of $D(\rho DU)$ at $M_1=5.8$. In contrast to the insulated-wall case, where, for $y<y_s$, $D(\rho DU)$ is always positive, decreasing the wall temperature T_w causes $D(\rho DU)$ to be negative from the wall to a second generalized inflection point. This second point, y_{s2}, occurs where $U<1-1/M_1$. Consequently, there can be an inflectional neutral solution associated with this point only for a sufficiently oblique wave. Further cooling moves y_{s2} outward. When it reaches y_s, $D(\rho DU)$ is negative everywhere and there are no longer either inflectional neutral solutions of any wave angle or the associated first-mode unstable solutions. Just before y_{s2} reaches y_s, U_s drops below $1-1/M_1$, and there are no 2D inflectional neutral solutions in spite of there being two generalized inflection points. However, as the noninflectional neutral solutions and the higher-mode unstable solutions depend only on a region of relative supersonic flow, these solutions are still present at all wall temperatures.

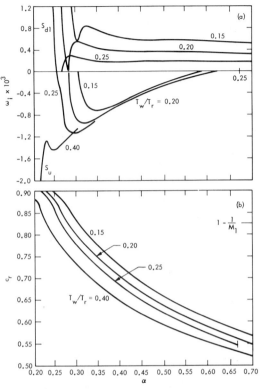

Fig. 11. Temporal amplification rates and phase velocities for cooled flat-plate boundary layers at $M_1=5.8$; detail showing unstable and damped supersonic solutions.

Some numerical results at $M_1=5.8$ are shown in Fig. 11 for $T_w/T_r=0.40$, 0.25, 0.20 and 0.15, where T_r is the recovery temperature. Only a portion of the eigenvalue diagrams are shown. At $T_w/T_r=0.40$, the unstable second-mode (not shown) still belongs to the S_{d1} family as in Fig. 8, and extends from α_{11} to α_{s2}. After passing through α_{s2}, the S_{d1} family connects to strongly damped supersonic solutions. The damped second-mode solutions of the S_u family, as shown, connect to a family of moderately damped supersonic solutions. At $T_w/T_r=0.25$, it is the unstable second mode of the S_{d1} family that, after passing through the subsonic neutral α_{s2} solution, connects to the moderately damped supersonic solutions. A new feature at this temperature ratio is that at an α just above α_{s2} there is a neutral wavenumber which is neither α_{s3} nor α_{12}. It is instead a supersonic neutral wave. Adjacent to this neutral solution at larger wavenumbers are supersonic unstable solutions with small amplification rates, and phase velocities that are almost identical to those shown in Fig. 11b. Both the supersonic neutral and unstable solutions are outgoing waves. In contrast, the supersonic damped solutions are incoming waves.

If the supersonic unstable solutions for $T_w/T_r=0.25$ are extended to larger wavenumbers than shown in Fig. 11a, they appear to continue almost indefinitely with monotonically decreasing amplification rates to an unknown upper wavenumber limit. The supersonic damped solutions, however, are terminated by a neutral solution at $\alpha=0.669$ (marked in Figs. 11a and 11b by a vertical line), as was verified with an eigenvalue search procedure for α and ω_r with $\omega_i=0$. As this neutral solution must represent an incoming wave in order to be contiguous with the damped incoming waves, it is necessary in calculating the eigenvalue to choose the lower sign in the exponential factor of the freestream solution, $\exp(\mp\lambda y)$. The usual choice of the upper sign gives all of the other possible eigensolutions. In agreement with the earlier discussion of necessary conditions, the critical layer of the outgoing-wave neutral solution occurs where $D(\rho DU)$ is positive, and of the incoming wave solution where $D(\rho DU)$ is negative.

At $T_w/T_r=0.20$, there are still two generalized inflection points, but $U_s<1-1/M_1$ at both of them; at $T_w/T_r=0.15$, $D(\rho DU)$ is negative everywhere in the boundary layer. Consequently, in neither case can there be 2D inflectional neutral solutions (α_{s2}). The S_{d1} family of unstable second-mode solutions joins the supersonic amplified solutions instead of the damped solutions as at $T_w/T_r=0.25$. The maximum amplification rate of the supersonic unstable solutions increases with decreasing temperature. The damped supersonic solutions form a separate branch, with the upper wavenumber limit an incoming wave neutral solution. The lower limit is uncertain at $T_w/T_r=0.20$. The damping rate becomes very small in the vicinity of $c=1-1/M_1$, but the existence of a neutral solution has not been established. Calculations made explicitly to reveal either an incoming-wave neutral solution or a sonic neutral solution with $\alpha>0$ were equally unsuccessful. However, at $T_w/T_r=0.15$, the lower limit is definitely an incoming-wave neutral solution, just as at the upper limit.

We may note in the three cooled-wall examples that the family of supersonic solutions, either amplified or damped, as the case may be, that is not attached to the S_{d1} family constitutes a solution family that does not exist at higher temperatures.

4. Viscous Theory

All of the inviscid solutions can be expected to have viscous counterparts, but the inverse is by no means true. We discuss first the S_u and S_{dn} families at finite Reynolds numbers, and then three classes of additional viscous solutions that have no inviscid counterparts.

Least Stable Discrete Mode

For incompressible flow, the viscous discrete eigenvalue spectrum is known to be finite (Mack [24]). It is the least stable discrete mode, or TS mode, whose limit as $R \to \infty$ is the inviscid solution. This mode is mode 1 at low R, and mode 3 at high R in the example given in [24]. In the compressible theory, the situation is more complicated because of the S_{dn} families of inviscid solutions which have viscous counterparts that are unrelated to the higher discrete modes of the incompressible theory. Until a complete spectrum has been worked out, these solutions may all be considered to belong to the least stable discrete mode. The viscous eigenvalue diagrams of the S_u and S_{dn} families of solutions are Reynolds number dependent and, except for the single example given below, will not be presented here. It will suffice to say that as R is varied at constant α, the solutions jump from one family to another, just as occurs with the inviscid solutions as a mean-flow parameter is varied.

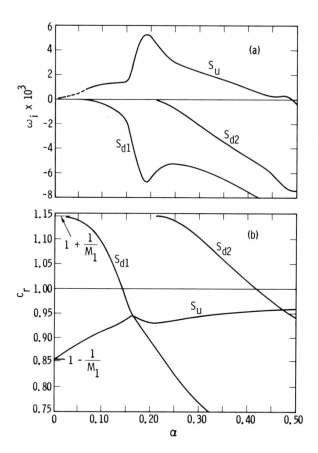

Fig. 12. Multiple viscous solutions at R=9000; M_1=6.8 insulated-wall, sharp-cone boundary layer.

Of the inviscid solutions we have discussed that are not generally considered in stability calculations, viscous counterparts have been obtained for the S_u family, and for the S_{d1} and S_{d2} families with c both greater than and less than 1.0. The solutions with c>1 are always damped at finite Reynolds numbers. Damped supersonic solutions at M_1=5.8 have also been found at R=5000 for T_w/T_r=0.15 that correspond to the supersonic solutions of Fig. 11. These solutions are connected to the second-mode unstable solutions via a subsonic neutral point in the same way as the inviscid solutions at T_w/T_r=0.25 in Fig. 11. For α>0.32, the damping rate at T_w/T_r=0.15 is about 2×10^{-3} and varies only slowly with α. No evidence has yet been found of a second supersonic solution to correspond with the two supersonic solutions of Fig. 11. It is probably only when R is greater than the Reynolds number where the neutral point becomes sonic that the inviscid pattern of Fig. 11 occurs.

An example of multiple viscous solutions is given in Fig. 12 for an insulated-wall boundary layer on a sharp cone at R=9000 (the equivalent flat-plate R is 5196). The Mangler transformation has been used for the mean boundary-layer flow. Comparison with Fig. 9, the inviscid eigenvalue diagram for the flat-plate boundary layer at M_1=7, shows a number of differences. At R=9000, a merger has taken place of the second and third-mode unstable regions. Both the first and second-mode regions with c>1 are damped at R=9000 instead of neutral as in the inviscid theory, and the lightly damped solutions of the S_{d1} family beyond the second-mode maximum have been replaced by strongly damped solutions.

Higher Discrete Modes

A few calculations have been made of the second and third discrete modes (following the classification of [24]) up to M_1=6. For α=0.1, β=0.2 at R=826, there is little effect of Mach number. The phase velocity of mode 2 increases slowly with increasing Mach number, and the damping rate hardly changes at all. The phase velocity of mode 3 at M_1=6 is about 30% below its incompressible value, and the damping rate first decreases and then increases such that at M_1=6 it is only 10% less than at M_1=0. These modes are only slightly dependent on β, and the dependence on α and R appears to be similar to the low-speed behavior shown in [24].

Compressible Squire Mode

Following eq. (5) in Section 2, we discussed the incompressible Squire mode that is responsible for the dominant mode of secondary instability at small, but finite, amplitudes of the primary TS wave. In order to investigate whether a similar mechanism has any possibility of being present in a compressible boundary layer, we have made calculations up to M_1=6 of the compressible counterpart to the Squire mode. First, we adopted a slightly different numerical procedure from the usual one in order to force the complete sixth-order incompressible equations to produce a Squire eigenmode instead of a TS eigenmode. We then applied the same procedure to the eighth-order compressible equations, and at low Mach numbers obtained the eigenvalues of the Squire mode. As the Mach number increases, the eigenvalues change even less than for discrete modes 2 and 3. At $M_1=5$, the damping rate is increased by only 5%, and at $M_1=6$ by another 5%. Just as at low speeds, the eigenvalues have little dependence on the spanwise wavenumber. As R is varied at constant α, the damping rate remains almost constant, but at constant frequency the damping rate increases rapidly with increasing R and the phase velocity decreases. At some low Reynolds number the phase velocity is equal to 1.0. Setting the coupling term a_{68} to zero has little effect on the eigenvalues.

A consequence of the weak dependence of the Squire mode on the Mach number is that the phase velocity remains low and becomes supersonic with respect to the free-stream at about $M_1=2$. As the phase velocities of the primary instability waves increase with Mach number, the primary and Squire modes quickly become detuned so that little possibility remains of a secondary instability as at low speeds.

The damping rate of the incompressible Squire mode with $\alpha=0.1$, $\beta=0.2$ at R=826 is only about 10% larger than for discrete mode 3 with the same wavenumber. The phase velocities differ by about 30%. As the Mach number increases, the Squire mode phase velocity increases, while that of discrete mode 3 decreases. Near $M_1=4.5$, the phase velocities are equal and the damping rates differ by about 25%. With $\beta=0$, this coincidence occurs near $M_1=6$.

Continuous Spectra

The c=1 continuous spectrum of the incompressible theory must also exist for the compressible boundary layer as well. Because of the viscous nature of this spectrum, a region of relative supersonic flow is not expected to have the importance it does for discrete mode 1. However, for supersonic Mach numbers there is also the continuous spectrum that was considered analytically by Lees and Lin [1] and calculated in some detail by Mack [9]. This spectrum represents the reflection of travelling Mach waves from the boundary layer. The reflected wave can be stronger or weaker than the incoming wave and has a small phase lag. It is only in the inviscid theory that examples have been found with a pure incoming or pure outgoing wave, as discussed in the preceding Section for $T_w/T_r < 0.30$ at $M_1=5.8$.

The incoming waves can be regarded as originating elsewhere than in the boundary layer under study. Another possibility is that they are produced by the reflection of outgoing waves at an oblique shock wave in hypersonic flow as has been considered by Petrov [25]. In the first instance, the condition of boundedness at $y \to \infty$ is no longer necessary, and incoming unstable waves can be allowed; in the second instance, a boundary condition appropriate to the reflection of a Mach wave from a shock wave must be imposed at a finite distance from the edge of the boundary layer.

5. Concluding Remarks

We have presented a number of examples of multiple solutions of the inviscid and viscous stability equations. It is entirely possible that additional solutions remain to be discovered. One thing that complicates a discussion of multiple solutions is the ambiguous meaning of the word 'mode'. The usual terminology of normal modes refers to the Fourier components such as eq. (1). These modes are the solutions of the dispersion relation for the least-stable waves. The higher neutral modes of the compressible theory are additional solutions with the same phase velocity, and the extension to contiguous amplified and damped solutions results in the eigenvalue diagrams of Figs. 7-9. Another group of modes are the compressible counterparts of the incompressible Squire modes. These modes, unlike the preceding, are purely viscous in nature. Finally, there are the modes that comprise the individual terms of an expansion of an arbitrary solution in eigenmodes. In the temporal theory, they all have the same wavenumber; in the spatial theory, the same frequency. The higher viscous modes can be discrete or continuous, and if discrete, can be finite or infinite in number depending on whether the solutions are or are not confined to a finite domain. In either case they can also be degenerate, and have multiple eigenfunctions for the same eigenvalue.

The continuum of normal modes is the ordinary subject matter of stability theory. The higher normal modes of the compressible theory are firmly established to the extent that the second mode has been measured experimentally, and in wind tunnels is the dominant unstable mode for hypersonic boundary layers. The neutral modes of the continuous spectrum which represents the reflection of a periodic Mach radiation field from a boundary layer also correspond to a measurable physical phenomenon. The higher discrete viscous modes must be excited in the immediate neighborhood of an artificial disturbance source, but their presence has yet to be clearly demonstrated. At low speeds, the highly damped Squire mode of the undisturbed boundary layer becomes strongly unstable when the base flow has added to it a periodic disturbance of finite, but small, amplitude, but it remains to be established whether this phenomenon persists beyond a very low Mach number.

A possible physical role has been found for the multiple solutions of Figs. 7-9, where the multiplicity refers to the same wavenumber but the additional solutions are not the higher discrete modes. This multiplicity requires a region of relative supersonic flow, and the number of solutions is wavenumber dependent: two solutions exist at low wavenumbers; three at somewhat higher wavenumbers; etc. An example was given previously [2] where the physical consequences of exciting the two solutions that exist at low wavenumbers were examined at $M_1=5.8$. First and higher-mode solutions respond differently to cooling; first-mode solutions are strongly stabilized by cooling; solutions dependent on the supersonic relative-flow region are slightly destabilized. When a fixed-frequency spatial wave with $c<1$ is excited at a low

Reynolds number at a fixed initial amplitude in a moderately cooled boundary layer, the downstream wave will first belong to the damped first mode, then to the unstable second mode. With more cooling, the <u>increased</u> damping of the first mode outweighs the amplification of the second mode, and the downstream amplitude <u>decreases</u>. At very low wall temperatures, the wave with the $c<1$ first-mode solutions is damped at all Reynolds numbers. It is the second wave, with $c>1$ at the initial Reynolds number, that leads to the unstable second-mode solutions at higher R. Thus, cooling <u>decreases</u> the damping of this wave in addition to destabilizing the unstable second mode, and the downstream amplitude <u>increases</u>.

What is missing from compressible stability theory is a firm connection with boundary-layer transition. There is little doubt that transition is preceded by linear instability in many instances, but just how the individual unstable waves act, alone or in combination, to trigger the transition process is not known. This is another way of saying that nothing yet exists in the compressible theory comparable to the secondary-instability and nonlinear theories that are such a prominent feature of incompressible stability theory. Unfortunately, there is little experimental guidance. An opportunity exists for theorists and numerical analysts to fill a large gap in present-day knowledge of how transition takes place at supersonic and hypersonic speeds.

Acknowledgement

The work described in this paper was carried out at the Jet Propulsion Laboratory, California Institute of Technology, under contract with the National Aeronautics and Space Administration (NASA). Support from the Aerodynamics Division of the Office of Aeronautics and Space Technology, NASA, is gratefully acknowledged.

References

1. Lees, L. and Lin, C.C., Investigation of the Stability of the Laminar Boundary Layer in a Compressible Fluid. NACA Technical Note No. 1115, 1946.

2. Mack, L.M., Boundary-Layer Stability Theory, Document No. 900-277, Rev. A, Jet Propulsion Laboratory, Pasadena, CA, 1969.

3. Mack, L.M., Boundary-Layer Linear Stability Theory, in "Special Course on Stability and Transition of Laminar Flow," AGARD Report No. 709, pp. 3-1 to 3-81, 1984.

4. Gapnov, S.A. and Maslov, A.A., Propagation of Disturbances in Compressible Fluids (in Russian), Akademia Nauk, Siberian Branch, Novosibirsk, USSR, 1980.

5. Lees, L. and Gold, H., Stability of Laminar Boundary Layers and Wakes at Hypersonic Speeds: Part I. Stability of Laminar Wakes, in "Proceedings of International Symposium on Fundamental Phenomenon in Hypersonic Flow," pp. 310-337, Cornell Univ. Press, 1964.

6. Brown, W.B., Exact Numerical Solutions of the Complete Linearized Equations for the Stability of Compressible Boundary Layers, Norair Report No. NOR-62-15, Northrop Aircraft Inc., Hawthorne, CA 1962.

7. Mack, L.M., Stability of the Compressible Laminar Boundary Layer According to a Direct Numerical Solution, in AGARDograph 97, Part I, pp. 329-362, 1965.

8. Gropengiesser, H., Beitrag zur Stabilitat freier Grenzschichten in kompressiblen Medien, Report DLR FB 69-25, Deutsche Luft- und Raumfahrt, 1969 (also available as NASA Technical Translation F-12,786, Washington, D.C., 1970).

9. Mack, L.M., Linear Stability and the Problem of Supersonic Boundary-Layer Transition, AIAA J., Vol. 13, pp. 278-289, 1975.

10. Mack, L.M., A Numerical Method for the Prediction of High-Speed Boundary-Layer Transition Using Linear Theory, in "Proceedings of Conference on Aerodynamic Analyses Requiring Advanced Computers," NASA SP-347, 1975.

11. Nayfeh, A.H. and El-Hady, N.M., Nonparallel Stability of Compressible Boundary-Layer Flows, Rep. No. VPI-E-79.13, Engineering Science and Mechanics Dept., Virginia Polytechnic and State Univ., Blacksburg, VA, 1980.

12. Gaponov, S.A., The Influence of Flow Non-parallelism on Disturbance Development in the Supersonic Boundary Layer in "Proceedings of the Eighth Canadian Congress of Applied Mechanics," pp. 673-674, 1981.

13. Drazin, P.G. and Davey, A., Shear Layer Instability of an Inviscid Compressible Fluid: Part 3, J. Fluid Mech., Vol. 82, Part 2, pp. 255-260, 1977.

14. Mack, L.M., On the Stability of the Boundary Layer on a Transonic Swept Wing, AIAA Paper No. 79-0264, 1979.

15. Lekoudis, S., Stability of Three-Dimensional Boundary Layers over Wings with Suction, AIAA Paper No. 79-0265, 1979.

16. Malik, M.R., COSAL - A Black-Box Compressible Stability Analysis Code for Transition Prediction in Three-Dimensional Boundary Layers, NASA CR-165925, 1982.

17. Herbert, Th., Subharmonic Three-Dimensional Disturbances in Unstable Plane Shear Flows, AIAA Paper No. 83-1759, 1983.

18. Herbert, Th., Analysis of the Subharmonic Route to Transition in Boundary Layers, AIAA Paper No. 84-0009, 1984.

19. Squire, H.B., On the Stability of Three-Dimensional Disturbance of Viscous Flow Between Parallel Walls, Proc. Roy. Soc. A, Vol. 142, pp. 621-628, 1933.

20. Tam, C.K.W. and Morris, P.J., The Radiation of Sound by the Instability Waves of a Compressible Plane Turbulent Shear Layer, J. Fluid Mech., Vol. 98, Part 2, pp. 349-381, 1980.

21. El-Hady, N.M., On the Effect of Boundary-Layer Growth on the Stability of Compressible Flows, unpublished, 1981.

22. Mack, L.M., Computation of the Stability of the Laminar Boundary Layer, in "Methods of Computational Physics" (B. Alder, S. Fernbach and M. Rotenberg, eds.), Vol. 4, pp. 247-299, Academic Press, NY, 1965.

23. Lees, L., The Stability of the Laminar Boundary Layer in a Compressible Fluid, NACA Technical Report No. 876, 1947.

24. Mack, L.M., A Numerical Study of the Temporal Eigenvalue Spectrum of the Blasius Boundary Layer, J. Fluid Mech., Vol. 79, pp. 497-520, 1976.

25. Petrov, G.V., Stability of Thin Viscous Shock Layer on a Wedge in Hypersonic Flow of a Perfect Gas, in "Laminar-Turbulent Transition" (V.V. Kozlov, ed.), Proceedings of 2nd IUTAM Symposium, pp. 487-493, Springer, Berlin, 1984.

Wave Phenomena in a High Reynolds Number Compressible Boundary Layer

A. BAYLISS, L. MAESTRELLO, P. PARIKH AND E. TURKEL

Introduction

This paper is a numerical study of the behavior of spatially unstable waves in a high Reynolds number, compressible laminar boundary layer. The numerical simulations are conducted by solving the laminar, two-dimensional, compressible Navier-Stokes equations over a flat plate with a fluctuating disturbance generated at the inflow. The primary objectives of this work are to study the nonlinear growth and distortion of the unstable waves and also to study techniques for the active control of these disturbances by time-periodic surface heating and cooling. The results presented here are an extension of the results presented in [1], [2].

An extensive experimental investigation of the evolution of linearly unstable waves in a boundary layer is described in [3]. The numerical simulations closely parallel the conditions of this experiment except for the Mach number of the mean flow which is considerably higher than in the experiment. The authors in [3] investigated the initial stage of the development of the disturbance, in particular the growth of the higher harmonics. It was shown that in this region three-dimensional effects were not important. Our results are in qualitative agreement with these experimental observations. Murdock [4] studied the growth of spatially unstable disturbances in an incompressible flow. The results presented here are similar to those obtained in [4]. However, our results are obtained for a much smaller initial disturbance level and bring out some additional features of the wave development. Furthermore, we account for compressibility effects.

The active control of unstable waves by surface heating was introduced in [5] and [6] for disturbances in water. The idea is to introduce a temperature disturbance by surface heating which is out of phase with the propagating disturbance. The growth of the unstable wave is reduced because the two waves cancel. In [5] and [6] a feedback control mechanism was used to generate the control signal. A feedback mechanism is needed since by an appropriate choice of phase the signals can be made to amplify rather than cancel. In the experiments of Maestrello [7], instantaneous transition in air was achieved by localized surface heating.

The use of active control techniques which attempt to modify the unstable wave, rather than the basic mean flow, offers considerable promise as an efficient method of delaying transition. It was shown in [8] that this technique should be considerably more difficult to apply in air than in water. There are three reasons for this. First, steady heating tends to destabilize the mean flow. Second, a much larger temperature disturbance is required to generate an equivalent change in the viscosity. Finally, there is the possibility that temperature disturbances can be transformed into acoustic disturbances, thereby losing effective phase control. Nevertheless, the numerical simulations presented in [2] and in the present paper demonstrate that this technique is feasible.

In section 2 we discuss the numerical model. The major feature of the numerical scheme is the use of fourth-order accurate finite differences for the inviscid terms of the Navier-Stokes equations. It is well known (see, for example, [9], [10]) that fourth-order accuracy is essential in preventing numerical dispersion and dissipation from significantly degrading the accuracy of wave propagation problems. Fourth-order accuracy is even more essential for the present problem, in order to prevent numerical errors on the inviscid terms, which act as an additional source of viscosity, from lowering the effective Reynolds number of the mean flow and hence, incorrectly stabilizing the flow.

In section 3 we present numerical results. The section is divided into two parts. In the first part the nonlinear evolution of the uncontrolled wave is described and compared with experimental observations. In the second part the simulation of localized time-periodic surface heating and cooling is discussed. Generally, the active control would be expected to be most effective in the linear regime before nonlinear growth and distortion becomes significant. When nonlinear effects predominate then periodic control is no longer possible as harmonics develop and the waves will not have a well-defined phase. One needs to develop control techniques which account for random amplitudes and phases. In section 4 we discuss our conclusions.

2. Numerical Model

In this section we describe the numerical model. An extensive discussion of the model is presented in [1]. The discussion here will be brief and the reader is referred to [1] for further details.

The laminar, compressible Navier-Stokes equations can be written in the form

$$w_t = F_x + G_y. \qquad (2.1)$$

Here, w is the vector $(\rho, \rho u, \rho v, E)^T$, ρ is the density, u and v are the x and y velocity components respectively, and E is the total energy. The functional forms of the functions F and G are standard and will be omitted for brevity.

The computational domain is the rectangle shown in figure 1. First, a basic steady flow, in this case a spreading boundary layer, is computed. Then an unsteady disturbance is specified at inflow and the development of the disturbance as it propagates through the steady flow is simulated by solving the system (2.1). Since the disturbance must be followed over a large number of wavelengths, it is essential to use a higher order accurate scheme. We therefore use a scheme which is second-order accurate in time and fourth-order accurate in space.

For the one-dimensional equation

$$W_t + F_x = 0,$$

we have

$$\overline{W}_i^{n+1} = W_i^n - \frac{\Delta t}{6 \Delta x} \; 7(F_{i+1}^n - F_i^n) - (F_{i+2}^n - F_{i+1}^n)$$

$$W_i^{n+1} = \frac{1}{2}\left(\overline{W}_i^{n+1} + W_i^n - \frac{\Delta t}{6\Delta x}\left\{7(\overline{F}_i^{n+1} - \overline{F}_{i-1}^{n+1}) - (\overline{F}_{i-1}^{n+1} - \overline{F}_{i-2}^{n+1})\right\}\right). \qquad (2.2)$$

The scheme (2.2) becomes fourth-order when F = F(u) if it is alternated with a symmetric variant in which there are backward differences in the predictor and forward differences in the corrector. It has a greatly-reduced truncation error compared with the second-order MacCormack scheme. Our experience has been that fourth-order accuracy is necessary to efficiently compute the class of problems considered here. Operator splitting is used so that the two-dimensional system (2.1) is solved by successive applications of one-dimensional solution operators of the form (2.2). This scheme is fourth-order accurate on the inviscid terms. The scheme is fourth-order on the viscous terms for a constant viscosity but is only second-order accurate on the viscous terms when the viscosity is spatially dependent. For this problem, due to the high Reynolds number the primary source of error is due to the inviscid terms and the scheme is very accurate. A sixth-order, in space, algorithm is presented in [2].

A detailed description of the boundary conditions is given in [1]. Radiation conditions are used at the outflow and upper boundaries. At the inflow we must specify three boundary conditions. These are the three incoming characteristic variables based on linearizing the function F in (2.1) and ignoring variations in the y direction. Let Q denote an incoming variable. We specify Q at inflow by

$$Q = Q_{steady} + \varepsilon\, e^{i\omega t}\, Q_{0S}(y)$$

where Q_{steady} is the steady state solution, and $Q_{0S}(y)$ is obtained from the linearized Orr-Sommerfeld equation (linearized about the inflow steady profile) for the (unstable) frequency ω. The Orr-Sommerfeld solutions were obtained from a program developed by J. R. Dagenhart at NASA Langley Research Center. This program neglects compressibility effects. Thus, there exists a discrepancy when comparing with linear theory at large flow velocities. However, for an inflow Mach number of 0.4 close agreement with linear theory is obtained.

3. Numerical Results

In this section we present numerical results for the model described in section 2. The section is divided into two parts. Part A is concerned with wave propagation through the mean flow without external active control. In Part B the effect of active surface heating and cooling is discussed.

A. Uncontrolled Wave Propagation

We first consider a boundary layer with a free stream Mach number of 0.4. The unit Reynolds number is 3.0×10^5. The computational domain is chosen so that at inflow $Re_{\delta *}$ (Reynolds number based on displacement thickness) is 990 and at outflow $Re_{\delta *}$ is 1730. Based on the inflow profile, the nondimensional frequency $F = (2\pi f \nu / U_\infty^2)$ of $.8 \times 10^{-4}$ is unstable. (Here f is the frequency in Hertz, ν the kinematic viscosity and U_∞ the free stream velocity.)

In figure 2 we plot the growth rates of the unstable disturbance as a function of $Re_{\delta *}$. The growth rates are computed by computing the RMS of $\rho u(t,y)$, integrating the result in y and normalizing by the value at inflow. The results in figure 2 are plotted for two different values of ε (amplitude of the inflow perturbation) and compared with results obtained from linear (incompressible) stability theory.

It is apparent that for small ε, the growth rates are very close to those predicted by linear theory. Differences can be attributed to nonparallel and possibly compressibility effects. In particular, the solution decays at roughly the same position as predicted by linear theory. For larger values of ε, the solution does not decay and exhibits a nonlinear growth. This behavior is similar to that observed by Thomas [11] using a vibrating ribbon in air.

In order to analyze the solution as a function of y we plot in figure 3 the RMS of ρu as a function of y at four different x locations. The results in figure 3 are obtained with $\varepsilon = 0.02$. In this case the profiles follow the basic shape of the inflow Tollmein-Schlichting profile. However, the amplitude increases as the disturbance grows downstream. The shape of Tollmein-Schlichting profile is preserved even though the solution is exhibiting a growth which is not predicted by linear theory (which, in fact, predicts decay for $Re_{\delta *} > 1500$).

In figure 4 the amplitude of the fundamental (F_1) and the first harmonic (F_2) are plotted as a function of y/δ where δ is the local boundary layer thickness for $Re_{\delta*} = 1579$. The data is normalized so that the peak of the fundamental is 1.0. It can be seen that the amplitude of the first harmonic has grown to roughly 30% of the peak of the fundamental. It is also apparent that the harmonic has a maximum near the wall and thus the nonlinearity is most pronounced there. This is in agreement with the experimental measurements of [3]. In figure 5 we show results using the data presented in [3] illustrating the experimental fluctuation level.

The experimental data clearly shows a much larger degree of nonlinearity than the computations. However, the flow velocity in [3] is much less than the present computation. The value of $Re_{\delta*}$ can not be determined from the information presented in [3]. The computation does not produce a double peak for the fundamental although there is some indication of a double peak for the harmonic.

It is apparent from both figures that nonlinearity is expected to be most noticeable in the solution near the wall and near the turning point where the fluctuation goes through zero. This can also be seen in figure 6 where ρu is plotted as a function of nondimensional time at several y locations for $Re_{\delta*} = 1579$. We observe that at this location the disturbance has grown so large that there is a cylic separation and reattachment near the wall.

The compressibility effects for this case are not large. In order to examine the effect of compressibility we compare in figure 7 the growth of the disturbance for free stream Mach numbers 0.4 and 0.7. The data is taken so that $Re_{\delta*}$ at inflow is 998.0 in both cases. The parameter ε is also the same in both cases. Both inflow Tollmien-Schlichting profiles are taken from an incompressible program and would therefore be expected to be slightly less accurate for $M = 0.7$.

The results demonstrate a very strong stabilizing effect due to increasing compressibility. In fact the $M = 0.7$ decays downstream following the stability curve while the $M = 0.4$ case does not decay due to nonlinear effects. Examination of the solution indicates that there is much less nonlinearity in the $M = 0.7$ case.

B. <u>Active Control by Surface Heating and Cooling</u>

We next consider the effect of active surface heating and cooling. All results are obtained for the $M = 0.4$ case described previously. The surface heating and cooling is effected by imposing a boundary condition of the form

$$\frac{T}{T_{ref}} = \frac{T_w}{T_{ref}} \pm \left(\alpha + \beta \sin(\frac{\omega t}{2} + \phi)\right)^2 \qquad (3.1)$$

where the + sign is for heating and the − sign is for cooling. T_w is the temperature at the wall ($520^\circ R$) and T_{ref} is a reference temperature. The form of (3.1) is chosen to model a combination of a DC current and an AC current. No attempt was made to optimize the coefficients α and β, however, the effect of varying the phase ϕ was studied.

It is well known that static heating is destabilizing in air and static cooling is stabilizing. The numerical results with $\alpha \gg \beta$ confirm this. In order to demonstrate the effect of phase we plot the growth rates in figure 8 for cooling with different values of ϕ and in figure 9 for heating. In all cases the control strip extends over roughly 20% of a wavelength and is centered at $Re_{\delta^*} = 1263$.

It is apparent from figures 8 and 9 that the response is very sensitive to the phase. In particular, cooling can destabilize with an appropriate choice of phase and heating can stabilize. These results clearly indicate that phase cancellation can be a viable mechanism for the control of unstable disturbances. By appropriate choice of phase the active heating and cooling can have effects exactly opposite from the static case.

In figures 10 and 11 we examine time traces for $\rho u(t)$ at different y locations. The traces are all taken at x = 1.0 ft., significantly downstream of the heating (or cooling) strip which was placed at 0.6 ft. The figures show that the residual effects of the active heating and cooling are an amplitude and phase change in the propagating wave. Very little distortion in the wave form is introduced by the active heating and cooling.

4. Conclusions

The behavior of unstable disturbances in a high Reynolds number flow can be effectively computed provided a fourth-order accurate finite difference scheme is used. The results show that significant nonlinear distortion is produced which is in qualitative agreement with experiment.

It is shown via a full Navier-Stokes solution that increasing compressibility can significantly stabilize the flow over a flat plate. In addition, it is shown that the mechanism of phase cancellation is a viable mechanism for the control of growing disturbances. The authors are currently extending these results to flows over curved surfaces thus accounting for non-zero pressure gradients.

Acknowledgment

Partial support for the first and fourth authors was provided by the National Aeronautics and Space Administration under NASA Contracts No. NAS1-17070 and NAS1-18107 while they were in residence at the Institute for Computer Applications in Science and Engineering, NASA Langley Research Center, Hampton, VA 23665.

Partial support was provided for the third author under NASA Contract No. NAS1-17919.

References

[1] Bayliss, A., Maestrello, L., Parikh, P., and Turkel, E., "Numerical Simulation of Boundary Layer Excitation by Surface Heating/Cooling," AIAA-85-0565, March 1985.

[2] Bayliss, A., Parikh, P., Maestrello, L., and Turkel, E., "A Fourth-Order Method for the Unsteady Compressible Navier-Stokes Equations," AIAA-85-1694, July 1985.

[3] Kachanov, Yu. S., Kozlov, V. V., and Levchenko, V. Ya., "Nonlinear Development of Wave in a Boundary Layer," Fluid Dynamics, Translation, Vol. 12, No. 1, January-February, 1977.

[4] Murdock, J., "A Numerical Study of Nonlinear Effects on Boundary Layer Stability," AIAA J., Vol. 15, No. 8, August 1977, pp. 1167-1173.

[5] Liepmann, H. W., Brown, G. L., and Nosenchuck, D. M., "Control of Laminar Instability Waves Using a New Technique," J. Fluid Mech., Vol. 118, 1982, pp. 187-200.

[6] Liepmann, H. W. and Nosenchuck, D. M., "Active Control of Laminar-Turbulent Transition," J. Fluid Mech., Vol. 118, 1982, pp. 201-204.

[7] Maestrello, L., "Active Transition Fixing and Control of the Boundary Layer in Air," AIAA-85-0564, March 1985.

[8] Maestrello, L. and Ting, L., "Analysis of Active Control by Surface Heating," AIAA J., Vol. 23, No. 7, July 1985, pp. 1038-1045.

[9] Kreiss, H. O. and Oliger, J., "Comparison of Accurate Methods for the Integration of Hyperbolic Systems, Tellus, Vol. 24, 1980, pp. 119-225.

[10] Turkel, E., "On the Practical Use of Higher Order Methods for Hyperbolic Systems," J. Comput. Phys., Vol. 35, 1980, pp. 319-340.

[11] Thomas, A. S. W., "The Control of Boundary Layer Transition Using a Wave-Superposition Principle," J. Fluid Mech., Vol. 137, 1983, pp. 233-250.

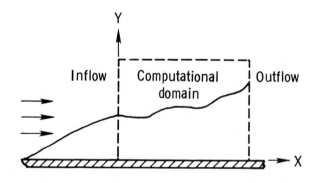

Figure 1. Schematic of computational domain.

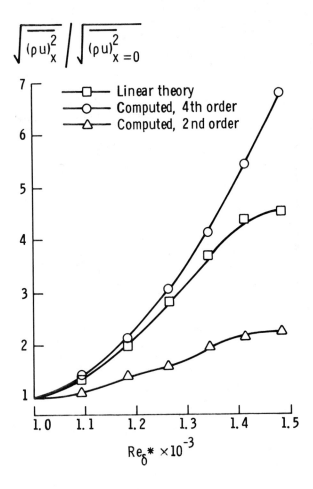

Figure 2. Comparison of amplitude growth with linear theory.

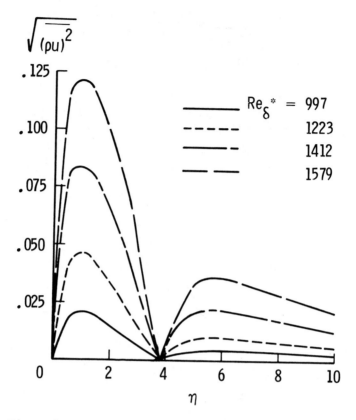

Figure 3. RMS ρu versus y at different x locations.

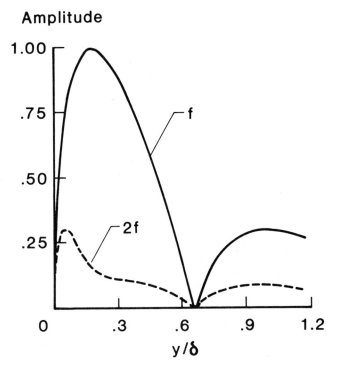

Figure 4. F_1 and F_2 versus y/δ at $Re_{\delta*}$ = 1579.

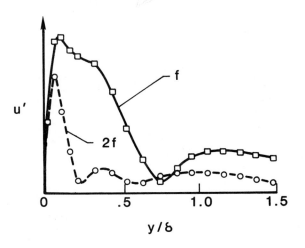

Figure 5. Experimental analysis of harmonic content of fluctuating disturbance (from [3]).

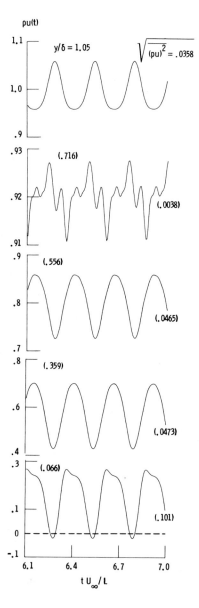

Figure 6. ρu versus t for selected values of y; $Re_{\delta*}$ = 1579.

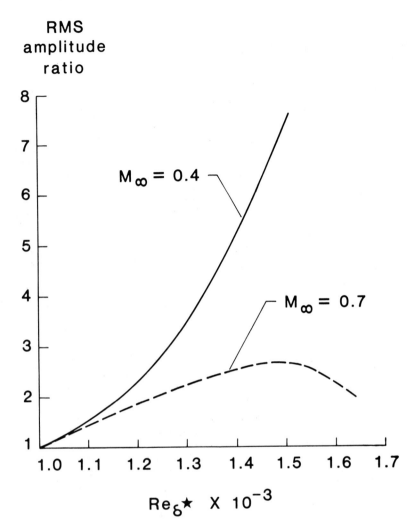

Figure 7. Growth rates for M = 0.7 and M = 0.4.

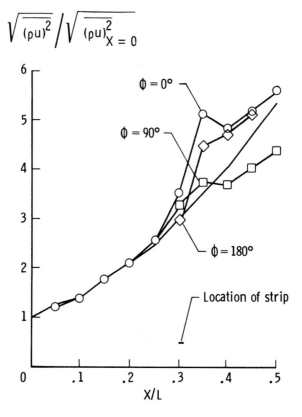

Figure 8. Effect of phase - cooling.

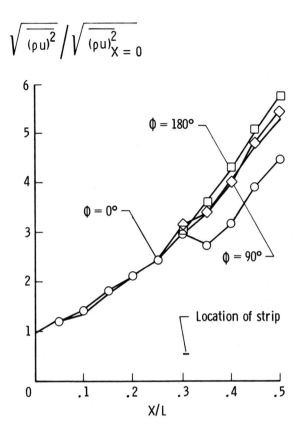

Figure 9. Effect of phase - heating.

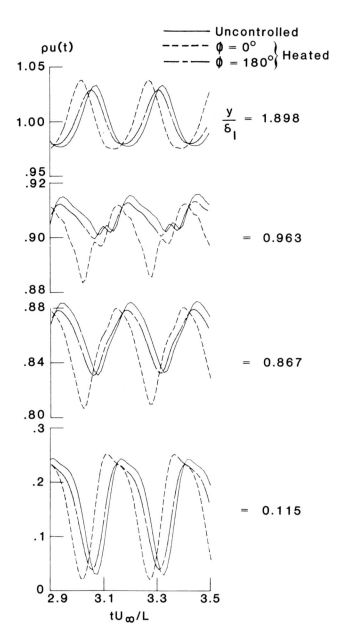

Figure 10. ρu versus t for different phases at x = 1.0 ft; heating at x = 0.6 ft.

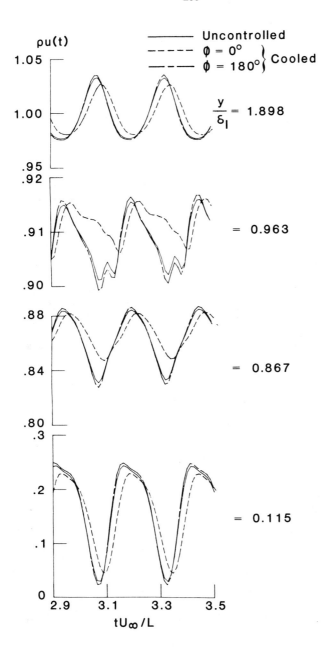

Figure 11. ρu versus t for different phases at x = 1.0 ft; cooling at x = 0.6 ft.

Instability of Time-Periodic Flows

Philip Hall

Abstract

The instabilities of some spatially and/or time-periodic flows are discussed, in particular flows with curved streamlines which can support Taylor-Görtler vortices are described in detail. The simplest flow where this type of instability can occur is that due to the torsional oscillations of an infinitely long circular cylinder. For more complicated spatially varying time-periodic flows a similar type of instability can occur and is spatially localized near the most unstable positions. When nonlinear effects are considered it is found that the instability modifies the steady streaming boundary layer induced by the oscillatory motion. It is shown that a rapidly rotating cylinder in a uniform flow is susceptible to a related type of instability; the appropriate stability equations are shown to be identical to those which govern the instability of a Boussinesq fluid of Prandtl number unity heated time periodically from below.

1. Introduction

Our main concern is with the nature of the centrifugal instability of time-periodic flows which interact with curved surfaces. It appears at this stage that theory and experiment here are in much closer agreement than is the case for the instability of flat Stokes layers. In the latter problem it seems that the instabilities which exist experimentally occur at such high Reynolds numbers that quasi-steady theory applies and the solution of the full time-dependent stability equations, being damped, are irrelevant. The reader is referred to the papers by Cowley and Kerczek, which appear in this volume, for a detailed discussion of the flat Stokes layer. In contrast to this situation, for curved Stokes layers the solutions of the time-dependent equation explain a great deal of the available experimental results, whilst the quasi-steady solutions of the equations are irrelevant except in the inviscid or high wavenumber viscous limits.

The existence of a Taylor-Görtler instability in the Stokes layers on a torsionally oscillating cylinder was demonstrated experimentally and theoretically by Seminara and Hall (1976). Since Rayleigh's criterion does not apply to time-dependent flows there is no simple way of determining whether a given time-dependent basic flow is inviscidly unstable. Indeed, since Floquet theory can only be used for equations with time-periodic coefficients, unless the basic state is varying periodically in time, it is not even clear how to define instability.

The onset of instability predicted theoretically is consistent with the available experimental observations, however, at high Taylor numbers it is found experimentally that a secondary instability which progressively doubles the axial wavelengths occurs. At some stage in this process the flow acquires an azimuthal dependence, there is as yet no adequate theoretical description of this regime and it seems likely that only a fully numerical investigation will reveal its structure.

More recently the instability mechanism found by Seminara and Hall (1976) has been shown experimentally and theoretically to occur in a wide range of flows of practical importance. These flows vary (periodically) in time and at least two spatial directions and exhibit the so-called steady streaming phenomenon associated with the work of, for example, Rayleigh (1883), Schlichting (1932), Stuart (1966), and Riley (1967). In these flows the Reynolds stresses in the boundary layer of the first-order oscillatory viscous flow drive a mean motion which may or may not be confined to the Stokes boundary layer. In fact, the structure of the steady streaming depends on the size of R_s, a Reynolds number associated with the mean motion. The importance of R_s was first explained by Stuart (1966) who showed that for $R_s \gg 1$ the steady streaming decays to zero in an outer boundary layer of relative thickness $R_s^{+1/2}$. It turns out that the instability of the basic oscillatory flow also occurs for $R_s \gg 1$ and that the steady streaming is then driven by both the Reynolds stresses of the oscillatory flow and the instability. It is likely that in some situations the instability might cause the premature separation of the steady streaming boundary layer.

The thermal convection analogue of the torsionally oscillatory cylinder problem has apparently not yet been investigated. The instability equations for an infinite layer of fluid heated sinusoidlly from above or below will be derived in this paper. It turns out that if the fluid has Prandtl number unity then the same equations govern the instability of the flow past a rapidly rotating cylinder in a uniform flow. The analogy between the problems is similar to that which is known to exist between the steady rapidly rotating Taylor vortex and steady Bénard problems. Some preliminary results for the solution of the eigenvalue problem governing the instability of these flows are given.

The procedure adopted in the rest of this paper is as follows: In Section 2 the instability of the flow on a torsionally oscillating cylinder is discussed. In section 3 the generalization of this instability to spatially varying time-periodic flows is discussed. In Section 4 we discuss the analogy between the instability problems for the flow around a rapidly rotating cylinder and that driven by the time-periodic heating of a fluid layer. Finally, in Section 5 we draw some conclusions.

2. The Torsionally Oscillating Cylinder Problem

Consider the viscous flow induced by an infinitely long circular cylinder of radius R rotating about its axis with angular velocity $\Omega \cos \omega t$. The time-periodic boundary layer on the cylinder is taken to be small compared to R so that,

$$\left(\frac{\nu}{\omega}\right)^{1/2} \ll R,$$

and the velocity field $\underset{\sim}{u}_B$ is then given by,

$$\underset{\sim}{u}_B = \Omega R(0, V(\eta, \tau), 0),$$

where

$$\tau = \omega t, \quad \eta = \{r - R\}\left\{\frac{\nu}{2\omega}\right\}^{-1/2}$$

and

$$\overline{V}(\eta, \tau) = \cos\{\tau - \eta\} e^{-\eta}. \qquad [2.1]$$

Thus $\overline{V}(\eta, \tau)$ is the usual Stokes velocity profile and it is convenient to define the Taylor-Görtler number T by

$$T = 2 \frac{\Omega^2 R}{\omega^{3/2} \nu^{1/2}}, \qquad [2.2]$$

and it is assumed from now on that $T \sim O(1)$.

The above time-periodic basic state is then perturbed to a disturbance which is periodic along the axis of the cylinder with wavelength $\frac{2\pi}{k}$ based on $\left(\frac{2\nu}{\omega}\right)^{1/2}$ the boundary layer thickness. After some manipulation we find that the radial and azimuthal velocity components U, V satisfy the coupled system:

$$\mathcal{L}\left\{\frac{\partial^2}{\partial \eta^2} - k^2\right\} U = 2k^2 \, T \overline{V}(\eta, \tau) V, \qquad [2.2a]$$

$$\mathcal{L} V = \sqrt{2} \frac{\partial \overline{V}}{\partial \eta} U, \qquad [2.2b]$$

where

$$\mathcal{L} \equiv \frac{\partial^2}{\partial \eta^2} - k^2 - 2 \frac{\partial}{\partial \tau}.$$

The system [2.2] must be solved subject to

$$U = V = \frac{\partial V}{\partial \eta} = 0, \quad \eta = 0, \quad [2.3]$$

and

$$U = V = 0, \quad \eta = \infty, \quad [2.4]$$

so that the no-slip condition is satisfied at the wall and the disturbance decays exponentially to zero at infinity. There is, of course, no justification for a quasi-steady approximation if T is $O(1)$ since the time dependence of the partial differential system is in no sense slow.

On the basis of Floquet theory we expect solutions of [2.2] of the form

$$U = e^{\sigma \tau} u(\eta, \tau), \quad V = e^{\sigma \tau} v(\eta, \tau)$$

where u, v are now periodic functions of τ. The Floquet exponent σ is a function of k and T and must be chosen such that the partial differential equations and boundary conditions are satisfied. If we write

$$u = \sum_{-\infty}^{\infty} e^{in\tau} u_n(\eta), \quad v = \sum_{-\infty}^{\infty} e^{in\tau} v_n(\eta),$$

then the eigenvalue σ is determined by the system

$$\left.\begin{array}{l}
[\frac{d^2}{d\eta^2} - k^2 - \sigma - 2in][\frac{d^2}{d\eta^2} - k^2]u_n = \frac{k^2 T}{2}[e^{-\eta(1+i)} v_{n-1} + e^{-\eta(1-i)} v_{n+1}], \\[6pt]
[\frac{d^2}{d\eta^2} - k^2 - \sigma - 2in]v_n = \frac{-1}{\sqrt{2}}[(1+i)e^{-\eta(1+i)} u_{n-1} + (1-i)e^{-\eta(1-i)} u_{n+1}], \\[6pt]
u_n = v_n = \frac{du_n}{d\eta} = 0, \quad \eta = 0, \\[6pt]
u_n = v_n = 0, \quad \eta = \infty,
\end{array}\right\} \quad [2.5]$$

for $n = 0, \pm 1, \pm 2, \ldots$.

The eigenvalue problem [2.5] can be solved numerically by truncating n such that $|n| \leq N$ and finding $\sigma = \sigma(N, K, T)$. Numerical calculations by Seminara and Hall (1976) showed that instability occurs for

$$T > 164.42$$

and the critical value of the wavenumber is $k = 0.859$. The neutral curve calculated by Seminara and Hall is shown in Figure 1. The calculations also showed that $u_n = 0$ for n odd and that $v_n = 0$ for n even. These results are in excellent agreement with the experiments performed by Seminara and Hall (1976) and Barenghi, Park, and Donnelly (1980) who visualized the instability using dye visualization.

However, in both sets of experiments a secondary instability at a Taylor number about 30% above the critical was observed. The linear theory of Seminara and Hall showed that when instability occurs the function $v_0(\eta)$ is zero so that there is no mean flow around the cylinder. When the secondary instability occurs the vortices initially interact with their neighbours to generate larger vortices. Subsequently this larger set of vortices interact to produce an even larger set and this process appears to continue without any equilibrium state ever being reached. At some stage in this process the azimuthal velocity field develops a mean component and there are signs of nonaxisymmetry.

At this time very little is understood theoretically about how and why this secondary instability takes place. Seminara and Hall (1977) showed that the first mode bifurcates supercritically at the critical Taylor number. Hall (1981) showed, using an approximate approach, that an axisymmetric mode with twice the wavelength of the first mode causes the first mode to lose stability very close to the experimentally observed critical Taylor number. However, it should be stressed that this approximate approach, which is similar to that recently used for mode interactions in spatially varying flows, cannot be rigorously justified, nevertheless it would appear that it describes very well the first stages in the secondary instability regime. It is of interest to note that in the steady Taylor and Görtler vortex instabilities the onset of the secondary wavy vortex modes leads to a new equilibrium state.

The possible role of nonaxisymmetric modes in the secondary instability problem is still not understood, certainly it is known from the work of Duck and Hall (1981) that θ-dependent modes occur quite close to the axisymmetric critical Taylor number and are more stable than the axisymmetric one. Moreover, we can show that when nonaxisymmetric modes are present a steady streaming phenomenon occurs and the instability can no longer be confined to the boundary layer. In this case the velocity field decays to zero in an outer layer of relative thickness $R_s^{1/2}$ where R_s is a steady streaming Reynolds number.

It would appear that little further theoretical work on the secondary instability stage is possible until more precise experimental results become available. The flow visualization methods used previously are not sensitive enough to determine the detailed flow structure after the secondary instability occurs.

3. Instabilities in Flows Exhibiting Steady Streaming

The prototype problem discussed in the previous section shows how a centrifugal instability mechanism can occur when a Stokes layer interacts with a curved surface. However, in general it is known that when such an interaction takes place a secondary steady streaming is set up. The nature of the steady streaming depends crucially on a steady streaming Reynolds number R_s whose importance was explained clearly by Stuart (1966). For large values of R_s the streaming decays to zero in a steady outer boundary layer of relative thickness $R_s^{1/2}$ whilst for small values of R_s the steady motion is confined to the Stokes layer. We shall see below that the instability mechanism described in Section 2 occurs for $R_s \gg 1$ and has an $O(1)$ effect on the steady streaming boundary layer.

The possibility that the instability described in Section 2 could occur in spatially varying oscillatory flows over curved walls was overlooked until Honji (1981) investigated the classical steady streaming flow induced by oscillating a cylinder transversely along its axis. In addition Honji investigated flows over wavy walls and over steps and found that a vortex instability of these time-periodic flows occurred at sufficiently large amplitudes of oscillation of the fluid velocity at infinity. We will discuss only the transversely oscillating cylinder problem in detail, related results for wavy walls, pipe flows, and spheres will be mentioned briefly later.

In his experiment Honji (1981) investigated the classical steady streaming problem of Schlichting (1932), Stuart (1966), and Riley (1967). For sufficiently small amplitude high frequency oscillations he found that the flow remained two-dimensional and was consistent with theoretical predictions. However, at a critical frequency dependent amplitude of oscillation the flow became three-dimensional in the neighbourhood of the positions on the cylinder where the tangent plane was parallel to the direction of motion of the cylinder. At a still higher critical amplitude of oscillation, Honji found that the flow became ´turbulent and separated.´ We will show below that a linear stability analysis predicts the first critical amplitude of oscillation whilst a nonlinear theory suggests that a finite amplitude instability might cause the basic steady streaming boundary layer to separate prematurely. The details of the calculation outlined below can be found in the paper by Hall (1984).

Suppose that a circular cylinder of radius a oscillates with velocity $U_0 \cos \omega t$ along a diameter in a fluid of viscosity ν. The parameters which govern the two-dimensional flow are

$$\beta = \frac{\omega a^2}{\nu}, \qquad [3.1a]$$

$$\lambda = \frac{U_0}{\omega a}, \qquad [3.1b]$$

$$R_s = \frac{U_0^2}{\omega\nu} = \lambda^2 \beta. \qquad [3.1c]$$

The frequency parameter β is taken to be large so that the boundary layer on the cylinder is thin compared to its radius whilst the amplitude parameter λ is taken to be small. Before specifying the size of the steady streaming Reynolds number R_s we write down the Taylor number T based on $(\nu/\omega)^{1/2}$ the boundary layer thickness. We obtain

$$T = \frac{2^{3/2} U_0^2}{a\nu^{1/2}\omega^{3/2}},$$

$$= 2^{3/2} R_s \beta^{-1/2},$$

so that if instability occurs at $O(1)$ values of T we must take R_s formally to be $O(\beta^{-1/2})$, thus we take the limit $\beta \to \infty$ with $R_s \beta^{-1/2}$ held fixed. We further note that in this limit $\lambda \sim \beta^{-1/4}$ so that the boundary layer on the cylinder is essentially a Stokes layer. In Figure 2 we have sketched the cylinder and the various regions of interest. The inviscid slip velocity for the basic flow has maxima at $\theta = \pm \pi/2$, and since the curvature of the boundary is constant we expect that these will be the least stable locations. A WKB approach to the instability problem quickly shows that $\theta = \pm \pi/2$ are turning points of the expansion. In fact, they are second-order turning points so that an inner region of angle $O(\beta^{-1/8})$ is needed at $\theta = \pm \pi/2$.

Suppose then that the basic flow is perturbed to a disturbance periodic along the axis of the cylinder with wavelength $\frac{2\pi}{k}$. The linear stability equations can be reduced to

$$L(\frac{\partial^2}{\partial \eta^2} - k^2)U = 2k^2 T \sin\theta \, \bar{v}_0 V - \frac{2^{5/4} \sin\theta \, \bar{v}_{0\eta\eta} T^{1/2}}{\beta^{1/4}} \frac{\partial U}{\partial \theta} + O(\beta^{-1/4}),$$

$$LV = 4 \sin\theta \, \frac{\partial \bar{v}_0}{\partial \eta} U + O(\beta^{-1/4}), \qquad [3.2]$$

where

$$L \equiv \frac{\partial^2}{\partial \eta^2} - k^2 - 2\frac{\partial}{\partial \tau} - \frac{2^{5/4} T^{1/2} \sin\theta \, \bar{v}_0}{\beta^{1/4}} \frac{\partial}{\partial \theta}$$

with

$$\bar{v}_0 = \cos\tau - \cos(\tau - \eta)e^{-\eta}. \qquad [3.3]$$

The Taylor number T is expanded as

$$T = T_0 + \beta^{-1/4} T_1 + \cdots$$

whilst near $\theta = \pi/2$ the velocity components U, and V expand as

$$U = U_0(\eta, \tau, \phi) + \beta^{-1/8} U_1(\eta, \tau, \phi) + \cdots$$

$$V = V_0(\eta, \tau, \phi) + \beta^{-1/8} V_1(\eta, \tau, \phi) + \cdots$$

where $\phi = \beta^{1/8}(\theta - \pi/2)$. The functions U_0, V_0 can be written

$$(U_0, V_0) = A(\phi)(u_0(\eta, \tau), v_0(\eta, \tau))$$

where A is an amplitude function to be determined whilst (u_0, v_0) satisfy

$$\{\frac{\partial^2}{\partial \eta^2} - k^2 - 2\frac{\partial}{\partial \tau}\}\{\frac{\partial^2}{\partial \eta^2} - k^2\}u_0 = 2k^2 T_0 \bar{v}_0 v_0,$$

$$\{\frac{\partial^2}{\partial \eta^2} - k^2 - 2\frac{\partial}{\partial \tau}\}v_0 = 4\frac{\partial \bar{v}_0}{\partial \eta} u_0,$$

$$u_0 = \frac{\partial u_0}{\partial \eta} = v_0 = 0, \quad \eta = 0,$$

[3.4]

$$u_0, v_0 \to 0, \eta \to \infty.$$

This eigenvalue problem was solved by Hall (1984) in exactly the same manner as that described in Section 2. The neutral curve $k = k(T_0)$ is shown in Figure 3, we see that instability is predicted locally near $\theta = \pi/2$ for

$$T_0 > 11.99. \qquad [3.5]$$

The amplitude function $A(\phi)$ is found at higher order to satisfy

$$\frac{d^2 A}{d\phi^2} + \mu[T_1 - \phi^2]A = 0. \qquad [3.6]$$

Here μ is a positive constant which must be calculated numerically, the solutions of this equation, which decay when $\phi \to \pm \pi/2$ are

$$A = U_n(-n - 1/2, 2\mu^{1/4} \phi)$$

where U_n is the nth parabolic cylinder function and

$$T_1 = T_{1n} = \frac{2[n + \frac{1}{2}]}{\sqrt{\mu}}.$$

The most unstable mode corresponds to n = 0 and gives

$$A_0 = e^{-\mu^{1/2}\phi^2/2},$$

with

$$T_{10} = \frac{1}{\sqrt{\mu}} = 5.51.$$

Thus the critical value of R_s is

$$R_s = R_{sc} = 4.24[\beta^{1/2} + .46\beta^{1/4} + \cdots], \qquad [3.7]$$

or in terms of λ the critical oscillation amplitude corresponds to

$$\lambda = \lambda_c = \frac{2.06}{\beta^{1/4}}[1 + \frac{.23}{\beta^{1/4}} + \cdots]. \qquad [3.8]$$

In Figure 4 we have plotted the results predicted by [3.8] and Honji's results, there seems little doubt that the instability mechanism is responsible for the onset of the three-dimensionality described by Honji.

The development of this instability into a weakly nonlinear region leads to some surprising results. We choose to write the Taylor number T in the form

$$T = T_{0c} + \beta^{-1/4}\overline{T}$$

where T_{0c} is the critical value of T_0. A Stuart-Watson expansion with the fundamental terms of order $\beta^{-1/8}$ shows that the generalization of [3.6] into the nonlinear stage is

$$\frac{d^2 A}{d\phi^2} + \mu[\overline{T} - \phi^2]A = \gamma A^3. \qquad [3.9]$$

Numerical calculations by Hall (1984) showed that γ is negative so that finite amplitude solutions bifurcate subcritically at the critical Taylor number. The determination of γ requires the solution of several partial differential systems so the possibility that a numerical error was made should not be overlooked. However, if γ is indeed negative then it suggests that experimentally three-dimensionality could be induced by sufficiently large subcritical perturbations. The nonlinear theory shows that the steady streaming boundary layer is, at least in an $O(\beta^{-1/8})$ neighbourhood of $\theta = \pi/2$, driven by both the instability and the basic oscillatory flow. If we define an outer boundary layer variable ξ by

$$\xi = (\frac{r-a}{a})\beta^{1/4}$$

then the outer steady boundary layer is given by

$$v = \frac{\nu}{a}\beta^{3/8}\psi_\xi, \quad u = \frac{-\nu}{a}\beta^{1/4}\psi_\phi$$

with

$$\psi_{\xi\xi\xi\xi} = \psi_\xi \psi_{\xi\xi\phi} - \psi_\phi \psi_{\xi\xi\xi}, \qquad [3.10]$$

which must be solved subject to

$$\psi = 0, \quad \xi = 0$$

$$\psi_\xi \to 0, \quad \xi \to \infty \qquad [3.11]$$

$$\psi_\xi \to T_{0c}[\frac{3\phi}{2^{3/2}} - 4.39 \cdot 2^{3/4} T_{0c}^{-1/2} A \frac{dA}{d\phi}], \quad \xi \to 0.$$

If we set $A = 0$ above we obtain the equations governing the attachment of a steady streaming boundary layer within a $\beta^{-1/8}$ neighbourhood of $\theta = \pi/2$. In that case [3.10] is solved subject to

$$\psi_\xi = 0, \quad \phi = 0, \qquad [3.12]$$

and the symmetry of [3.9] about $\phi = 0$ means that [3.12] can still be applied since either A or $\frac{dA}{d\phi} = 0$ at $\phi = 0$. We note that for large ϕ the condition [3.11] reduces to $\psi_\xi \to T_{0c} \frac{3\phi}{2^{3/2}}$, $\xi \to 0$, so that, assuming that the boundary layer remains attached for finite values of ϕ, the extra term proportional to $A\frac{dA}{d\phi}$ in [3.11] merely produces an origin shift in the large ϕ asymptotic solution of [3.10]. The linear eigenfunctions $A_n(\phi)$ for $n \neq 0$ all have intervals where $A_n(\phi)A_n^1(\phi)$ is positive so that the possibility exists that in the nonlinear regime the slip velocity in [3.11] might change sign at some value of ϕ. If the magnitude of the inviscid slip velocity is sufficiently large where this occurs then the attached flow strategy fails and the steady streaming boundary layer will detach prematurely from the cylinder. This possibility does not occur for more general flows where the point of attachment of the steady streaming layer and the most unstable position do not coincide. In this case the steady streaming driven by the instability is weak compared to that of the basic flow.

For more general steady streaming flows Papageorgiu (1985) has shown that both concave and convex curvature lead to this local vortex type of instability. Consider then an oscillatory viscous flow adjacent to the wall $y = 0$ induced by a outer potential flow with slip velocity $U_0 U(x) \cos \omega t$. If the wall has radius of curvature $aR(x)$ we write down the local Taylor number T_ℓ defined by

$$T_\ell = \frac{2^{3/2} U_0^2 U^2(x)}{a \nu^{1/2} \omega^{3/2} R(x)}.$$

The mechanism described above for the circular cylinder problem occurs near $x = x_m$ if

$$T_\ell(x_m) \sim 11.99,$$

$$\frac{d}{dx} T_\ell = 0, \quad x = x_m.$$

Hence with variable curvature the most unstable locations do not necessarily occur where the inviscid slip velocity has a maximum, this produces some interesting results. Thus for an ellipse oscillating transversly there can be two or six local regions of instability depending on the angle of attack and the eccentricity of the ellipse.

Papageorgiu has shown that for concave curvature the corresponding condition is

$$|T_\ell| = 7.104,$$

$$\frac{d}{dx} |T_\ell| = 0, \quad x = x_m$$

so that a locally concave wall is more susceptible to instability than is a convex one. The result that either a concave or a convex wall lead to instability is quite different than the classical results for steady flows, however, for a time-dependent flow Rayleighs criterion does not apply so there is no reason why this should not be the case.

The implications of Papageorgiu's calculations for flows in curved pipes and over wavy walls are important. For a curved pipe he has shown that an oscillating azimuthal pressure gradient leads to instability at <u>both</u> the inner and outer bends, however, the outer bend is the most unstable location. This instability leads to rapid local variations in the shear stress at the wall which might have important consequences for aortic blood flow. It should also be pointed out that in this problem the turning point structure changes from that found for the circular cylinder problem, the technical problems associated with this change were overcome using the expansion procedure devised by Soward and Jones (1981). A similar difficulty arises when the instability of the flow due to a torsionally oscillating sphere is considered, in this problem the instability is localized near the equation. As yet the generalization of the expansion procedure of Soward and Jones into the nonlinear regime has not been given.

4. The Linear Stability Equations for the Flow Around a Rapidly Rotating Cylinder

Consider the flow of a viscous fluid of kinematic viscosity ν around a circular cylinder of radius a. The cylinder rotates with angular velocity $\frac{V}{a}$ whilst the flow a long way from the cylinder has speed U in the x direction. We follow the notation of Moore (1957) and define the parameter ε by

$$\varepsilon = \frac{U}{V}, \qquad [4.1]$$

and a Reynolds number R by

$$R = \frac{Va}{\nu}. \qquad [4.2]$$

Moore has discussed the above flow in the limit $\varepsilon \to 0$, here we will also assume that $R \gg 1$. For large values of R it is known from the work of Glauert (1957), Moore (1957), and Wood (1957) that a boundary layer of thickness $aR^{-1/2}$ is set up on the surface of the cylinder. Here we will consider the possible instability of this boundary layer to a Taylor-Görtler vortex type of perturbation. An examination of the effective Taylor-Görtler number for such a flow suggests that instability is likely for $O(1)$ values of the parameter

$$T = \varepsilon R^{1/2}, \qquad [4.3]$$

hence we limit our investigation to the limit $\varepsilon \to 0$ with T, which we will refer to as the Taylor number, held fixed. On the assumption that $\varepsilon \sim R^{-1/2}$ we can show from the work of Moore (1957) that in the boundary layer u and v, the radial and azimuthal velocity components, are given by

$$u = V\varepsilon R^{-1/2}\{\overline{u}(\eta, \theta) + O(R^{-1/2})\}, \qquad [4.4a]$$

$$v = V\{1 + \varepsilon v(\eta, \theta) + O(R^{-1})\}. \qquad [4.4b]$$

Here η is a boundary layer variable defined by

$$\eta = \frac{(r-a)}{a}\left(\frac{R}{2}\right)^{1/2} \qquad [4.5]$$

whilst $\bar{u}$ and $\bar{v}$ can be obtained from Moore's work, in fact we shall only need an explicit form for $\bar{v}$ which is given by

$$\bar{v} = ie^{i\theta - \eta(1+i)} + \text{complex conjugate.} \qquad [4.6]$$

Thus $\bar{v}$ is just the Stokes layer velocity profile with θ taking the place of the time variable. It is this correction to the irrotational flow $v = 1/r$ which leads to instability. This result follows from the fact that the irrotational flow $v \sim 1/r$ is neutrally stable according to Rayleigh's criterion.

Suppose that we define the dimensionless variables τ and Z by

$$Z = \sqrt{\frac{R}{2}} \frac{z}{a}, \quad \tau = \frac{Vt}{a},$$

where z represents distance along the axis of the cylinder and t denotes time. The flow given by [4.4] is perturbed by writing

$$\frac{u}{V} = R^{-1/2}[\bar{u} + O(R^{-1/2})] + U(\eta, \theta, \tau)e^{ikZ}, \qquad [4.7a]$$

$$\frac{v}{V} = [1 + \bar{v}(\eta, \theta) - \sqrt{\frac{2}{R}}\eta + O(R^{-1})] + V(\eta, \theta, \tau)e^{ikZ}, \qquad [4.7b]$$

$$\frac{w}{V} = W(\eta, \theta, z)e^{ikZ}. \qquad [4.7c]$$

Thus k is the wavenumber of the vortex type of instability and, unlike the usual case with a centrifugal instability mechanism the three components of the distrubance velocity field are of comparable size. The disturbed velocity field [4.7] can be substituted into the Navier-Stokes equations and after some manipulation we obtain the linear stability equations:

$$\{\frac{\partial^2}{\partial \eta^2} - k^2 - 2\frac{\partial}{\partial \theta} - 2\frac{\partial}{\partial \tau}\}U = \sqrt{2}\frac{\partial P}{\partial \eta} - 4v, \qquad [4.8a]$$

$$\{\frac{\partial^2}{\partial \eta^2} - k^2 - 2\frac{\partial}{\partial \theta} - 2\frac{\partial}{\partial \tau}\}V = \sqrt{2}\, TU \frac{\partial \bar{v}}{\partial \eta}, \qquad [4.8b]$$

$$\{\frac{\partial^2}{\partial \eta^2} - k^2 - 2\frac{\partial}{\partial \theta} - 2\frac{\partial}{\partial \tau}\}W = ikP\sqrt{2}. \qquad [4.8c]$$

Here P is the pressure perturbation scaled on $\rho \nu R^{-1/2} V^2$ and terms of relative order $R^{-1/2}$ have been neglected. The continuity equation corresponding to [4.8] is

$$\frac{\partial U}{\partial \eta} + ikW = 0. \qquad [4.9]$$

It is convenient at this stage to eliminate P and W from [4.8] and [4.9] to give the coupled pair of equations:

$$\{\frac{\partial^2}{\partial \eta^2} - k^2 - 2\frac{\partial}{\partial \ell} - 2\frac{\partial}{\partial \tau}\}\{\frac{\partial^2}{\partial \eta^2} - k^2\}U = 4k^2 V,$$

[4.10]

$$\{\frac{\partial^2}{\partial \eta^2} - k^2 - 2\frac{\partial}{\partial \ell} - 2\frac{\partial}{\partial \tau}\}V = \sqrt{2}\ TU\frac{\partial \bar{v}}{\partial \eta},$$

which must be solved subject to

$$V = U = \frac{\partial U}{\partial \eta} = 0, \quad \eta = 0, \infty.$$

Thus we assume that the instability is confined to the boundary layer, this is to be expected so long as we ignore nonlinear effects.

Suppose next that a viscous fluid of Prandtl number unity and viscosity ν occupies the region $z \geq 0$ and that the plane $z = 0$ has temperature $2T_0 \cos \omega t$. We define dimensionless variables τ and η by

$$\tau = \omega t,$$ [4.11a]

$$\eta = \{\frac{\omega}{2\nu}\}^{1/2} z$$ [4.11b]

so that the motionless state has temperature field T given by

$$T = T_0\ \bar{v}(\tau, \eta)$$ [4.12]

where $\bar{v}$ is as defined by [4.6] with θ replaced by τ. Thus the temperature is zero except for a thin region of thickness $\Delta = \{\frac{\omega}{2\nu}\}^{-1/2}$ whilst in the boundary layer the temperature is given by the Stokes layer velocity profile. In the usual way the basic state is now perturbed by a disturbance with horizontal wavenumber k scaled on Δ^{-1}. If the Boussinesq approximation is made then, following Chandrasehkar (1963), Chapter II, we obtain

$$\{\frac{\partial^2}{\partial \eta^2} - k^2\}\{\frac{\partial^2}{\partial \eta^2} - k^2 - 2\frac{\partial}{\partial \tau}\}W = R\Theta,$$ [4.13a]

$$\{\frac{\partial^2}{\partial \eta^2} - k^2 - 2\frac{\partial}{\partial \tau}\}\Theta = \frac{\partial \bar{v}}{\partial \eta} W,$$ [4.13b]

where Θ is the temperature perturbation and W the z velocity component of the perturbation. The parameter R appearing in [4.13] is the Rayleigh number defined by

$$R = g\alpha\Delta^3 T_0 \nu^{-2}$$ [4.14]

where α is the coefficient of volume expansion. If the plane $\eta = 0$ is rigid then (4.13) are to be solved subject to

$$0 = W = \frac{\partial W}{\partial \eta} = 0, \quad \eta = 0, \infty. \qquad [4.15]$$

Thus if we seek neutral solutions of [4.10] with $\frac{\partial}{\partial \tau} = 0$ we see that the resulting eigenvalue problem is identical to that given by [4.13], [4.15] if we associate T with $R/4\sqrt{2}$ and θ with τ respectively. Thus there exists an analogy between the neutral stability problems for the flow past a rapidly rotating cylinder and a time-periodically heated fluid layer. The analogy corresponds to that between the steady Taylor vortex instability between circular cylinders rotating at almost the same speed and Bénard convection between rigid walls. The eigenvalue problem [4.13]-[4.14] can be solved using the method outlined in Section 2, so far we have found one mode of instability with a minimum value of $R \sim 54$. However, other modes exist but it is not yet clear which is the most unstable.

5. Conclusion

The instability mechanism found several years ago for the flow around a torsionally oscillating cylinder exists in many flows of practical importance. When the basic oscillatory flow which supports the instability is spatially varying the instability becomes spatially concentrated near the most unstable positions. Moreover the instability drives the secondary steady streaming boundary layer at the same order as does the basic oscillatory flow. Recently it has been shown by Hall (1985) that this mechanism can lead to the instability of high frequency Tollmien-Schlichting waves interacting with a wall of either convex or concave curvature.

There are many flows where the basic state has a steady component of the same order as the oscillatory part, in such flows there exists the possibility of interactions between the instability mechanisms associated with the steady and unsteady components. An obvious example of this is the Taylor vortex problem with a fixed outer cylinder and an inner cylinder rotating with angular velocity $\Omega[1 + \cos \omega t]$. For $\varepsilon \ll 1$, the flow is susceptible to steady Taylor vortices at sufficiently large vaules of Ω whilst for $\varepsilon \gg 1$, and $\omega \gg 1$ the Stokes layer mode is possible. For intermediate values of ε the modes interact in an as yet undetermined manner, the ratio of the axial lengthscales for the different modes is large so that some progress could probably be made asymptotically.

Another interaction problem yet to be investigated is that between the Tollmien-Schlichting and Taylor-Görtler modes of instability. Experimentally it is known that local transition in the flat Stokes layer takes place as Reynolds numbers as low as 300. It follows from [2.2] that for Stokes layers on curved walls with local radius of curvature R such that

$$\frac{(\frac{\nu}{\omega})^{1/2}}{R} > \sim 10^{-3}$$

the centrifugal mode will occur first. This effectively means that in most situations the mode is likely to be dominant.

The preliminary results, which we have obtained so far for the sinsuiodally heated fluid layer, suggest that a convection mode of instability similar to that found by Seminara and Hall (1976) exists for that flow. As yet no experimental investigations of this problem have been made, it will be interesting to see how this flow evolves at high Rayleigh numbers.

Acknowledgement

Research was supported by the National Aeronautics and Space Administration under NASA Contract No. NAS1-17070 while the author was in residence at the Institute for Computer Applications in Science and Engineering, NASA Langely Research Center, Hampton, VA 23665.

References

[1] Chandrasehkar, S., 1963, <u>Hydrodynamic and Hydromagnetic Stability</u>, Dover.

[2] Duck, P. W. and Hall, P., 1981, ZAMP, **32**, p. 102.

[3] Glauert, M. B., 1957, J. Fluid Mech., **2**, p. 89.

[4] Hall, P., 1981, J. Fluid Mech., **105**, p. 523.

[5] Hall, P., 1984, J. Fluid Mech., **146**, p. 347.

[6] Honji, H., 1981, J. Fluid Mech., **107**, p. 509.

[7] Moore, D. W., 1957, J. Fluid Mech., **2**, p. 541.

[8] Park, K., Barenghi, C., and Donnelly, R. J., 1980, Physics Letters, **78a**, p. 152.

[9] Raleigh, Lord, 1883, Phil. Trans., <u>A</u>, **175**, p. 1.

[10] Riley, N., 1967, J. Inst. Math. Appls., **3**, p. 419.

[11] Schlichting, H., 1932, Phys. Z., **33**, p. 327.

[12] Seminara, G. and Hall, P., 1976, PRS(A), **350**, p. 299.

[13] Seminara, G. and Hall, P., 1977, PRS(A), **354**, p. 119.

[14] Soward, A. and Jones, C., 1983, QJMAM, **36**, p. 19.

[15] Stuart, J. T., 1966, J. Fluid Mech., **24**, p. 673.

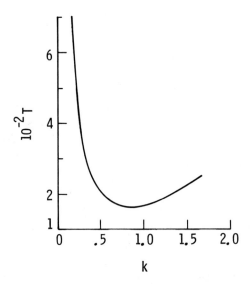

Figure 1. The neutral curve for the torsionally oscillating cylinder problem.

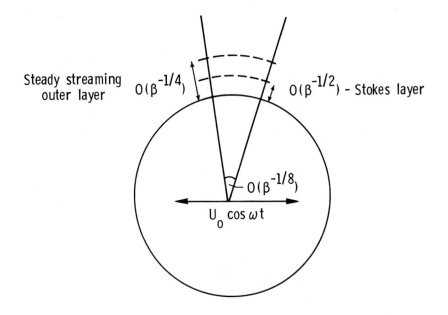

Figure 2. The unstable region for the transversely oscillating cylinder.

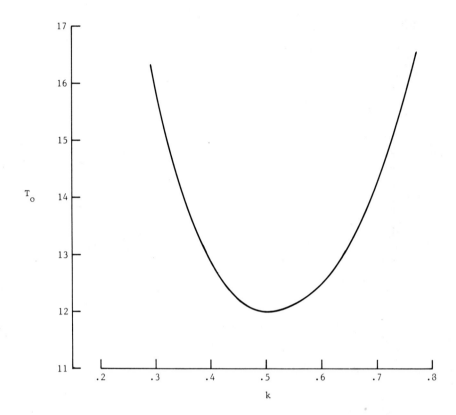

Figure 3. The eigenrelation for (3.4).

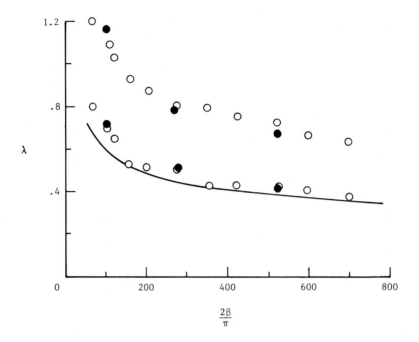

Figure 4. Comparison of theory with Honji's experiments.

Stability Characteristics of Some Oscillatory Flows-Poiseuille, Ekman and Films

CHRISTIAN VON KERCZEK

I. INTRODUCTION

This paper describes some results of a research project to study the stability characteristics of elementary time periodic parallel flows. The purpose of this study is to learn how time periodic oscillations modify the stability characteristics of steady flows. It is useful to understand such effects because small amplitude time periodic modulations of a flow may be a way to either enhance or delay mixing processes if the modulations destabilize or stabilize the underlying steady flow.

The study of the stability theory of time periodic flow was reviewed by Davis (1976). He gives a conceptual framework for comparing stability characteristics of time periodic flows to those of the corresponding steady flows. The present paper emphasizes the modifications of the stability characteristics of some well known steady flows when small amplitude sinusoidal parallel flow oscillations are superimposed on them.

The flows described here are sinusoidally time modulated Hagen-Poiseuille flow, Ekman layer flow and the flow of a liquid film down a vertical plate. The results of these studies were obtained and reported as Ph.D. thesis topics by John T. Tozzi (Hagen-Poiseuille flow, 1982), Pabitra N. Majumdar (Ekman layer flow, 1983) and Ronald J. Bauer (film flow, 1986). These three flows are briefly described in Sections II-IV. The appendix gives a brief description of the method of solution of the stability problems posed by these studies.

II. OSCILLATORY HAGEN-POISEUILLE FLOW

Hagen-Poiseuille flow is the pressure driven flow down a straight circular cross-section pipe. If z is the coordinate down the axis of the pipe, r is the radial coordinate and t is time then the basic oscillatory velocity field in the pipe is

$$\vec{V} = (0, 0, U(r,t)) \tag{1}$$

$$P = -z(1 + \Lambda \cos \Omega t) \tag{2}$$

where

$$U(r,t) = 1 - r^2 + \Delta \text{Re}\left\{ i \left[\frac{J_o(\hat{\beta}r)}{J_o(\hat{\beta})} - 1 \right] e^{i\Omega t} \right\} \tag{3}$$

$$\Lambda = Q_o/P_o \qquad \hat{\beta} = i^{3/2} \beta$$
$$\Omega = a\omega/U_o \qquad \Delta = 4\Lambda/\beta$$
$$\beta = a(\omega/\nu)^{1/2}$$

and J_o is the zeroeth order Bessel function. These equations are in the dimensionless form in which the pipe radius a is the length scale, the steady pressure velocity $U_o = P_o a^2/4\nu$ is the velocity scale and a/U_o is the time scale. P_o is the steady pressure gradient amplitude and Q_o is the oscillatory pressure gradient amplitude. The radian frequency of the imposed sinusoidal pressure gradient is ω.

The mean flow is characterized by the Reynolds number $R_E = U_o a/\upsilon$. The oscillatory flow is characterized by the radius to Stokes layer thickness ratio β and velocity amplitude ratio Δ (Tozzi and von Kerczek, 1985).

Figure 1 shows the graphs of the instantaneous velocity profile shapes, $U(r,t)$ at various instants of time for a typical case. One of the main features of these profiles is that the steady flow Rayleigh instability criterion,

$$\frac{\partial^2 U}{\partial r^2} - \frac{1}{r} \frac{\partial U}{\partial r} = 0 \tag{4}$$

is instantaneously satisfied for some value of r for more than three fourths of the oscillation cycle. Intuitions developed on the basis of steady flow stability theory would suggest the oscillatory flow is unstable, or at least less stable than the steady flow.

Only axisymmetric disturbances were examined. These disturbances can be represented by the meridianal velocity fluctuations (u,v) in terms of a stream function $\phi e^{i\alpha z}$ in such a way that

$$(u,v) = \text{Re} \left(-\frac{i\alpha}{r} \phi e^{i\alpha z}, \frac{1}{r} \frac{\partial \phi}{\partial r} e^{i\alpha z}\right) \tag{5}$$

where α is the axial wave number. Then ϕ satisfies the unsteady Orr-Sommerfeld equation

$$L \frac{\partial \phi}{\partial t} = \frac{1}{Re} L^2 \phi - i\alpha U L \phi + i\alpha \left[\phi L U + \alpha^2 U \phi\right] \tag{6}$$

with boundary conditions

$$\phi(0,t) = \frac{\partial \phi}{\partial r}(0,t) = 0$$

$$\phi(1,t) = \frac{\partial \phi}{\partial r}(1,t) = 0 \qquad (7)$$

where

$$L = \frac{\partial^2}{\partial r^2} - \frac{1}{r}\frac{\partial}{\partial r} - \alpha^2$$

The streamfunction $\phi(r,t)$ has the form $f(r,t)e^{(\lambda+i\gamma)t}$ where $f(r,t)$ is $2\pi/\Omega$ periodic in time.

The main results of solving this stabiity problem are shown in Figures 2 and 3. Figure 2 shows the relative change σ of the decay rate λ of the most relevant wall mode disturbance (see Tozzi and von Kerczek, 1985). Here σ is defined by

$$\sigma = \frac{\lambda_s - \lambda}{\lambda_s} \qquad (8)$$

where λ denotes the decay rate of the oscillatory flow and λ_s is the decay rate of the corresponding disturbance in the steady flow at the same mean Reynolds number. Note that except at the irrelevantly large wave number of $\alpha=11$, the oscillations stabilize Hagen-Poiseuille flow. However, this stabilization is very weak.

Of even more importance, Figure 3 shows that the intantaneous growth rates (i.e., negative decay rate) are mostly much smaller than the growth rate of the steady flow. Thus, the Rayleigh inflection point criterion, Eq. 4, is not at all relevant to this flow even though these oscillations are very low frequency when $\beta = 5.6$.

Although the stabilization of steady Hagen-Poiseuille flow by super-imposed sinusoidal oscillations is in accord with the experiments of Sarpkaya (1966), the stability theory says nothing of quantitative relevance to such experiments. In particular Sarpkaya's (1966) transition results indicate that Hagen-Poiseuille flow is stabilized, at $\beta=5.6$, for values of Δ increasing from zero up to 0.31. For values of $\Delta > 0.31$ the stabilization decreases. However, stability theory for axisymmetric disturbances indicates monotonic stabilization of the flow for $0 \leq \Delta < 0.44$.

III. THE OSCILLATORY EKMAN LAYER

The viscous boundary layer above a horizontal plane in a rotating system with a geostrophic flow infinitely far above the plane is called the Ekman layer. This parallel viscous flow is a simple model of certain flows in the atmospheric boundary layer and in the upper parts of the ocean. Furthermore, Ekman layers often occur at solid boundaries in rotating systems of fluids.

The stability of the steady Ekman layer has been well studied. Lilly (1966) showed that the Ekman layer is unstable, at a Reynolds number of 55, to disturbances propagating in the direction of 65° clockwise from the direction of the geostrophic wind. This disturbance mode is associated with a coriolis-viscous force balance (see Lilly, 1966) and is called a class A disturbance. At a Reynolds number of about 110 a second mode, class B disturbance, that is associated with a velocity profile inflection point becomes unstable. This mode propagates in a direction of about 85° counterclockwise from the geostrophic wind.

There are many situations in which the geostrophic wind driving the Ekman layer is time periodic and this time variation can have a variety of forms. Majumdar (1983) investigated two types of unsteadiness. One unsteadiness consisted of the geostrophic flow having its amplitude modulated in time. The second unsteadiness consisted of the geostrophic wind having both its amplitude and direction modulated in time. Only results from the second case are discussed here.

The oscillatory Ekman layer forms above the x-y plane. The geostrophic wind at $z \to \infty$ is given by

$$\vec{V}_g = U_g \vec{i} + W_g (\vec{i} \cos \omega t + \vec{j} \sin \omega t) \qquad (9)$$

where U_g is the amplitude of the steady component and W_g is the amplitude of the oscillatory component of the geostrophic wind. In equation (9) $\vec{i}$ and $\vec{j}$ are unit vectors tangent to the x and y axes respectively, t is time and ω is the oscillatory angular velocity. The flow is made dimensionless by the velocity U_g, the steady Ekman layer thickness $\delta_E = \sqrt{\nu/\Omega}$, where Ω is the angular velocity of rotation of the entire system, and time scale δ_E/U_g. The following dimensionless parameters emerge from this scaling: The mean Reynolds number $R_E = U_g \delta_E/\nu$, the frequency ratio $\sigma = \omega/\Omega$, the amplitude ratio $\Delta = W_g/U_g$ and the reduced frequency $k = \omega/\Omega R_E$.

The basic steady Ekman layer velocity distribution is

$$\vec{v}_s(z) = \vec{i}(1 - e^{-z} \cos z) + \vec{j} e^{-z} \sin z \qquad (10)$$

where z is the dimensionless vertical coordinate. The superimposed oscillatory velocity profile is

$$\vec{v}_o(z,t) = \vec{i} u_o(z,\sigma) \cos kt + \vec{j} v_o(z,\sigma) \sin kt \qquad (11)$$

where $u_o(z,\varepsilon)$ and $v_o(z,\varepsilon)$ are too complicated to be reproduced here but can be found in Majumdar (1983). The entire flow whose stability was investigated is given by the basic velocity

$$\vec{V} = \vec{V}_s + \Delta\vec{v}_o \quad . \tag{12}$$

Plane wave disturbances of wave number α and propagation direction $\varepsilon - 90°$ were investigated. The angle ε is the direction of the wave crests and is measured counterclockwise from the direction of the geostrophic wind (the x-axis). It is convenient to define a new set of axes (x',y',z) in which x' is in the direction of ε and now y' is in the disturbance propagation direction. The disturbance velocity is (u',v',w) in this new coordinate system. The velocity components (v',w) can be derived from the streamfunction $\phi(z,t)e^{i\alpha y'}$ by the formulas

$$v' = \text{Re}\left[-\frac{\partial \phi}{\partial z} e^{i\alpha y'}\right], \quad w = \text{Re}\left[i\alpha\phi e^{i\alpha y'}\right] \quad . \tag{13}$$

The velocity u' is given by the formula

$$u' = \text{Re}\left[\mu(z,t) e^{i\alpha y'}\right] \tag{14}$$

Then the stability equations describing the growth of the disturbance are

$$\frac{\partial}{\partial t} L\phi = \frac{1}{R_E} L^2\phi - i\alpha\left[VL - \frac{\partial^2 V}{\partial z^2}\right]\phi + 2\frac{\partial \mu}{\partial z} \tag{15a}$$

$$\frac{\partial \mu}{\partial t} = \frac{1}{R_E} L\mu - i\alpha\left[V_\mu + \frac{\partial U}{\partial z}\phi\right] - 2\frac{\partial \phi}{\partial z} \tag{15b}$$

where

$$L = \frac{\partial^2}{\partial z^2} - \alpha^2 \qquad \begin{aligned} V &= \vec{v} \cdot \vec{j}' \\ U &= \vec{v} \cdot \vec{i}' \end{aligned}$$

and $(\vec{i}', \vec{j}')$ are the unit vectors in the directions of (x', y') respectively. The boundary conditions are that $\phi = \phi_z = \mu = 0$ at $z=0$, and ϕ, ϕ_z and μ remain bounded as $z \to \infty$.

Equation (15) was solved by the method outlined in the appendix with the special modifications outlined by Majumdar (1982). The disturbance mode (ϕ, μ) can be represented by Floquet theory as

$$(\phi, \mu) = (f, g) \, e^{(\lambda + i\gamma)t} \tag{16}$$

where f and g are $2\pi/k$ time periodic. Thus the growth rate λ determines the stability of the flow. Majumdar calculates the growth rate λ as a series expansion in oscillation amplitude Δ about the most unstable steady flow mode. Then λ has the form

$$\lambda = \lambda_o + \Delta^2 \lambda_2 + \ldots \tag{17}$$

Majumder calculated at least three terms of this expansion. Only the most important of his results will be shown here.

Figure 4 shows the graph of λ_2 versus frequency ratio σ for the case of $R_E = 55$, $\alpha = 0.3151$ and $\varepsilon = -23.15°$. This is the neutrally stable class A disturbance in the steady Ekman layer. The second abscissa of Figure 4 corresponds to the relative oscillation period if the global rotation rate Ω is measured in revolutions per day. It is noteworthy that oscillations of frequency larger than the global rotation rate have only a small stabilizing effect. However, oscillation frequencies of the same order or smaller than the rotation rate have a more substantial stabilizing effect. The break in the λ_2 curve at $\sigma=2$ occurs because no periodic basic state solution exists for this value of σ (see Majumdar, 1983).

Figure 5 shows overall growth rate λ as a function of Reynolds number R_E for $\alpha = 0.3151$ and $\varepsilon = -23.15°$. The upper curve is the growth rate for no oscillations ($\Delta=0$) and the lower curve is for oscillation with frequency ratio $\sigma=1$ and amplitude ratio $\Delta = 0.106$. This small amplitude oscillation stabilizes the Ekman layer for values of R_E up to at least 70. Thus, it would seem that an actual atmospheric Ekman boundary layer might be more stable than the theoretical steady Ekman layer. This is because the atmospheric Ekman layer experiences the diurnal variations of the geostrophic wind.

IV. OSCILLATORY FILM FLOW

The flow of a liquid film on a flat surface is a model of certain coating processes. The stability of such films is sometimes the governing factor limiting the speed of coating. It would seem that if sinusoidal modulation of such a film flow stabilizes it, then important modifications to the coating process may be possible. This section describes the work of Bauer (1986) which aims to examine the effects of sinusoidal plate oscillation on the stability of a liquid film.

In Bauer's problem a liquid film of density ρ, kinematic viscosity υ, mean thickness h and zero surface tension flows down a vertical plate aligned with the downward pointing (+ x-direction) gravitational vector $\vec{g}$. The y axis is perpendicular to the plate which is located at $y = -h$. The plate executes sinusoidal oscillations in the x-direction with speed $\hat{v} \cos \omega t$. This flow is described in terms of the length scale h, velocity scale $\hat{u} = h^2 g/2\nu$ and time scale $h/\hat{u}$. As a result of this scaling the following parameters describe the state of flow: The mean flow Reynolds $R_E = \hat{u}h/\nu$, the velocity amplitude ratio $\Delta = \hat{v}/\hat{u}$, the frequency parameter $\beta = h^2/\delta_s^2$, where δ_s is the Stokes layer thickness $\sqrt{\nu/\omega}$, and the dimensionless frequency $\Omega = h\omega/\hat{u}$.

The basic fluid velocity field is given by

$$\vec{V} = (U(y,t), 0) \tag{18}$$

where

$$U(y,t) = U_s + \Delta U_o . \tag{19a}$$

$$U_s = 1 - y^2 \tag{19b}$$

$$U_o(y,t) = \mathrm{Re}\left\{\frac{\cosh(1+i)y\beta}{\cosh(1+i)\beta} e^{1\Omega t}\right\} . \tag{19c}$$

Two-dimensional disturbances (u,v) of streamwise wave number α are considered. These disturbances result in a free surface disturbance described by the deflection $\eta(x,t)$. On the basis of linear small disturbance theory the velocity perturbations (u,v) can be described by the streamfunction $\phi(y,t)e^{i\alpha x}$ such that

$$u = \mathrm{Re}\left[\frac{\partial \phi}{\partial y} e^{i\alpha x}\right], \quad v = \mathrm{Re}\left[-i\alpha\phi e^{i\alpha x}\right] . \tag{20}$$

The free surface deflection can be described by the mode function $f(t)$ such that

$$\eta(x,t) = \text{Re}\left[f(t)e^{i\alpha x}\right] . \qquad (21)$$

The resulting equations governing ϕ and f are

$$\frac{\partial}{\partial t} L\phi = \frac{1}{R_E} L^2\phi - i\alpha\left[UL\phi - \frac{\partial^2}{\partial y^2}\phi\right] \qquad (22)$$

$$\frac{\partial^2 \phi}{\partial y \partial t} + i\alpha\,(U\frac{\partial \phi}{\partial y} - \phi\frac{\partial u}{\partial y}) - \frac{1}{R_E}(\frac{\partial^3 \phi}{\partial y^3} - 3\alpha^2\frac{\partial \phi}{\partial y}) = 0 \qquad (23)$$

at $y=0$,

$$\frac{\partial f}{\partial t} = -i\alpha(Uf + \phi) \qquad y=0 \qquad (24)$$

$$\frac{\partial^2 \phi}{\partial y^2} + \frac{\partial^2 u}{\partial y^2} f + \alpha^2 \phi = 0 \qquad y=0 \qquad (25)$$

The wall boundary conditions are

$$\phi = \phi_y = 0 \text{ at } y = -1 .$$

For the derivation (see Bauer, 1986 or, using slightly different notation, Benjamin, 1957).

Since the basic state U is time periodic, the solution (ϕ,f) of equations (22-25) can be expressed in the Floquet form

$$(\phi,f) = (\psi,g)\,e^{(\lambda+i\gamma)t} \qquad (26)$$

where ψ and g are $2\pi/\Omega$ periodic in time.

A first step to examine this stability problem is to solve the system of equations (22-25) by a small wave number α expansion. Equation (26) is substituted into equations (22-25) and ψ, g, λ and γ are expanded as

$$\psi = \psi_o(y,t) + \alpha\psi_1(y,t) + \alpha^2\psi_2(y,t) + \ldots \qquad (27)$$

etc. This small wave number expansion follows the method of Yih (1963, 1968) and the details can be found in Bauer (1986). Bauer finds that to second order in α the growth rate λ is just the combined results of Yih's steady flow solution (1963) and Yih's purely oscillatory solution (1968). Thus

$$\lambda = \lambda_o + \alpha^2(\lambda_{s2} + \lambda_{p2}) + \ldots \qquad (28)$$

where

$$\lambda_{s2} = \frac{8R_E}{15} \qquad (29a)$$

$$\lambda_{p2} = -\left\{3R_E\Delta^2/16\rho^2\left[\sinh^2\beta + \cos^2\beta\right]^2\right\}$$
$$\cdot (\beta\cosh2\beta\sin2\beta + \beta\sinh2\beta\cos2\beta - \sinh2\beta\sin2\beta) . \qquad (29b)$$

Figure 6 shows the neutral curves $\lambda_{s2} = 0$ as a function of β and amplitude Δ. Inside the horizontal u shaped curves $\lambda_{s2} > 0$. Figure 7 gives the stablity diagram, i.e., curves of constant growth rate λ_o (denoted by γ in the figure) as a function of wave number α and Reynolds number R_E for steady vertical film flow. Thus the combination of Figures 6 and 7 can be used to determine the stability of vertical oscillatory film flow to small wave number disturbances.

The steady vertical film flow is unstable to very long wave length disturbances as Reynolds number decreases. However, in a practical situation the finite length of the film imposes a low wave number cutoff on the possible disturbances modes. This low wave number cutoff effectively chops off the modal growth rates below a certain value of α and thus gives a lower bound on the critical value of R_E. In a practical situation then, the results in Figure 7 would thus be applied in the vicinity of the cutoff values of α_c and R_{Ec}. Figure 7 shows that with the judicious choice of β and Δ substantial stabilization to larger values of α and R_E than α_c and R_{Ec} might be obtained.

REFERENCES

1. Bauer, R.J. (1986), "Effect of Oscillatory Excitation on Film Flow Stability," Forthcoming Ph.D. Thesis, The Catholic Univ. of America, Washington, D.C.

2. Benjamin, T.B. (1957), "Wave Formation in Laminar Flow Down an Inclined Plane," J. Fluid Mech., Vol. 2, pp. 554-574.

3. Davis, S.H. (1976), "The Stability of Time-Periodic Flows," Annual Review of Fluid Mechanics, Vol. 8, pp. 57-74.

4. Lilly, D.K. (1966), "On the Instability of Ekman Boundary Flow," J. Atmos. Sci., Vol. 23, pp. 481-494.

5. Majumdar, P.N. (1983), "The Instability of Oscillatory Ekman Boundary Layer Flow," Ph.D. Dissertation, The Catholic Univ. of America, Washington, D.C.

6. Orszag, S.A. (1971), "Accurate Solution of the Orr-Sommerfeld Stability Equation," J. Fluid Mech., Vol. 50, pp. 689-703.

7. Tozzi, J.T. (1982), "The Stability of Pulsatile Flow in a Conduit of Circular Cross Section," Ph.D. Dissertation, The Catholic Univ. of America, Washington, D.C.

8. Tozzi, J.T. and C.H. von Kerczek (1985), "The Stability of Oscillatory Hagen-Poiseuille Flow," ASME J. Appl. Mech., to appear.

9. Sarpkaya, T. (1966), "Experimental Determination of the Critical Reynolds Number for Pulsating Poiseuille Flow," ASME J. Basic Eng., pp. 589-598.

10. von Kerczek, C.H. (1982), "The Instability of Oscillatory Plane

Poiseuille Flow," <u>J. Fluid Mech.</u>, Vol. 116, pp. 91-114.

11. Yih, C.S. (1963), "Stability of Liquid Flow Down an Inclined Plane," <u>Physics Fluids</u>, Vol. 6, pp. 321-334.

12. Yih, C.S. (1968), "Instability of a Horizontal Liquid Layer on an Oscillating Plane," <u>J. Fluid Mechs.</u>, Vol. 31, pp. 737-751.

Acknowledgement: The author would like to take this opportunity to thank John Tozzi, Pete Majumdar and Ron Bauer with whom he had the pleasure to work on these problems. This research was in part supported by NSF Grant No. CME-7900929.

APPENDIX

A general solution procedure for solving linear stability problems for parallel time periodic flows with a nonzero mean is outlined here. The method was first developed by von Kerczek (1982) and was used by Tozzi, Majumdar and Bauer in their studies.

The linear stability theory for a parallel time periodic flow results in a partial differential equation in the cross-stream coordinate y and time t of the form

$$\frac{\partial}{\partial t} L\phi = M\phi + \Delta N\phi \qquad (A1)$$

where L, M and N are linear differential operators with respect to y and have coefficients depending on y and t. In the case of Ekman flow ϕ is a 2-vector and in the case of film flow one of the boundary conditions is an ordinary differential equation in t and thus can be dealt with as if ϕ is a 2-vector as in Ekman flow.

The coefficients of equation (A1) are time periodic with period T because of the periodicity of the basic flow. In the present method for solving systems of equations such as (A1) the unknown vector ϕ is expanded in a Chebyshev polynomial series as developed by Orszag (1971) as follows:

$$\phi = \sum_{n=0}^{N} f_n(t)\, T_n(y) \qquad (A2)$$

By substituting Eq. (A2) into (A1) and using the boundary conditions the stability problem is converted to a system of the following ordinary differential equations with time periodic coefficients:

$$\frac{d\vec{f}}{dt} = P\cdot\vec{f} + \Delta Q\cdot\vec{f} \qquad (A3)$$

where P is a constant matrix and Q is time periodic matrix. The vector $\vec{f}$ denotes the vector of Fourier coefficients f_n;

$$\vec{f} = (f_1, \ldots, f_N)^+ .$$

The Floquet theorem says that $\vec{f}$ has the form

$$\vec{f} = \vec{g}(t)\, e^{(\lambda+i\gamma)t} \qquad (A4)$$

where $\vec{g}(t+T) = \vec{g}(t)$. Hence stability is determined by value of λ.

The mean flow by itself has a stable or unstable mode $\vec{f}_o = \vec{g}_o\, e^{(\lambda_o+\gamma_o)t}$. Thus disturbance modes $\vec{f}$ can be calculated as a series expansion in Δ about $\vec{f}_o$ as follows:

$$\left. \begin{array}{l} \vec{g} = \vec{g}_o + \Delta \vec{g}_1 + \ldots \\ (\lambda + i\gamma) = (\lambda_o + i\gamma_o) + \Delta(\lambda_1 + i\gamma_1) + \ldots \end{array} \right\} \quad (A5)$$

It is fairly easy to calculate many terms of the expansion (A5). Full details of the method are described by von Kerczek (1982).

It is noteworthy that because Q(t) is sinusoidal and all steady flow modes $\lambda_o + i\gamma_o$ are simple (for the flows considered here) the expansion term

$$\lambda_1 + i\gamma_1 = 0$$

but $\vec{g}_1 = 0$. Thus the long term modifications to the stability of the mean flow by the modulation occurs only to order Δ^2.

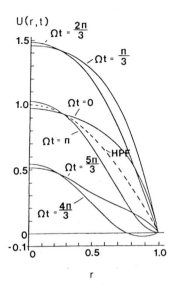
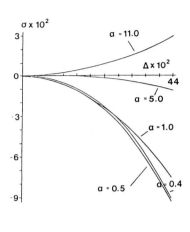

Fig. 1. Basic flow velocity profiles for oscillatory Hagen-Poiseuille flow. HPF denotes the steady-flow profile. For the unsteady profiles, $\beta = 5.6$, $\Delta = 0.5$.

Fig. 2. Relative chance of least stable wall mode versus for $R_E = 3000$, $\beta = 5.6$, in oscillatory Hagen-Poiseuille flow.

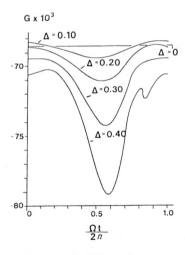

Fig. 3. Instantaneous growth rate versus time for least stable wall mode. $R_E = 3000$, $\beta = 5.6$, in oscillatory Hagen-Poiseuille flow.

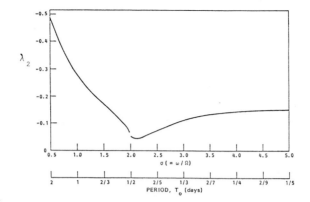

Fig. 4. Values of λ_2 modifying the growth-rate of the most unstable steady Ekman layer at $R_E = 55$.

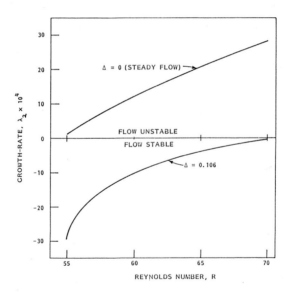

Fig. 5. Stabilization of the most unstable steady Ekman layer disturbance. (Oscillatory component of geostrophic flow: $\sigma = 1$, $T^o = 1$ day; $\Delta = 0.106$.

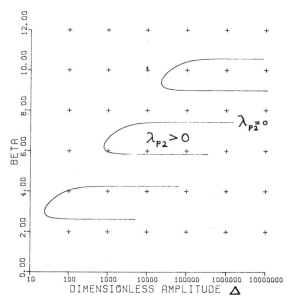

Fig. 6. Neutral curves for the effects of flow oscillation on the growth rate of steady film flow.

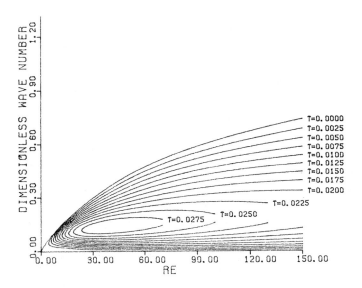

Fig. 7. Neutral and disturbance constant growth rate curves for steady film flow.

The Stability of the Stokes Layer: Visual Observations and Some Theoretical Considerations

P.A. Monkewitz and A. Bunster

Summary

The Stokes boundary layer produced by fluid oscillating parallel to a flat plate is studied by hydrogen bubble visualization. No significant departure from the laminar velocity profiles is found up to a Reynolds number, based on boundary layer thickness and freestream velocity amplitude, of about 600. The first visible finite amplitude disturbances appear shortly before and around flow reversal at the edge of the boundary layer. A theoretical model consistent with these observations is proposed. It is based on a quasisteady approach, valid for large Reynolds numbers, and involves nonlinear interaction between modes which possibly leads to a net disturbance growth over one freestream cycle.

1. Introduction

The Stokes boundary layer which is generated by a fluid in time-harmonic motion parallel to an infinite stationary wall - or conversely by the wall moving harmonically in a stationary fluid - is a prototypical time-dependent viscous flow and has as such received considerable attention. It falls into the class of parallel shear flows and is an exact solution of the incompressible Navier-Stokes equations in the laminar regime. The Stokes layer is relevant to the study of many unsteady phenomena such as the damping of acoustic waves in ducts and of shallow water waves and, for instance, the flow over oscillating bodies. For these applications the mechanism of transition to turbulence, which requires knowledge about the stability characteristics of the flow, is of particular interest. In the following we will focus our attention on the stability of the Stokes layer. The work on this subject notably by Kerczek and Davis [1,2] has been summarized by Davis [3]. As discussed in this review, the definition of stability boundaries in time-periodic flows is no longer clear cut. The following three regimes have to be distinguished: monotone stability when all disturbances decay monotonically to zero, transient stability when some disturbances temporarily grow but all of them experience a net decay over one basic flow period and finally instability when some disturbances grow over one cycle. We note that, apart from instability, transient stability can also lead to transition if the temporary amplification of a disturbance is large enough. In this light it might be more suggestive to call it transient instability. Nevertheless the two regimes should be clearly distinguished as they imply quite different excitations. In case of instability the introduction of one disturbance pulse will lead, after a sufficient number of flow periods, to a finite amplitude disturbance and possibly to turbulence. With transient stability, on the other hand, a continuous excitation (by free stream disturbances for instance) has to be provided to "restart" a temporarily growing disturbance every cycle. This question which is potentially important in applications will be further discussed in Section 5.

The results of stability analyses are somewhat puzzling and summarized below. Kerczek and Davis find, by an energy method for arbitrary disturbances, critical Reynolds numbers R_E and R_T with monotone stability if $R < R_E$ and transient stability if $R < R_T$. R is based as usual on the boundary layer thickness δ, involving the freestream oscillation frequency Ω and the kinematic viscosity ν, and on the freestream velocity amplitude $\hat{U}$.

$$\delta \equiv (2\nu/\Omega)^{\frac{1}{2}} \quad ; \quad R \equiv \hat{U}\delta/\nu = 2\hat{X}/\delta \qquad (1)$$

For this flow R has a particularly simple physical interpretation as twice the ratio of displacement amplitude $\hat{X} = \Omega^{-1}\hat{U}$ and δ. As an aside, experimental facilities for the study of the Stokes layer therefore have to be large since the smallness of δ is limited by probe resolution and Reynolds numbers of interest are of the order of several hundred. The values for two-dimensional (2D) and three-dimensional (3D) disturbances are found to be $R_E(3D) = 19.0$, $R_E(2D) = 38.9$, $R_T(3D) = 24.2$ and $R_T(2D) = 46.6$. Although these results are not rigorous for the infinite Stokes layer (see footnote in [3]), they differ very little from the rigorous results for finite channels with β as small as eight, where β is the ratio of wall separation to boundary layer thickness. A linear Floquet analysis by v. Kerczek and Davis [2] for $\beta = 8$ on the other hand shows transient stability in the range

0<R<800 for wavelengths between 4.8δ and 21δ. It is noted here that, although monotone stability is suggested by Figure 5 of [2], the amplitude of the periodic function multiplying the decaying exponential (with negative Floquet exponent) is arbitrary such that only transient stability can be inferred. Later, Hall [4] showed that the least stable mode in the above study does not correspond to the least stable mode when β = ∞, but it is generally believed that the infinite Stokes layer also displays linear transient stability probably for all R of practical interest. A different approach taken by Kerczek and Davis [2] consists in freezing the velocity profile at different times during a cycle and in performing an Orr-Sommerfeld calculation on the now steady flow. With this "quasisteady" approach the lowest critical Reynolds numbers R_Q is found to be 86 for a profile at flow reversal. By a multiple scale analysis, M. Monkewitz [5] has shown that substantial amplifications of quasisteady modes can be achieved over part of a cycle. One problem with this type of analysis is that the normal modes are often "Rayleigh" modes related to inflection points of the profile. If tracked over one cycle they therefore completely change character and make a comparison with results from Floquet theory difficult. This question will be further discussed in Section 4 and 5.

The experimental findings regarding transition in Stokes layers are not less ambiguous than the theoretical results. Experiments on purely oscillatory flows have been recently reported by, among others, Hino et al. [6], Ramaprian and Mueller [7], Hayashi and Ohashi [8] and have been reviewed by Carr [9]. The occurrence of "weak turbulence" reported by Hino and others can probably be identified with the incidence of a transient instability which in this flow does not manifest itself as a single frequency velocity disturbance at a fixed location and therefore may be difficult to distinguish from turbulence. It is clear though that turbulence, if only "transient turbulence", does occur in the Stokes layer but transition Reynolds numbers reported in the literature vary between 280 and about 700. These wide discrepancies indicate that the instability is possibly subcritical, making transition strongly dependent upon the disturbance level in the different facilities.

To fix the notation, the laminar velocity profile (2) is listed below in nondimensional form. The standard nondimensionalization of the streamwise coordinate x_*, the coordinate normal to the wall y_* and the time t_* with the Stokes boundary layer thickness δ, and $\hat{U}$ is used (a hat denoting the amplitude of a time-harmonic quantity throughout the paper). In addition we define the "slow time" $\tilde{t}$, also referred to as the Stokes phase, which is appropriately scaled to describe the time evolution of the laminar flow (2).

$$U = \cos \tilde{t} - \exp(-y) \cos(\tilde{t}-y) , \quad V = 0;$$
$$x = x_*/\delta , \quad y = x_*/\delta , \quad t = \hat{U}t_*/\delta , \quad \tilde{t} = 2R^{-1}t = \Omega t_* \qquad (2)$$

This profile is plotted on Figure 1 at various phases $\tilde{t}$. Also included in the figure are the trajectories of two inflection points of (2), labeled I and II for future reference. Of course the profile (2) has an infinity of inflection points at $(\tilde{t}-y)=n\pi$ which become exponentially weak at large y.

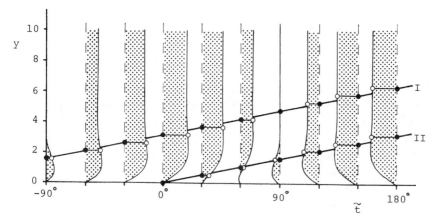

Figure 1. The velocity profile in the laminar Stokes layer at different phases. O , Inflection points; ——●——,Trajectories of inflection points.

2. Experimental apparatus

The experiment was carried out in a U-channel with water as working fluid. The facility is shown schematically on Figure 2 with a horizontal testsection of 240 cm length and a rectangular cross section of 89 mm height x 305 mm width. The length of the horizontal testsection was dictated by the requirement of reaching Reynold's numbers beyond transition while avoiding that any distortions ori-

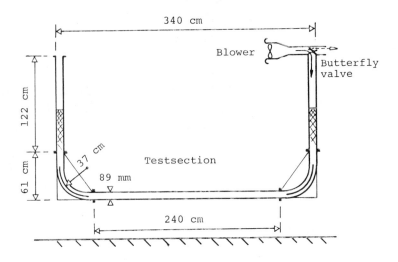

Figure 2. Schematic of the U-channel facility.

ginating at the bends could affect the uniformity of the free stream velocity at the center of the horizontal section where measurements were taken. Joined to the testsection are two bends with turning vanes and two upright channels to complete the U. The water column in the channel is made to oscillate at the natural U-tube frequency by periodically pressurizing one of the vertical legs. This is achieved by diverting the exhaust of a continuously running blower down into one of the upright legs with a crank-driven butterfly valve. The valve frequency is thereby matched to the natural frequency of the U-tube with the aid of a motor controller and the equilibrium "sloshing amplitude" is adjusted by changing the blower discharge. The operating parameters for the experiment are summarized in Table 1.

Ω [rad/sec]	δ [mm]	$\hat{X}$ [mm]	R
2.09	1.1	152	277
2.09	1.1	203	369
2.09	1.1	254	462
2.09	1.1	305	554
2.09	1.1	356	647

Table 1. Operating parameters of the U-channel facility.

The Stokes boundary layer was made visible by the hydrogen bubble technique [10]. A 0.05 mm diameter stainless steel wire was stretched vertically between the top and bottom center of the horizontal section as shown in Figure 3. This diameter was selected to give sufficient mechanical strength while keeping the Reynolds number R_D, based on the wire diameter D, below 40 to avoid vortex shedding (At the highest R of Table 1 where the velocity amplitude $\hat{U}$ is 0.74 m/sec, $\hat{R}_D = 30$). Flush mounted copper anodes were incorporated into the top and bottom surfaces of the channel. After adding a small amount of hydrochloric acid to the filtered working fluid (water), the voltage applied between the wire and the anodes could be held as low as 20 Volts which produced the smallest, most uniform bubbles. To facilitate identification of the bubbles, the wire was pulsed at a rate of 20 - 50 Hz with pulses of 10 ms duration. The bubbles were illuminated by a sheet of light from a 600W tungsten-quartz bulb and filmed with a high-speed camera at 400 frames per second through an optical glass window. By using a macro lens a field of view of 20 mm in the streamwise direction by 12 mm in the direction normal to the top wall (i.e. along the wire) was obtained, which was adequate to resolve the boundary layer profile. For each amplitude, slightly more than half a basic period was filmed, including two consecutive flow reversals to provide on film the necessary phase information.

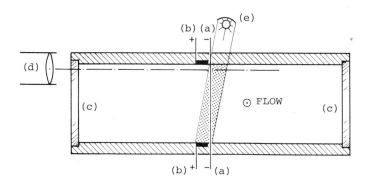

Figure 3. Cut through the center of the testsection with: (a) 0.05 mm stainless steel hydrogen bubble wire: (b) Copper anodes: (c) Optical glass windows; (d) 16 mm high speed movie camera (—·—·, optical axis); (e) 600W tungsten-quartz light source.

3. Experimental results

The movie was analyzed by projecting consecutive frames and determining the displacement of individual bubbles. Care was taken to only consider bubbles far enough from the wire (typically 100D = 5 mm according to [10] and [11]) to avoid systematic errors in excess of 10% due to the wire wake. With this technique instantaneous velocity profiles between Stokes phases of $-90°$ and $0°$ were obtained for R from 277 to 554. They are compared on Figure 4 to the laminar profile (2) and it is obvious that, given statistical errors estimated at 10 - 20% of the freestream velocity (depending on phase), the laminar Stokes profile shows no significant distortions up to approximately R = 500. With the exception of $\tilde{t} = -90°$ (Figure 4a) no clear trend of the deviations can be discerned with increasing R. This good correspondence between measurement and laminar profile is in agreement with the data of Ramaprian and Mueller [7], obtained with laser velocimetry.

The next question to ask was when and where disturbances of the laminar Stokes profile would first become visible, i.e. would first attain finite amplitude. From the movie this instance was determined to be shortly before and at flow reversal where a vortical disturbance appeared very clearly at the highest R of 647. Instantaneous velocity vector diagrams (shown as Figure 5) were obtained at $\tilde{t} = 60°$ and $90°$ in the same manner as the instantaneous profiles of Figure 4. The striking feature of these diagrams is that the disturbances appear to be centered at or beyond the edge of the boundary layer (compare with Figure 1). Although the field of view is somewhat too small for an experimental confirmation, the disturbance appears to be spatially periodic at least on Figure 5a. It's wavelength can be roughly estimated at 10δ.

By using the terminology of quasisteady stability analysis, the key features of the above observations may be interpreted as follows. The fact that disturbances first become visible around flow reversal may be due to the fact that there phase velocities of quasisteady

Figure 4. Comparison of measured instantaneous velocity profiles with the laminar profile (2) for different Reynolds numbers:● R = 277, ○ R = 369, △ R = 462, □ R = 554. The Stokes phases are: (a) $\tilde{t} = -90°$, (b1) $\tilde{t} = -72.5°$, (b2) $\tilde{t} = -66°$, (b3) $\tilde{t} = -70.5°$, (c) $\tilde{t} = -55°$, (d) $\tilde{t} = -36°$, (e) $\tilde{t} = -18°$, (f) $\tilde{t} = 0°$.

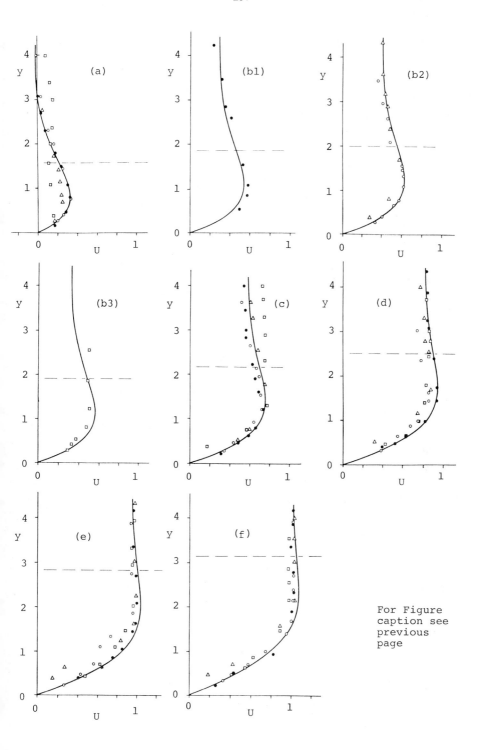

For Figure caption see previous page

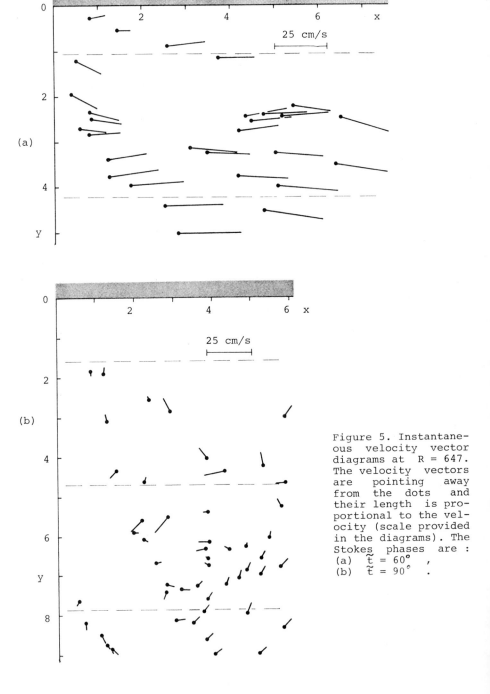

Figure 5. Instantaneous velocity vector diagrams at $R = 647$. The velocity vectors are pointing away from the dots and their length is proportional to the velocity (scale provided in the diagrams). The Stokes phases are:
(a) $\tilde{t} = 60°$,
(b) $\tilde{t} = 90°$.

normal modes become small. As a consequence a disturbance of fixed wavenumber will have a small frequency which makes it easier to spot on movie. Disturbances near maximum free stream velocity, on the other hand, are expected to have frequencies of the order of the Reynolds number (see next section) and would therefore be difficult to detect at a "sampling rate" of 400 frames per second. If they were of sufficient amplitude though, the high frequency, folded back onto the interval between zero and the Nyquist frequency of 200 Hz, could still be detected. The observation of the Stokes phase where finite disturbances first appear is therefore thought to be reliable and not an artifact of the recording method. The second observation of the wall distance at which these disturbances are centered means that the modes which first attain a visible amplitude of a few percent of U are the ones associated with an outer inflection point - the second or possibly the third from the wall. If it is assumed that disturbances travel with inflection points away from the wall, this would mean that it takes a mode about half a cycle after inception of "its" inflection point to reach an observable amplitude (see Figure 1).

4. A weakly nonlinear model for mode interaction

As discussed in the introduction, one of the questions in connection with the stability of the Stokes layer is whether it can become unstable in the strict sense of the word. It is clear that on a linear basis this is not possible at least for the Reynolds numbers of practical interest such that nonlinearities have to be included. In this paper we will not pursue the approach proposed by Davis [12] but propose a model based on a quasisteady leading order approximation. It allows a mode, say mode A associated with the inflection point I of Figure 1, to pass its energy on to a mode closer to the wall, say mode B associated with the inflection point II having the same wavenumber k as mode A, before it gets damped out in the free stream where the inflection points become exponentially weak. Such a mechanism would achieve two things: first, disturbance energy would be transmitted much more "efficiently" over a cycle by always concentrating it on the most unstable or the least stable mode at any point within the cycle. Second, the stability boundary could be determined directly from mode amplitudes instead of integrated disturbance energy since, for instance, the mode shape I (associated with the inflection point I) at $\tilde{t} = 0°$ is the same as the mode shape II at $\tilde{t} = 180°$ (see Figure 1). Hence, if mode I would transmit its entire energy to mode II between $\tilde{t}$

= 0° and 180° then the stability boundary would be simply defined by amplitude of II at $\tilde{t} = 180°$ equal to amplitude of I at $\tilde{t} = 0°$. As is well known, a necessary condition for the efficient (resonant) transfer of energy between two modes is the equality or near equality of their real phase speeds. It has been shown by M. Monkewitz [5] that the phase speed of quasisteady modes quickly approaches the speed of "their" inflection point. This is shown on Figure 6 where the velocity of the inflection points I and II are plotted together with the real phase speeds c_r of the corresponding instability modes. It is obvious that, as the inflection points approach the freestream, the difference of the two c_r becomes exponentially small. Another condition for interaction is that the two modes have some spatial overlap which is the case for the two modes I and II of interest as shown on Figure 7. Of course this description is oversimplified and one will have to consider a linear combination of modes which, at critical conditions, is reproduced after half a cycle.

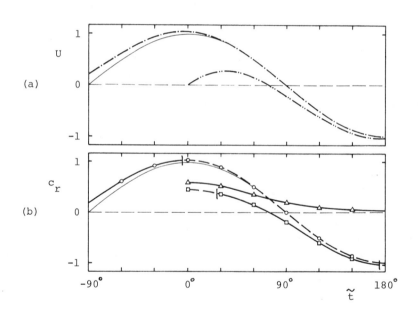

Figure 6. (a) Freestream velocity (———) and the velocities of inflection points I (—.—) and II (—..—) (see Figure 1). (b) Real phase speeds c_r of the quasisteady modes associated with the inflection points I and II for R = 500, k = 0.45 (from [5] with kind permission). $c_{r,I}$: —○— when mode is amplified and — -○- — when it is damped. $c_{r,II}$: —□— when mode is amplified and — -□- — when it is damped. —△—: $c_{r,I} - c_{r,II}$.

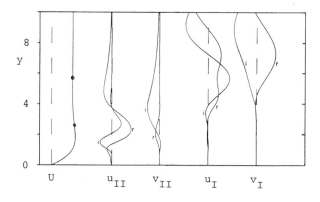

Figure 7. Mode shapes associated with inflection points I and II at R = 500, k = 0.45. Real (r) and imaginary (i) parts of velocity components parallel (u) and normal (v) to the wall (from [5] with kind permission).

The formal analysis of an interaction between two dominant modes is now sketched. It is based on a multiple scale approach where the fast time t defined in (2) is the natural variable to describe the development of the disturbance (if its wavelength scales on δ) and is related to the timescale of the freestream Ω^{-1} through the Reynolds number. The small parameter is therefore

$$\varepsilon^2 = 2R^{-1} \tag{3}$$

Since in a standard weakly nonlinear formulation the interaction between two modes of equal wavenumber k takes place at the order of amplitude cube and since we are only interested in interactions which take place on a time scale Ω^{-1} or faster (interactions over several cycles do not appear to be relevant) the disturbance amplitude has to be $O(\varepsilon)$. The stream function Ψ is thus expanded in powers of ε according to

$$\Psi(x,y,t,\tilde{t}...) = \Psi_0(y,\tilde{t}) + \sum_1^\infty \varepsilon^n \Psi_n(x,y,t,\tilde{t}...) \tag{4}$$

Ψ_0 is thereby the streamfunction corresponding to the velocity profile (2). Insertion of (4) into the vorticity equation (5) yields the succession of inhomogeneous Rayleigh equations (6).

$$\{ \frac{\partial}{\partial t} + \Psi_y \frac{\partial}{\partial x} - \Psi_x \frac{\partial}{\partial y} \} \nabla^2 \Psi = \frac{1}{2} \varepsilon^2 \nabla^4 \Psi \tag{5}$$

$$L(\Psi_1) \equiv \nabla^2 \Psi_{1t} + U \nabla^2 \Psi_{1x} - U_{yy} \Psi_{1x} = 0 \qquad (6)$$

$$L(\Psi_2) = \left[\tfrac{1}{2}\Psi_{o yy} - \Psi_{o\tilde{t}}\right]_{yy} + J\left[\Psi_1, \nabla^2 \Psi_1\right]$$

$$L(\Psi_3) = -\nabla^2 \Psi_{1\tilde{t}} + \tfrac{1}{2}\nabla^4 \Psi_1 + J\left[\Psi_1, \nabla^2 \Psi_2\right] + J\left[\Psi_2, \nabla^2 \Psi_1\right]$$

Above, L is the Rayleigh operator and J stands for the Jacobian. We note that in this problem the Rayleigh equation with the slow time $\tilde{t}$ as parameter arises naturally at leading order which is not as annoying as in the boundary layer since the basic velocity profiles contains inflection points and satisfies Rayleigh's criterion for instability. The linear solution is now set up in WKB form as a superposition of two modes A and B with c.c. standing for the complex conjugate.

$$\Psi_1 = e^{ikx} \{A(\tilde{t})\phi_A(y;\tilde{t}) \exp(-ik \int c_A(\tilde{t})dt) \qquad (7a)$$

$$+ B(\tilde{t})\phi_B(y;\tilde{t}) \exp(-ik \int c_B(\tilde{t})dt)\} + \text{c.c.}$$

$$c_A = c_r + \varepsilon^2 c_A \quad , \quad c_B = c_r + \varepsilon^2 c_B \qquad (7b)$$

In addition, the analysis is restricted to the region where an interaction is strong, that is where the difference of phase speeds and growth rates are small. Both are assumed here to be $O(\varepsilon^2)$ which makes the approach somewhat heuristic since growth rates do not go to zero in the limit of infinite Reynolds number. This is of course related to the fact that no single critical Reynolds number exists in this problem which would allow the rigorous implementation of the Stuart-Watson scheme. Introducing (7a) together with the assumption (7b) into the inhomogeneous equation for Ψ_2 then yields the "mean flow correction" $\Psi_2^{(o)}$ and the harmonic $\Psi_2^{(2k)}$. We note that the first term on the right hand side involving Ψ_o is the equation governing the basic flow and therefore identical to zero. Using (7b) it can also be easily shown that the mean flow correction $\Psi_2^{(o)}$ does not depend on the fast time. In addition we note that for the present time the no-slip condition at y=0 has been neglected, i.e. no viscous wall layer of thickness ε has been incorporated into the formulation. One obtains thus formally with the coefficients F being functions of y and $\tilde{t}$:

$$\psi_2 = \psi_2^{(o)}(y,\tilde{t}) + e^{2ik(x-\int c_r dt)} \phi_2^{(2k)}(y;\tilde{t}) \tag{8a}$$

$$\psi_2^{(o)}{}_{\tilde{y}\tilde{t}} = |A|^2 F_{A\bar{A}} + |B|^2 F_{B\bar{B}} + \mathrm{Re}\{A\bar{B}\, F_{A\bar{B}}\}$$

$$\phi_2^{(2k)} = A^2 F_{AA} + B^2 F_{BB} + AB\, F_{AB} \tag{8b}$$

At $O(\varepsilon^3)$ now the right hand side in (6) has to be orthogonal to the adjoint solution A and B to avoid secular terms. This yields, after algebraic manipulations, to two first order equations for $A_{\tilde{t}}$ and $B_{\tilde{t}}$ with coefficients G and H which may depend on the slow time.

$$A_{\tilde{t}} = A\, G_A + B\, G_B + A|A|^2 G_{AA\bar{A}} + A^2\bar{B}\, G_{AA\bar{B}} \tag{9a}$$
$$+ |A|^2 B\, G_{A\bar{A}B} + A|B|^2 G_{AB\bar{B}} + \bar{A}\, B^2 G_{\bar{A}BB} + B|B|^2 G_{BB\bar{B}}$$

$$B_{\tilde{t}} = A\, H_A + B\, H_B + A|A|^2 H_{AA\bar{A}} + A^2\bar{B}\, H_{AA\bar{B}} \tag{9b}$$
$$+ |A|^2 B\, H_{A\bar{A}B} + A|B|^2 H_{AB\bar{B}} + \bar{A}\, B^2 H_{\bar{A}BB} + B|B|^2 H_{BB\bar{B}}$$

These two equations in which some coefficients may be zero describe the interaction of the two modes. If A is identified with the inflectional mode I, the case of particular interest is when the initial amplitude of B is zero. The question is then whether this mode associated, say, with the inflection point II can be excited by A. It is surprising to find that this is possible even without the cubic terms since there is a linear "diffraction" term with a coefficient H_A which is easily shown to be nonzero. Physically this term can be understood if the unsteadiness is thought to be a succession of infinitesimal impulsive changes of the velocity profile. At each change a new initial value problem has to be solved where the initial value (the "old" mode shape) does not quite match the "new" mode shape such that other modes and possibly transients are excited at infinitesimal amplitudes in a somewhat analogous fashion as in the spatial problem considered by Goldstein [13]. If these other modes are generated in phase at each successive flow change (i.e. $c_A \cong c_B$) constructive interference occurs which is reflected by the $O(1)$ coefficient H_A. As soon as this mechanism has produced a nonzero B any of the cubic terms can greatly enhance the energy transfer from A to B. Without evaluating the interaction coefficients, one can only speculate how strong the nonlinear effect will be. The size of a typical amplitude $|A|$, though, can be estimated from experiments. Taking a typical disturbance of 2% of $\hat{U}$ at say R = 500, the corresponding $|A|$ is indeed of order unity.

To find the stability boundary the procedure would consist of choosing the correct initial $|A|$ ($|B| = 0$) in order to produce an identical $|B|$ after half a cycle while ensuring at the same time that $\psi_2^{(0)}$ is also periodic with the same period as the freestream. Considering (9b) there is not much doubt that this is possible for a sufficiently large $|A|$.

5. Conclusions

In the last section a mechanism has been sketched which is compatible with the visual observations of Section 3 and by which the stability boundary could be reached in a Stokes layer. According to the Floquet analysis in [2] the nonlinearity is thereby essential to reach this boundary at Reynolds numbers of interest, i.e. of the order of several hundred. The linear transfer of energy between modes on the other hand possibly provides a different interpretation of Kerczek and Davis' results [2] who conclude that instability modes are probably not able to follow inflection points as they move away from the wall and hence cannot "take advantage" of the powerful Rayleigh mechanism to grow. The present analysis suggests the alternate possibility that the instability modes might very well follow the inflection points but that the linear transfer of energy is just too weak to avoid the decay of disturbance energy over one cycle at Reynolds numbers below 800. The ideas put forward in this paper may be summarized on Figure 8 which shows the subcritical instability suggested in section 4. The amplitude axis is labeled A_i to indicte that the boundaries may depend on the size of several modes and the boundaries themselves are drawn in a completely arbitrary fashion. In particular their "noses" do not necessarily reach R_T and R_E respectively.

Of course this sketch leaves many questions unanswered. First there is the question of mathematical consistency which possibly requires the introduction of a second small parameter to characterize growth rates which are small of the order of a few hundredth (see [5]), virtually independent of R for large R. Then the question also

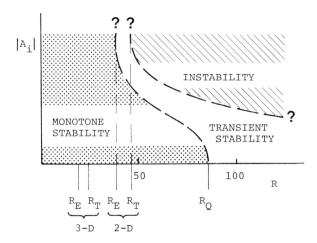

Figure 8. Schematic of proposed stability diagram for the Stokes layer.

arises whether higher order nonuniformities, i.e. the dependence of A and B on times slower that $\tilde{t}$, automatically disappear on the stability boundary (if it can be reached). Finally and most importantly the question of relevance has to be raised: if $|A|$, in order to reach the stability boundary, has to be so large that the associated disturbance experiences a nonlinear breakdown with transition to turbulence <u>before</u> it can interact with B then the proposed mechanism is clearly not relevant. To decide this question, the program outlined in the last section will have to be fully implemented.

This work has been carried out with the support of the NSF Fluid Mechanics Program under grant MEA 82-14719 which is gratefully acknowledged.

References

1. C.v. Kerczek and S.H. Davis. The Stability of Oscillatory Stokes Layers, Studies in Appl. Math. <u>51</u> (1972), p.239.

2. C.v. Kerczek and S.H. Davis. Linear Stability Theory of Oscillatory Stokes Layers. J. Fluid Mech <u>62</u> (1974), p.753.

3. S.H. Davis. The Stability of Time-Periodic Flows. Ann. Rev. Fluid Mech. <u>8</u> (1976), p.57.

4. P. Hall. The Linear Stability of Flat Stokes Layers. Proc. R. Soc. Lond. A <u>359</u> (1978), p.151.

5. M.A. Monkewitz, Lineare Stabilitats - untersuchungen an den Oszillierenden Grenzschichten von Stokes. Ph.D. thesis No. 7297, Federal Institute of Technology, Zurich Switzerland (1983).

6. M. Hino, M. Sawamoto and S. Takasu. Experiments on Transition to Turbulence in an Oscillatory Pipe Flow. J. Fluid Mech. 75 (1976), p.193.

7. B.R. Ramaprian and A. Mueller. Transitional Periodic Boundary Layer Study. J. of the Hydraulics Div. ASCE 106 (1980), Proc. paper #15908, p.1959.

8. T. Hayashi and M. Ohashi. A Dynamical and Visual Study on the Oscillatory Turbulent Boundary Layer. Proc. of 3rd Symposium on Turb. Shear Flows, Univ. of Calif., Davis (Sept. 9-11, 1981), p.8.1.

9. L.W. Carr. A Review of Unsteady Turbulent Boundary-Layer Experiments. Proc. IUTAM Symp. on Unsteady Turbulent Shear Flows, Toulouse France 1981, p.3, Springer.

10. F.A. Schraub, S.J. Kline, J. Henry, P.W. Runstadler, A. Littel. Use of Hydrogen Bubbles for Quantitative Determination of Time-Dependent Velocity Fields in Low-Speed Water Flows. J. Basic Engr., Trans. ASME ser. D, 87 (1965), p.429.

11. L.S.G. Kovasznay. Hot Wire Investigation of the Wake Behind Cylinders at Low Reynolds Numbers. Proc. Roy. Soc. A 198 (1949), p. 174.

12. S.H. Davis. Finite Amplitude Instability of Time-Dependent Flows. J. Fluid Mech. 45 (1970), p.33.

13. M.E. Goldstein. Sound Generation and Upstream Influence Due to Instability Waves Interacting with Nonuniform Mean Flows. J. Fluid Mech. 149 (1984), p.161.

High Frequency Rayleigh Instability of Stokes Layers

STEPHEN J. COWLEY

1. INTRODUCTION

A knowledge of the conditions under which unsteady flows become turbulent is important in many applications, e.g. the transport of sediment along the ocean bed (Li 1954), physiological investigations of the larger blood vessels (e.g. Pedley 1980), and various pipe-line and aerodynamic problems.

This study will be concerned with the stability of sinusoidally oscillating flow above a plane stationary wall. Fluid of kinematic viscosity ν, and oscillations of radian frequency ω and velocity amplitude U are assumed. The resulting *Stokes-layer* flow has a boundary-layer thickness $\delta = (2\nu/\omega)^{\frac{1}{2}}$, and is theoretically equivalent to the flow generated by a plate oscillating under fluid which is stationary at infinity. It is one of the simplest unsteady flows possible, and is an exact solution of the Navier-Stokes equations. As well as being a prototype problem, from which conclusions can be inferred for more complicated non-periodic, non-parallel flows, an understanding of the stability of Stokes layers is of relevance to many high-Reynolds-number oscillating flows which include Stokes layers as the viscous boundary-layer correction (e.g. Riley 1965, Lyne 1971, Bodonyi & Smith 1981).

There have been a number of previous theoretical studies of this stability problem. Von Kerczek & Davis (1974) (subsequently referred to by vKD) introduced an upper stationary boundary a distance $\beta\delta$ above an oscillating plate, and sought a critical Reynolds number above which there exist *linear* disturbances that grow over a cycle. Using Floquet theory and a numerical Galerkin technique they could find no unstable disturbances (the largest Reynolds number, $R = U\delta/\nu$, examined was 800, and $\beta = 8$ was taken). Hall (1978) presented an improved version of this theory in which there was no need for an upper boundary, but he too found that the flow was stable for all Reynolds numbers investigated, i.e. $R < 320$. Oscillatory flow through circular pipes of radius $\beta\delta$ has also been studied. For a restricted range of R, Yang & Yih (1977) found only decaying disturbances, while Pelissier (1979) discovered growing disturbances with $\beta = 11.4$ for $R \geq 90$ (however, these results seem anomalous and may be a numerical artefact).

These theoretical conclusions are not in obvious accord with the experimental evidence. For example, in a channel with an oscillating bottom and $\beta \approx 175$, Li (1954) observed turbulence for $R \geqslant 565$. There have also been a number of experiments performed in circular pipes. Turbulent bursts have been reported above Reynolds numbers of 500 ($\beta \geqslant 5$; Sergeev 1966), 550 ($\beta \geqslant 2$; Hino, Sawamoto & Takasu 1976, Ohmi et al. 1982a, Iguchi et al. 1982), 280 ($\beta \geqslant 40$; Merkli & Thomann 1975), and 640 ($\beta \geqslant 4.8$; Clarion & Pelissier 1975). The last authors also observe disturbances 'which have the appearance of a Tollmien-Schlichting instability' for $R \geqslant 250$, while Hino et al. note 'weak' instabilities for $R \geqslant 380$ ($\beta = 2.7$) and $R \geqslant 180$ ($\beta = 3.9$). Clamen & Minton (1977) report laminar flow for $R \leqslant 280$ ($\beta \approx 20$), while Ramaprian & Mueller (1980) found 'transitional' flow at $R = 370$. For a flow with a mean-flow component, but which includes Stokes layers close to the boundary, Nerem, Seed & Wood (1972) have recorded turbulent bursts for $R \geqslant 350$ in the descending canine aorta, and for $R \geqslant 210$ in the ascending aorta.

These varied results have led a number of authors (e.g. Hino et al. 1976) to identify four broad types of flow: (I) laminar flow, (II) disturbed flow, (III) flows with turbulent bursts which relaminarise over part of the oscillation cycle, and (IV) fully turbulent flow. Critical Reynolds numbers of approximately 280 and 550 have been suggested for the boundary between the first two types (e.g. Ohmi et al. 1982a, Iguchi et al. 1982). There also appears to be a consensus that the turbulence is generated at times soon after the maximum velocity amplitude has been attained (Nerem et al. 1972, Merkli et al. 1975, Hino et al. 1976, Kiser et al. 1976, Ohmi et al. 1982a,b, Monkewitz & Bunster 1985), although Clarion et al. (1975) suggest that the disturbances appear soon after the wall shear reverses.

A comparison between theory and experiment suggests that there is at least a range of Reynolds numbers where linear disturbances decay over a period, but where Stokes layers appear unstable. A possible resolution is that although all disturbances experience net decay over a cycle, in experiments certain of them grow to such an extent over part of the period that the flow can become nonlinearly unstable (see for example Rosenblat 1968, Rosenblat & Herbert 1970, Finucane & Kelly 1976, Davis & Rosenblat 1977, von Kerczek 1982, and Hall 1983, for other applications of this idea). This explanation is consistent with the observed indeterminacy in the critical Reynolds number, because this would tend to vary between experiments according to the level of extraneous disturbances (cf. Hagen-Poiseuille pipe flow). The difficulty with this explanation is that figure 5 of vKD suggests that the kinetic energy of disturbance modes decays monotonically (see also von Kerczek 1973). The question also arises why Floquet theory apparently fails to predict a critical Reynolds number for two-dimensional Stokes layers, while there is excellent agreement between theory and experiment for the stability of Stokes layers modified by centrifugal and stratification effects (Seminara & Hall 1976, von Kerczek & Davis 1976).

The aim here is to demonstrate that for sufficiently large Reynolds numbers disturbances can grow significantly over part, if not all, of the oscillation cycle. The analysis is based on Benney & Rosenblat's (1964) approach of studying high-frequency disturbances superimposed upon the underlying time-dependent flow. The multiple scales method used is valid for asymptotically large Reynolds numbers, and has previously been applied to this problem by Tromans (1977), whose analysis we follow, and by P. Hall and P. Blennerhassett (private communication). This approach results, at leading order, in an inviscid Rayleigh eigenvalue problem for a succession of quasi-steady velocity profiles with a parametric dependence on time.

In the past objections have been raised that a quasi-steady Rayleigh theory is inapplicable because ω^{-1} is the only time scale (e.g. vKD). However, in section 2 it is shown that for disturbances with wavelengths comparable with δ, the relevant time scale for the growth/decay of inflectional instabilities is $(R\omega)^{-1}$, which is much smaller than ω^{-1} if $R \gg 1$.

It should also be emphasised that the present quasi-steady approach is different from that of Collins (1963), Obremski & Morkovin (1969) and Monkewitz (1983). Their method is based on a heuristic approximation in which the Reynolds number is first assumed sufficiently large that a multiple scales analysis can be performed in time, but then taken to be of order one so that the viscous terms can be retained in the Orr-Sommerfeld equation (see Smith, 1979, for a criticism of a similar approach often made in the stability analysis of steady non-parallel flows[+], e.g. Bouthier, 1973). The present analysis assumes that the Reynolds number is asymptotically large throughout, although as a consequence this means that a critical Reynolds number for instability cannot be predicted.

2. FORMULATION

The flow will be described by Cartesian coordinates $\delta(x,y)$, the corresponding velocity components $U(u,v)$, and a time coordinate $\omega^{-1}t$. At $y = 0$ there is a rigid wall, and as $y \to \infty$, $u \to \cos(t)$. The solution for the unperturbed flow is

$$u = u_0(y,t) \equiv \cos(t) - \exp(-y)\cos(t-y), \quad v = 0 \ . \tag{2.1}$$

The stability of this flow to linear disturbances will be examined by means of a normal mode analysis. Attention will be focused on two-dimensional perturbations because any three-dimensional disturbance of fixed wavenumber is related to a two-dimensional one at a lower Reynolds number (e.g. see vKD). For a disturbance of

---

[+] In fact Allmen & Eagles (1984) have shown recently that this heuristic approximation can sometimes be surprisingly good.

wavenumber α we write

$$(u-u_0, v, p-x\sin(t)) = (\tilde{u}, \tilde{v}, R\tilde{p}) \exp(i\alpha x) \quad , \tag{2.2}$$

where $\rho\omega\delta U p$ is the pressure, and ρ is the fluid density. Substitution of (2.2) into the Navier-Stokes and continuity equations, and neglect of all nonlinear perturbation quantities, yield

$$2\tilde{u}_t + R(i\alpha u_0 \tilde{u} + \tilde{v} u_{0y} + 2i\alpha \tilde{p}) = \tilde{u}_{yy} - \alpha^2 \tilde{u} \quad , \tag{2.3a}$$

$$2\tilde{v}_t + R(i\alpha u_0 \tilde{v} + 2\tilde{p}_y) = \tilde{v}_{yy} - \alpha^2 \tilde{v}, \quad i\alpha \tilde{u} + \tilde{v}_y = 0 \quad . \tag{2.3b,c}$$

Following Tromans (1977) we assume that the disturbances of interest have high frequencies when the Reynolds number is large. An examination of the terms in (2.3) then suggests the introduction of the fast time scale $\tau = Rt$, and a perturbation expansion of the form

$$(\tilde{u}, \tilde{v}, \tilde{p}, \alpha) = [(u_1, v_1, p_1, \alpha_1) + R^{-\frac{1}{2}}(u_2, v_2, p_2, \alpha_2) + \ldots] \exp(-i\vartheta(\tau)) \quad , \tag{2.4a}$$

where the u_i, v_i, p_i are functions of y and t, and

$$\frac{d\vartheta}{d\tau} = \Omega_1(t) + R^{-\frac{1}{2}}\Omega_2(t) + \ldots \quad . \tag{2.4b}$$

The $R^{-\frac{1}{2}}$ terms in (2.4) are present because the largely inviscid disturbance has a slip velocity at the wall and so generates a perturbation boundary layer on the plate. On substitution of (2.4) into (2.3), at leading order Rayleigh's equation is obtained:

$$(u_0 - c_1)(v_{1yy} - \alpha_1^2 v_1) - u_{0yy} v_1 = 0 \quad , \tag{2.5a}$$

where $\alpha_1 c_1 = 2\Omega_1$. The appropriate boundary conditions for no normal flow into the wall, and zero disturbance at infinity, are

$$v_1(0,t) = 0, \quad v_1 \to 0 \text{ as } y \to \infty \quad . \tag{2.5b}$$

The nature of (2.5) allows a solution to be found of the form $v_1 = A(t)V(y,t)$, where $V(y,t)$ satisfies a normalization condition, e.g. $\int_0^\infty |V|^2 dy = 1$.

It is a relatively straightforward numerical matter to solve (2.5) for the complex wavespeed c_1 at any specified time t (see section 3). The leading order instantaneous growth-rate of a quantity Q may then be calculated from the definition:

$$\text{local growth-rate} = \text{Re}(Q^{-1} \partial Q/\partial t) \quad . \tag{2.6}$$

Although at higher orders this growth-rate will usually depend on the disturbance quantity in question (e.g. see Bouthier 1973), the largest contribution to the growth of linear properties of the perturbation is independent of the quantity, and over the interval t_0 to t is given by

$$\exp(-i \int_{\tau_0}^{\tau} \Omega_1 \, d\tau) = \exp(-\tfrac{1}{2} i \alpha_1 R \int_{t_0}^{t} c_1 \, dt) \quad . \tag{2.7}$$

The next order correction to (2.7) is a multiplicative factor of order $\exp(R^{\frac{1}{2}})$, which although large in magnitude is small compared to (2.7) provided that $t-t_0 > 0(R^{-\frac{1}{2}})$. It may be found in principle by substitution of the higher-order terms of (2.4) into the governing equations. However, our aim is to ascertain if and when disturbances can grow, and to this end it is sufficient to study the leading-order solution.

3. NEUTRAL MODES AND GROWTH-RATES

The symmetry of the Stokes-layer velocity profile (2.1) in time means that it is sufficient to consider the times $0 \leq t < \pi$ only; the eigenvalues for times $\pi \leq t < 2\pi$ are given by $c_1(t) = -\bar{c}_1(t-\pi)$, where an overbar denotes a complex conjugate. We note also for future reference that: (i) at $t = 0$ there is an inflection point on the boundary which moves away from the wall as time increases, (ii) at $t=\pi/4$ and $t=\pi/2$ the wall shear and outer flow, respectively, reverse direction, and (iii) there are an infinite number of inflection points satisfying Fjortoft's criterion for all values of t; hence it is probable, but not certain, that unstable modes can be found at all times (e.g. Tollmien 1936).

Equation (2.5) was solved numerically using the orthonormalization shooting method provided in the subroutine package SUPORT. The eigenvalue, c_1, was then determined by a secant iteration on the lower boundary condition. As a check certain of the results were also obtained by Newton iteration of a fourth order finite-difference scheme. Graphs of growth-rate against wavenumber for various times have been presented previously by Tromans (1977), and will not be repeated here. Our numerical results are largely in agreement with his where they overlap, although we will place a slightly different emphasis on their interpretation. In particular, the distinction between a 'boundary-layer mode' and a 'free-shear-layer mode' is possibly not as clear cut as made out by him, since for certain wavenumbers they merge continuously into each other. The eigenvalue problem for $t = \pi/2$ has also been studied by Gill & Davey (1969); again our results agree with theirs for the mode reported there.

In figure 1 the numerically found neutral curves for two of the modes have been plotted. For convenience, in this figure and throughout the rest of the paper, the subscript 1 on α_1, c_1, and v_1 in (2.5), etc. is omitted. Also, we will differentiate between the two modes illustrated by referring to them by A or B according as whether their *left* neutral branches lie on curves A or B. No difficulty was encountered in finding mode A for times $0 < t < 1.35$. However, for $1.35 < t < \pi$ convergence problems sometimes did arise, and there seemed to be two reasons for this. First,

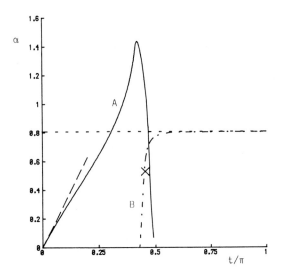

Figure 1: Graphs of neutral curves plotted as parametric functions of time. ———: Mode A, —·—·—: Mode B. : Asymptotic neutral curve for small wavenumber. ······: Limiting asymptotic neutral curve for modes centred far from the wall (although not illustrated, an infinite number of neutral curves should be clustered near this asymptote, see section 3). X: Mode crossing point.

inviscid modes are often associated with inflection points (cf. Howard 1964), and the strength of such points exponentially decreases away from the wall. Because the growth-rate of the mode and the 'strength' of the associated inflection point are related (see below), $\text{Im}(c)$ becomes very small and an accurate guess of eigenvalue and eigenfunction is necessary if the iteration is to converge. Second, for $1.35 \lesssim t \lesssim 1.57$, the growth-rates of both modes A and B are comparable, and at $\alpha = \alpha_m \approx 0.53$, $t = t_m \approx 1.42+$ the two modes seem to coalesce so that there are two eigenfunctions for the same eigenvalue c. These two difficulties will be considered in turn.

Inflection points far from the wall

As mentioned above, inviscid modes are often associated with inflection points, and up to the time that is the first root of $\cos(t) - \exp(-t) = 0$, i.e. $t \approx 1.293$, the wavespeed of neutral mode A is given by the fluid velocity, u_s, at the inflection point closest to the wall (this wavespeed condition ensures that the solution to (2.5) is not singular where $u_0 = c$, e.g. see Drazin & Reid 1981). For $t \gtrsim 1.293$ the existence of other positions with velocity u_s means that critical layers develop and the wavespeed differs from u_s. However, the modes still seem to remain associated with inflection points, so suggesting an asymptotic analysis for the case when these are far from the wall.

To investigate this possibility we consider a two region solution with $y = O(1)$ and $y = O(\log(\epsilon^{-1}))$, where $\epsilon \ll 1$. The relevant inflection point is assumed to be in the outer region, which suggests from (2.1) that if the mode wavespeed is close to the local fluid velocity, then it will expand as

$$c = \cos(t) + \epsilon C + \ldots \quad . \tag{3.1}$$

Substitution of (3.1) into (2.5) yields the leading order equation for $y = O(1)$:

$$v_{yy} + (2\tan(t-y) - \alpha^2)v = 0 \quad . \tag{3.2}$$

The boundary condition $v(0) = 0$ is to be applied, and at the critical layers where $u_0 = c$, i.e. $y_c = t + n\pi + \pi/2$, the contour of integration in the complex y-plane needs to be deformed above/below the pole according to $u_{0y}(y_c) \gtrless 0$ (see e.g. Drazin & Reid 1981).

Floquet theory implies that the general solution of (3.2) has the form

$$v = k_1 \exp(\lambda_1 z)w_1(z) + k_2\exp(\lambda_2 z)w_2(z) \quad , \tag{3.3}$$

where the λ_i and k_i are constants, $z = y-t$, and the w_i are π-periodic functions of z. The boundary condition can be satisfied by choosing k_1 and k_2 appropriately, and without loss of generality we assume that $k_1 = 1$ and $\text{Re}(\lambda_1) > \text{Re}(\lambda_2)$ (the strict inequality proves to be correct). Hence as $y \to \infty$, $v \sim \exp(\lambda_1 z)w_1(z)$.

Far from the wall we set

$$y = \log(\epsilon^{-1}) + Y, \quad \log(\epsilon^{-1}) = 2n\pi + Y_0, \quad v = \exp(2n\pi\lambda_1)V \quad , \tag{3.4}$$

where $0 \leq Y_0 < 2\pi$, and n is a large integer. The governing equation and boundary conditions then become from (2.5), (3.1) and (3.4)

$$(C+\exp(-Y)\cos(t-Y-Y_0))(V_{yy}-\alpha^2 V) + 2\exp(-Y)\sin(t-Y-Y_0)V = 0 \quad , \tag{3.5a}$$

$$V \to 0 \text{ as } Y \to \infty, \quad V \to \exp(\lambda_1(Y+Y_0-t))w_1(Y+Y_0-t) \text{ as } Y \to -\infty \quad . \tag{3.5b}$$

Equations (3.5a,b) specify an eigenvalue problem for $C(t,Y_0)$, which was solved numerically. Note that the problem can be simplified by the transformations $Y+Y_0-t \to Y$ and $C\exp(t-Y_0) \to C_0$ which yield (3.5) with Y_0 and t set equal to zero. If due account is also taken of the modes obtained by the transformation $t \to t+\pi$, $C \to -\bar{C}$, the asymptotic expansion (3.1) can be expressed as

$$c = \cos(t) + C_0 \exp(-t-2n\pi) + \ldots \qquad (3.6a)$$
$$c = \cos(t) - \bar{C}_0 \exp(-t-2n\pi-\pi) + \ldots \qquad (3.6b)$$
$$\left. \right\} \quad 2n\pi+t \gg 1, \quad 0 \leq t < \pi.$$

The exponential decay of Im(c) postulated by Tromans (1977), and which is also present in our numerical solutions of (2.5), is thus confirmed.

Equation (3.3) was solved numerically by deforming the integration path into the complex y-plane. The value of λ_1 so obtained was then used as a boundary condition in the solution of (3.5), which was solved by the same numerical techniques as for (2.5). λ_1 was found to be positive, and hence these asymptotic modes have their maximum amplitude in the outer region. The dotted line in figure 1 is the wavenumber, $\alpha = 0.807$, of the neutral curve predicted from (3.5). A very rapid approach to this asymptote is suggested; in fact for $\alpha = 0.42$ and $t \geq 5\pi/8$, (3.6a) with $n = 0$ is within 1% of numerical solutions of (2.5).

The nature of the above asymptotic solutions means that strictly an infinite number of unstable modes exist at all times. Therefore, although not illustrated, curve B in figure 1 should run up to $t = \pi$, wrap around to $t = 0$, and then repeat this procedure indefinitely, in this way representing an infinite number of neutral modes clustered near the dotted asymptote. In observable flows an infinite number of *growing* modes are unlikely to exist. This is because (3.6) shows that the growth-rate of many modes is exponentially small; thus for the higher modes it is not possible to ignore the small viscous effects which are always present at finite Reynolds numbers, and which will probably lead to the modes' decay (see also Monkewitz 1983). Elementary scaling arguments suggest that the above solution will need modification when $\varepsilon = O(R^{-1})$; hence for a mode generated when $t = O(1)$, viscous effects will become important when $t = O(\log(R))$.

The existence of a mode crossing point

The numerical results suggest that there exists a time, t_m, and wavenumber, α_m, for which modes A and B have the same eigenvalue, c_m. The consequences of this are perhaps best illustrated by considering the evolution of one of the modes as the parameters t and α are varied so as to trace out a closed curve around t_m, α_m. On completion of one circuit the initial mode will have changed continuously into the other, while on completion of another circuit, the original mode will be recovered (cf. the branch point of a square root). It therefore appears difficult to distinguish definitively between the two modes by, for instance, associating one with the inflection point. Indeed, for $\alpha > \alpha_m$ the mode related to the inflection point seems to change as time increases past t_m.

In order to assess the relevance of the mode crossing point to experiments, we recall that at moderate Reynolds numbers the flow is observed to *relaminarise* after turbulent bursts. This suggests a thought experiment in which random disturbances

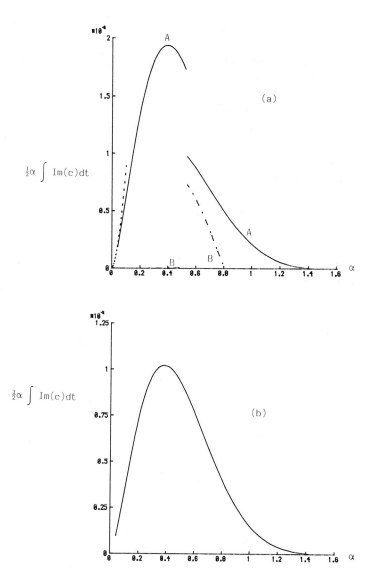

Figure 2: Integrated growth-rate $\frac{1}{2}\alpha \int \mathrm{Im}(c)\,dt$. The lower limit of integration lies on the left branch of the relevant neutral curve. (a) The upper integration limit either lies on the right branch, or is infinite, as appropriate. The discontinuity at $\alpha \approx 0.53$ is a result of the mode crossing. ———: Mode A, —·—·: Mode B. -----: Asymptotic approximation for small α (main contribution from $t > \pi/4$). (b) The upper integration limit is $t=1.35$, which is just prior to the inception of mode B.

with all wavenumbers are introduced for $t \geq 0$ into an initially unperturbed flow, i.e. disturbances are introduced for times just after the outer flow begins to decelerate. Some of these disturbances will decay algebraically, others will generate unstable modes.

First, we note that those modes associated with inflection points well away from the wall are probably not of the greatest importance because: (i) they do not grow significantly, (ii) they are ones most likely to be affected by finite Reynolds number effects, and (iii) they can be viewed as the continuation in time of modes which had larger growth-rates in previous half-cycles.

Next we consider mode A for wavenumbers $\alpha < \alpha_m$. This mode exists for times $t > t_A$, where $t_A(\alpha)$ specifies the left branch of neutral curve A in figure 1. We find that it continues to exist and grow for all time, although its growth-rate decays exponentially. The increase in amplitude of a mode excited by t_A is given by the integral (2.7) with $t_0 = t_A$; as $t \to \infty$, this integral tends to the *finite* limiting value plotted in figure 2a. Hence, at leading order, a disturbance represented by this mode grows only by a finite amount, i.e. there is no unbounded growth.

Suppose now that mode A is excited at $t_A(\alpha)$ but with $\alpha > \alpha_m$. It then grows only until the time specified by the right branch of curve A, after which it disappears and the disturbance will decay. The integrated growth-rate of the mode over this finite period is also plotted in figure 2a. The discontinuity at $\alpha = \alpha_m$ is a direct result of the mode crossing.

Similarly we can consider mode B. For $\alpha < \alpha_m$, this mode can be excited only between $t = t_B$, where $t_B(\alpha)$ lies on curve B in figure 1, and the times specified by the right branch of curve A. This is a short period, and the growth-rates are so small that in figure 2a the integrated growth-rates over this interval are almost indistinguishable from the α-axis. For $\alpha > \alpha_m$, a mode B disturbance generated at t_B will continue to exist for all time (although the inclusion of viscous effects would probably lead to its decay). Again the integrated growth-rate is plotted in figure 2a.

From this figure we see that the wavenumber of the disturbance which grows by the greatest amount is $\alpha \approx 0.39$. However, because it takes an infinite time to achieve this increase, a more relevant quantity to consider is possibly the integrated growth-rate upto a specified time; for instance in figure 2b this quantity has been plotted for $t = 1.35$ (mode A only). Again there is a range of wavenumbers with significant growth, in this case centred on $\alpha \approx 0.38$. Next, in figures 3a,b, the wavenumber and integrated growth-rate of the mode which has grown by the largest amount since inception has been plotted against time (the flow is assumed disturbance free at $t = 0$). The most rapid increase in perturbation amplitude occurs in the period after the wall shear reverses at $t = \pi/4$, suggesting that this is when disturbances are most likely to become observable. This is in agreement with Clarion & Pelissier (1975) who report disturbances in their Stokes layers soon after flow reversal. It is less clear why in most of the other experiments reviewed in section 1 disturbances tend to be observed soon after the outer flow begins to decrease at t=0, although a fresh instability mode (i.e. mode A) does of course develop during this period.

The only direct indication in our inviscid theory of the flow relaminarization

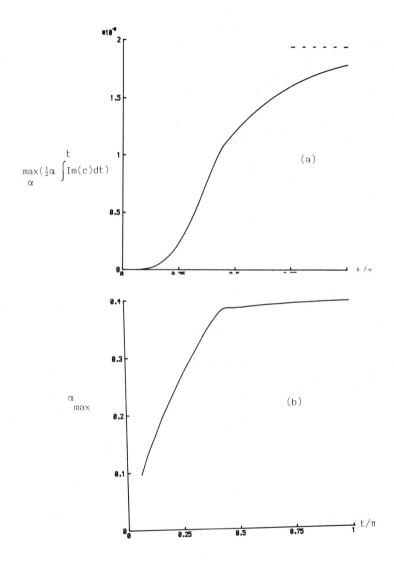

Figure 3: (a) Integrated growth rate of the mode which has increased by the greatest amount since inception assuming no modes exist at $t = 0$, i.e. $\max_\alpha \{\frac{1}{2}\alpha \int^t \text{Im}(c)dt\}$, where the lower integration limit lies on the left branch of the neutral curve. ----: Asymptotic limit for $t = \infty$. (b) Wavenumber of mode of greatest increase.

observed at moderately large Reynolds numbers, is the behaviour of modes with wavenumbers $\alpha > \alpha_m$ excited in the first half of the oscillation cycle; for these modes cease to exist before the outer flow reverses (see figure 1). In order to explain completely the relaminarization of the flow as the outer velocity increases some finite-Reynolds-number effects will have to be introduced. However, we note that during the second half of the oscillation cycle the growth-rates of the modes rapidly decrease; it is therefore feasible that small viscous effects can lead to the modes' decay and the relaminarization of the flow.

Finally, it would be interesting to see whether the consequence of mode crossing could be observed in either carefully controlled experiments, or finite-Reynolds-number numerical calculations. In particular, for modes excited early in the cycle, there should be a sharp difference in behaviour according as the wavenumber is greater or less than α_m; for $\alpha < \alpha_m$ the modes should exist over the whole cycle (and longer), while for $\alpha > \alpha_m$ they should last for half a cycle or less.

4. DISCUSSION

A multiple scales analysis has shown that at sufficiently large Reynolds numbers high-frequency perturbations to Stokes layers can grow. The most significant growth-rates tend to occur as the outer flow is decreasing, with the largest values developing not long after the wall shear has reversed direction at $t = \pi/4$. This suggests that disturbances are most likely to be observed as the outer flow decreases, which is in qualitative agreement with experiment.

The predicted change in disturbance amplitude over an oscillation cycle or part is $O(\exp(R))$, hence a slight increase in Reynolds number can lead to a relatively larger increase in amplitude for initial perturbations of the same size. If the generation of turbulent bursts is triggered by disturbances reaching a certain magnitude, this suggests that the critical Reynolds number for transition to turbulence could be dependent on the level of extraneous disturbances in an experiment. A careful investigation in which perturbations of given wavelength, amplitude, etc. are introduced might therefore prove productive.

None of the modes found evolves through a succession of quasi-steady states in such a way as to return, after one period, to its original state multiplied by a constant factor (if such a factor existed it would be one of the characteristic roots of Floquet theory). Consequently, the present results are not necessarily incompatible with either previous Floquet theory calculations at large finite Reynolds numbers or the hypothesis that most linear disturbances experience net decay over a period. In fact it may be possible to combine the above inviscid modes with others which include small viscous effects, so as to produce only Floquet characteristic forms with *net decay* over a period. However, the results of sections 2 & 3 strongly suggest that at large Reynolds numbers some of the characteristic forms will grow at least at some point in the oscillation cycle. It may be worthwhile, therefore, to extend vKD's calculations to higher Reynolds numbers to see if the disturbance kinetic energy, etc., does indeed cease to be a monotonic decreasing function of time.

There still remains the quandary as to why a Floquet theory approach does not give good agreement with experiment for a Stokes layer on a plane wall, whereas it does for Stokes layers modified by centrifugal effects, *inter alia*. An insight can be obtained by performing a high-frequency stability analysis similar to the above for the Stokes layer generated on a circular cylinder oscillating about its axis. The governing equations for this centrifugal stability problem are (Seminara & Hall 1976):

$$(\partial^2/\partial y^2 - \alpha^2 - 2\partial/\partial t)(\partial^2/\partial y^2 - \alpha^2)\tilde{u} - 2\alpha^2 T \hat{u}_0 \tilde{v} = 0 ,$$

$$(\partial^2/\partial y^2 - \alpha^2 - 2\partial/\partial t)\tilde{v} - \sqrt{2}\hat{u}_{0y}\tilde{u} = 0 , \qquad (4.1)$$

$$\tilde{u} = \tilde{v} = \tilde{u}_y = 0 \text{ on } y = 0 , \quad \tilde{u}, \tilde{v}, \tilde{u}_y \to 0 \text{ as } y \to \infty$$

where T is the Taylor number, and $\hat{u}_0 = \exp(-y)\cos(t-y) = \cos(t) - u_0$ is the Stokes layer velocity profile. Seminara & Hall found that according to Floquet theory, the critical Taylor number for instability is 164. For large values of T an asymptotic analysis is possible by introducing the fast time variable $\hat{\tau} = T^{\frac{1}{2}}t$, and expanding $\tilde{u}$, etc. as in (2.4). On substitution into (4.1), an inviscid eigenvalue problem for the centrifugal instability of a succession of quasi-steady Stokes-layer velocity profiles is obtained at leading order, viz.:

$$u_{1yy} - (\alpha^2 - \lambda \hat{u}_0 \hat{u}_{0y})u_1 = 0 , \qquad (4.2)$$

where $\lambda = \alpha^2/\sqrt{2}\Omega^2$, $u_1(0) = 0$ and $u_1 \to 0$ as $y \to \infty$ (cf. (2.5)). Numerical solution of (4.2) shows that the most unstable mode has a *periodic* parametric dependence on time, and that the same mode remains the most unstable throughout the oscillation period. As a result the large-T limits of the Floquet characteristic roots can be readily calculated from the inviscid asymptotic solution, unlike the case of a Stokes layer on a plane wall studied here. Hence in stability studies of other time periodic flows an investigation of the inviscid quasi-steady problem may be a useful preliminary. If such a theory readily yields asymptotic characteristic roots, and further, the same mode remains the 'most dangerous' throughout an oscillation cycle, then a stability test based on net decay/growth over a cycle may be appropriate.

Although the asymptotic forms of the Stokes-layer modes for small wavenumber are not presented here, they are relatively straightforward to derive except at $t = \pi/4$ (e.g. Drazin & Howard 1962, Gill & Davey 1969). Two of these asymptotic expansions have been used in drawing the figures. For sufficiently small wavenumbers viscous effects become important, and for certain of the modes lower neutral branches can be identified (the curves in figure 1 are upper branches). For instance, for

$0 < t < \pi/4$ the lower branch to mode A is on the $\alpha = O(R^{-\frac{1}{4}})$ triple-deck scale (e.g. see Smith 1979). However, some of the modes, in particular those which are connected with points of zero shear in the velocity profile, are unstable for wavenumbers down to $\alpha = O(R^{-1})$. For this scaling the quasi-steady assumption breaks down, and the relevant governing equations for the disturbances are just the linearised classical unsteady boundary-layer equations. The consequences of this instability existing within the classical boundary-layer formulation have been discussed by Cowley, Hocking & Tutty (1985). Small wavenumber asymptotic analysis also suggests that as $\alpha \to 0$ the curves A & B in figure 1 join onto neutral curves with the upper branch Tollmien-Schlichting scale.

Finally we note that since the analysis has been based on a quasisteady approximation it is also applicable to non-periodic unsteady flows. Further, since the relevant wavelengths are comparable with the width of the boundary layer, the extension to non-parallel boundary-layer flows is straightforward (cf. Cowley et al. 1985, Tutty & Cowley 1985).

The author is grateful to Dr T.W. Ng and Ms. S.T. Lam for performing some of the calculations, and to Professor S.H. Davis for his interest and comments. This work was funded in part by the National Aeronautics and Space Administration, Materials-Processing-in-Space Program.

REFERENCES

Allmen, M.J. & Eagles, P.M. (1984) Stability of divergent channel flows: a numerical approach. Proc. R. Soc. Lond. A392, 359-372.
Benney, D.J. & Rosenblat, S. (1964) Stability of spatially varying and time-dependent flows. Phys. Fluids 7, 1385-1386.
Bodonyi, R.J. & Smith, F.T. (1981) The upper branch stability of the Blasius boundary layer, including non-parallel flow effects. Proc. R. Soc. Lond. A375, 65-92.
Bouthier, M. (1973) Stabilite lineaire des ecoulements presque paralleles. J. de Mecanique 11, 599-621.
Clamen, M. & Minton, P. (1977) An experimental investigation of flow in an oscillating pipe. J. Fluid. Mech. 81, 421-431.
Clarion, C. & Pelissier, R. (1975) A theoretical and experimental study of the velocity distribution and transition to turbulence in free oscillatory flow. J. Fluid Mech. 70, 59-79.
Collins, J.I. (1963) Inception of turbulence at the bed under periodic gravity waves. J. Geophysical Res. 18, 6007-6014.
Cowley, S.J., Hocking, L.H. and Tutty, O.R. (1985) On the stability of solutions of the classical unsteady boundary-layer equation. Phys. Fluids 28, 441-443.
Davis, S.H. & Rosenblat, S. (1977) On bifurcating periodic solutions at low frequency Stud. Appl.Math. 57, 59-76.
Drazin, P.G. & Howard, L.N. (1962) The instability to long waves of unbounded parallel inviscid flow. J. Fluid Mech., 14, 257-283.
Drazin, P.G. & Reid, W.H. (1981) Hydrodynamic Stability. Cambridge University Press
Finucane, R.G. & Kelly, R.E. (1976) Onset of instability in a fluid layer heated sinusoidally from below. Int. J. Heat Mass Transfer 19, 71-85.
Gill, A.E. & Davey, A. (1969) Instabilities of a buoyancy-driven system. J. Fluid Mech. 35, 775-798.
Hall, P. (1978) The linear stability of flat Stokes layers. Proc. R. Soc. Lond. A359, 151-166.

Hall, P. (1983) On the nonlinear stability of slowly varying time-dependent viscous flows. J. Fluid Mech., 126, 357-368.
Hino, M., Sawamoto, M. & Takasu, S. (1976) Experiments on transition to turbulence in an oscillatory pipe flow. J. Fluid Mech. 75, 193-207.
Howard, L.N. (1964) The number of unstable modes in hydrodynamic stability problems. J. de Mecanique 3, 433-443.
Iguchi, M., Ohmi, M. & Megawa, K. (1982) Analysis of free oscillating flow in a U-shaped tube. Bull. JSME 25, 1398-1405.
Kiser, K.M., Falsetti, H.L., Yu, K.H., Resitarits, M.R., Francis, G.P. & Carroll, R.J. (1976) Measurements of velocity wave forms in the dog aorta. J. Fluids. Engng. 6, 297-304.
Li, H. (1954) Stability of oscillatory laminar flow along a wall. Beach Erosion Board, US Army Corps Eng., Washington, D.C. Tech. Memo No. 47.
Lyne, W.H. (1971) Unsteady viscous flow in a curved pipe. J. Fluid Mech. 45, 13-31.
Merkli, P. & Thomann, H. (1975) Transition to turbulence in oscillating pipe flow. J. Fluid Mech. 68, 567-575.
Monkewitz, M. (1983) Ph.D. Thesis No. 7297, Federal Institute of Technology, Zurich, Switzerland.
Monkewitz, P.A. & Bunster, A. (1985) The stability of the Stokes layer: visual observations and some theoretical considerations. This volume.
Nerem, R., Seed, W.A. & Wood, N.B. (1972) An experimental study of the velocity distribution and transition to turbulence in the aorta. J. Fluid Mech. 52, 137-160.
Obremski, H.J. & Morkovin, M.V. (1969) Application of a quasi-steady stability model to periodic boundary-layer flow. AIAA J. 7, 1298-1301.
Ohmi, M., Iguchi, M., Kakehashi, K. & Masuda, T. (1982a) Transition to turbulence and velocity distribution in an oscillating pipe flow. Bull. JSME 25, 365-371.
Ohmi, M., Iguchi, M. & Urahata, I. (1982b) Transition to turbulence in a pulsatile pipe flow, part 1, wave forms and distribution of pulsatile velocities near transition region. Bull. JSME 25, 182-189.
Pedley, T.J. (1980) The fluid mechanics of large blood vessels. Cambridge University Press.
Pelissier, R. (1979) Stability of a pseudo-periodic flow. J. Appl. Math. Phys. (ZAMP) 30, 577-585.
Ramaprian, B.R. & Mueller, A. (1980) Transitional periodic boundary layer study. J. Hydr. Div. ASCE 106, 1959-1971.
Riley, N. (1965) Oscillating viscous flows. Mathematika 12, 161-175.
Rosenblat, S. & Herbert, D.M. (1970) Low-frequency modulation of thermal instability. J. Fluid Mech. 43, 385-398.
Rosenblat, S. (1968) Centrifugal instability of time-dependent flows. Part 1. Inviscid, periodic flows. J. Fluid Mech. 33, 321-336.
Seminara, G. & Hall, P. (1976) Centrifugal instability of a Stokes layer: linear theory. Proc. R. Soc. Lond. A350, 299-316.
Sergeev, S.I. (1966) Fluid oscillations in pipes at moderate Reynolds numbers. Mekh. Zhidk. Gaza. 1, 168-170 (Transl. in Fluid Dyn. 1, 121-122).
Smith, F.T. (1979) On the non-parallel flow stability of the Blasius boundary layer. Proc. Roy. Soc. Lond. A336, 91-109.
Tollmien, W. (1936) General instability criterion of laminar velocity distributions. Tech. Memor. Nat. Adv. Comm. Aero. Wash. No. 792.
Tromans, P. (1977) Ph.D Thesis, University of Cambridge.
Tutty, O.R. & Cowley, S.J. (1985) On the stability and the numerical solution of the unsteady interactive boundary-layer equation. Submitted to J. Fluid Mech.
von Kerczek, C. (1973) Ph.D. Thesis, The Johns Hopkins University.
von Kerczek, C. (1982) The instability of oscillatory plane Poiseuille flow. J. Fluid Mech. 116, 91-114.
von Kerczek, C. & Davis, S.H. (1974) Linear stability theory of oscillatory Stokes layers, J. Fluid Mech. 62, 753-773.
von Kerczek, C. & Davis, S.H. (1976) The instability of a stratified periodic boundary layer. J. Fluid Mech. 75 287-303.
Yang, W.H. & Yih, C.-S. (1977) Stability of time-periodic flows in a circular pipe. J. Fluid Mech. 82, 497-505.

Ginzburg–Landau Equation: Stability and Bifurcations

L. Sirovich and P.K. Newton

1. Introduction

The Ginzburg-Landau (G-L) equation is a prototypical amplitude evolution equation governing the envelope of supercritical wave disturbances. We now demonstrate this with a brief heuristic derivation of the G-L equation. Consider Fig. 1, which shows a typical neutral stability curve separating a base flow and the bifurcated flow that results after instability has set in. The ordinate is a control parameter such as Rayleigh or Reynolds number while the abscissa is a wave number. To arrive at this curve one considers the linear equations governing the perturbation, $\mathscr{A}$, to the base flow. These may be written as an eigenvalue equation

$$L\mathscr{A} = \sigma \mathscr{A}$$

plus boundary conditions. The eigenvalue

$$\sigma = \sigma(k, R)$$

is in general complex, and the condition

$$\text{Re } \sigma = 0$$

yields the neutral stability curve. As indicated in the figure near the critical point (k_c, R_c) the curve is fit by a parabola

$$R - R_c \propto (k - k_c)^2.$$

More generally the eigenvalue σ has the expansion

(1) $\quad \sigma = i\omega_c + \sigma_0(R - R_c) + iv(k - k_c) - \lambda(k - k_c)^2 + \ldots$

where ω_c, σ_0 and the group speed v are real; but λ is complex. If we write

(2) $\quad \mathscr{A} = A(x,t)\exp[i(\omega_c t + k_c x)]$

then the lead terms of the dispersion relation explicitly given in (1) also arise if we seek plane wave solutions to $\mathscr{A}$ with A such that

(3) $\quad \dfrac{\partial}{\partial t} A - v \dfrac{\partial}{\partial x} A = \sigma_0(R - R_c)A + \lambda A_{xx}\left[-\beta|A|^2 A\right].$

The non-linear term in parenthesis was introduced on heuristic grounds by Landau [1] to account for the bifurcation. Stuart [2] and Watson [3] then gave formal developments leading to the Landau term for the case of parallel flows. In later developments [4-15] this equation was shown to govern amplitude evolution near a critical point for a wide variety of physical phenomena, e.g. Poiseuille flow, Bernard convection, internal waves, surface waves, chemical reactions, semi-conductor waves, etc.

The G-L equation, (3), may be simplified by transforming to a frame of reference moving with the group speed v and then after a suitable normalization put in the form [16],

(4) $\quad iA_t + (1 - ic_0)A_{xx} = i\rho A - (1 + i\rho)|A|^2 A,$

with

(5) $\quad \rho = c_0/c_1, \quad 0 \leqslant c_0^2 < c_1.$

Numerical experiments performed on (4) have shown that it gives rise to a rich variety of phenomena [17-23]. These include instability, bifurcation, motion on low dimensional torii and chaos. These calculations have proven to be quite lengthy and as a result a relatively sparse sampling of cases are available. In what follows we further investigate these experiments by mainly analytical methods. Using relatively simple tools we are able to treat wide parameter ranges of the G-L equation. In certain instances this leads to a reinterpretation of previous results.

2. Background [16]

Equation (4) is satisfied by the Stokes' solution,

(6) $\quad A_0 = e^{it}.$

This is a particular instance of a spatially periodic solution to (4) having wavelength

(7) $\quad L = 2\pi/q$.

A linear stability analysis easily shows that the Stokes solution is stable unless

(8) $\quad q^2 < \dfrac{2(1 - c_0^2/c_1)}{1 + c_0^2} = q_0^2$.

Thus when the *box* exceeds $2\pi/q_0$ in size the state of uniform oscillation given by (6) becomes unstable. If

(9) $\quad q^2 < q_0^2/n^2$

at least n harmonics are linearly unstable. A simple estimate shows that the growth rate of a harmonic is order

(10) $\quad \epsilon^2 = q_0^2 - q^2$.

The stability relation (8) is a surface in (q_0^2, c_0, c_1)-space and typical sections are shown in Figure 2.

Although linear theory exhibits exponential growth one may show that solutions are pointwise bounded in time [16]. Specifically solutions to (4) are Lagrange stable;

(11) $\quad \max\limits_{0<x<L} |A(x,t)|^2 < K(L)$,

where as indicated the constant depends on the size of the box, L.

In the numerical experiments of Moon et al. [18] and Keefe [21] which we will refer to frequently, the following approach was adopted. The amplitude A is expanded in harmonics

(12) $\quad A = \sum\limits_{n} A_n(t)\cos nqx$

which when substituted into (4) yields an infinite system of coupled ordinary equations

(13) $\quad \dfrac{\partial}{\partial t} A_n = F_n(A_0, A_1, A_2, ...); \quad n = 0,1, ...\ .$

Moon et al. [18] and Keefe [21] typically truncate this system at 32 harmonics and integrate forward in time using the initial data,

(14) $\quad A(t = 0) = 1 - .02 \cos qx$.

In certain instances extremely lengthy computation was required in order to ascertain the character of the time asymptotic motion.

3. Steady State Oscillations

Implicit in the results of Moon et al. [18] is the fact that at primary bifurcation the G-L equation, (4), produces a limit cycle solution in the form of a steady state oscillation

(15) $\quad A = e^{i\Omega t}\Phi(x).$

This was noticed by us [24] and also independently by Keefe [21] who did not fully pursue this form.

In [16] we show that (15) can be used as the basis of a straightforward perturbation analysis. The small parameter in this expansion is given by (10) which measures the departure from the neutral stability curve. We write

$$\Omega(\epsilon) = \sum_{n=0} \epsilon^{2n}\Omega_n; \quad \Omega_0 = 1,$$
$$\Phi(x;\epsilon) = \sum_{n=0} \epsilon^n \Phi_n(x); \quad \Phi_0 = 1.$$

This is substituted in (4) and the resulting perturbation solution is carried through $O(\epsilon^4)$.

When these perturbation solutions were compared with the exact limit cycle solution (see next section), the agreement was remarkable even for $\epsilon = O(1)$. An example of this is shown in Figure 3. For this reason the perturbation solution just described was used as the basis for the stability analysis of the limit cycle solution and hence for the determination of secondary bifurcation [25]. We only mention this in passing since in the next section we describe the results of an exact treatment.

4. Exact Limit Cycle Solution [26]

Next the steady state oscillation form for A, (15), is substituted into (4) to obtain

(16) $\quad (1 - ic_0)\Phi_{xx} = (\Omega + i\rho)\Phi - (1 + i\rho)\Phi|\Phi|^2.$

If this is put into first order form under the transformation

(17) $\quad \Phi = \hat{U} + i\hat{V}, \quad \Phi_x = \hat{u} + i\hat{v}$

then it may be shown that the resulting flow in $(\hat{U}, \hat{V}, \hat{u}, \hat{v})$-space is volume preserving. From (16) we observe that if Φ solves (16) then $\Phi \exp(i\theta_0)$ for any real constant θ_0 also solves this equation. From this it follows that the order of (16) can be reduced by one. To accomplish this we write

(18) $\quad \Phi = \sqrt{r(x)} \, \exp\left[i \int_0^x v(s)ds\right]$

which when substituted into (16) leads to

(19) $\quad \dfrac{d}{dx}\begin{bmatrix} r \\ u \\ v \end{bmatrix} = \begin{Bmatrix} 2ur \\ \dfrac{c_0\Omega + \rho}{1 + c_0} - \dfrac{c_0 + \rho}{1 + c_0^2}r - 2uv \\ \dfrac{\Omega - c_0\rho}{1 + c_0^2} - \dfrac{1 - c_0\rho}{1 + c_0^2}r - u^2 + v^2 \end{Bmatrix}.$

One may show that the flow this induces in (r^2, u, v)-space is also volume preserving.

Returning to the calculations discussed in the previous section the limit cycle solution corresponds to an *even*, *L*-periodic solution of (16). In terms of the system (19) this means that $u(x)$ and $v(x)$ are odd functions while $r(x)$ is even. Hence, to find the limit cycle solution for a given value of Ω we can forward integrate (19) starting with initial data

(20) $\quad u(0) = 0, \ v(0) = 0, \ r(0) = R_0,$

if R_0 is the appropriate initial value. This was made the basis of a Newton iteration scheme for determining the correct value of R_0. In particular we examine the trajectory generated by (19) and (20) as it crosses the Poincare section $u = 0$, after a half cycle. If the value of v at the Poincare section is $\hat{v}(R)$, with R the guessed value of R_0, then the iterated value is given by

(21) $\quad R_0 = R - \hat{v}(R)/\hat{v}_R(R).$

The derivative $\hat{v}_R$ satisfies the linear variational equation gotten from (19)

(22) $\quad \dfrac{d}{dx}\begin{bmatrix} r_R \\ v_R \\ u_R \end{bmatrix} = \begin{bmatrix} 2u & 0 & 2r \\ \dfrac{-(c_0 + \rho)}{1 + c_0^2} & -2u & -2v \\ -\dfrac{1 - c_0\rho}{1 + c_0^2} & 2v & -2u \end{bmatrix}\begin{bmatrix} r_R \\ v_R \\ u_R \end{bmatrix}.$

This is then simultaneously integrated with (19). Since a Newton scheme possesses quadratic convergence only a few iterations are necessary to achieve high accuracy in R_0. A typical limit cycle solution generated in this way was shown in Figure 3.

As a result of the speed of this method many cases could be determined in a short time and in Figure 4 we show some curves taken from the non-linear *dispersion relation*

(23) $\quad \Omega = \Omega(q; c_0, c_1)$.

5. Secondary Bifurcation [26]

From the numerical experiments [18,21] we know that the second stage in the route to chaos is the two torus motion which results at the instability of the one torus motion. To examine the stability of the solution just obtained we write

(24) $\quad A = [\phi(x) + \psi(x,t)]\exp(i\Omega t)$.

Then in the limit of infinitesimal perturbation ψ, we find the variational equation,

(25) $\quad \partial_t \psi = \mathcal{L}\psi = z_1 \psi + z_2 \psi_{xx} + z_3(2|\phi|^2 \psi + \phi^2 \psi^*)$
with
$\quad\quad z_1 = \rho - i\Omega, \quad z_2 = c_0 + i, \quad z_3 = i - \rho$.

It is therefore of interest to consider the eigenvalue problem for the operator $\mathcal{L}$,

(26) $\quad \mathcal{L}\psi = \lambda \psi$

with ψ an L-periodic and even function. For our purposes it may also be assumed that λ is real. An immediate result is that corresponding to the zero eigenvalue two eigenfunctions are known, viz.

(27) $\quad \lambda = 0: \quad \psi_1 = i\phi; \quad \psi_2 = \phi_x$.

The first ψ_1 is obtained from the time derivative and the second ψ_2 from the spatial derivative of the limit cycle solution (15) when substituted into (4). For future reference observe that ψ_1 is an even and ψ_2 an odd function of x.

It is convenient to reformulate (26) in first order form. To do this we write

(28) $\quad u + iv = \psi, \quad \mu + i\nu = \psi_x$.

The vector

(29) $\quad \mathbf{u} = (u, v, \mu, \nu)$

then satisfies

(30) $\quad \lambda \mathbf{M}\mathbf{u} + \left[\dfrac{d}{dx} - \mathbf{F}\right]\mathbf{u} = 0$

where the matrix **F**, specified in the appendix, is L-periodic

(31) $F(x + 2\pi/q) = F(x)$.

The constant matrix **M** is also given in the appendix.

One may show that the condition for instability, and hence secondary bifurcation, is $\lambda = 0$. Thus we want to find a solution to

(32) $\left[\dfrac{d}{dx} - F\right] u = 0$

which is L-periodic and an even function. To treat this problem we introduce the Jacobi matrix

(33) $U = \left[\dfrac{\partial u}{\partial a}, \dfrac{\partial u}{\partial b}, \dfrac{\partial u}{\partial c}, \dfrac{\partial u}{\partial d}\right]$

where
$(a, b, c, d) = u(0) = u_0$

represents the initial data. Clearly **U** satisfies

(34) $\left[\dfrac{d}{dx} - F\right] U = 0$

and

(35) $U(x = 0) = 1$.

It is a consequence of our earlier remark that (17) leads to a volume preserving map so that

(36) $\det U = 1$.

In particular the Floquet matrix

(37) $\mathscr{F} = U(x = 2\pi/q)$

fulfills this condition,

(38) $\det \mathscr{F} = 1$.

The solution to the problem (32) can be represented by

(39) $u(x) = U(x) u_0$

and the periodicity condition then may be written as

(40) $\mathbf{u}(2\pi/q) = \mathscr{F}\mathbf{u}_0 = \mathbf{u}_0.$

Thus $\mathbf{u}_0$ is an eigenvector of the Floquet matrix corresponding to a unit eigenvalue. Two such eigenvectors are known from (27). If we write

(41) $\psi_1 = i\Phi \rightarrow \mathbf{u}_1, \quad \psi_2 = \Phi_x \rightarrow \mathbf{u}_2$

then clearly

(42) $\mathscr{F}\mathbf{u}_1(0) = \mathbf{u}_1(0), \quad \mathscr{F}\mathbf{u}_2(0) = \mathbf{u}_2(0)$

To summarize: *From (42) $\mathscr{F}$ always has at least two unit eigenvalues. From (38) it follows that the product of the remaining two eigenvalues is unity.* Secondary bifurcation will be demonstrated if we find an Ω for which $\mathscr{F}$ has an additional unit eigenvalue. *Direct calulation has shown that this does not occur*!

If we return to the original eigenvalue problem (30) (or equivalently (26)) we can say that *the geometric multiplicity of $\lambda = 0$ is two.* However for secondary bifurcation to occur we must have the algebraic multiplicity of $\lambda = 0$ to be at least three. In order for this to be so we must have the following scenario: Denote the eigenfunction which goes unstable by $\Psi(x;\Omega)$ where the dependence of Ψ on Ω has been made explicit. For Ω greater than its critical value Ω_0, the corresponding value of λ is negative. *As Ω approaches Ω_0 from above; $\lambda \rightarrow 0$ and $\Psi \rightarrow \psi_1$* (since it must be even). If this set of circumstances is translated into analytical terms the condition on Ω_0 is that

(43) $\mathscr{L}(\Omega_0)p = i\Phi$

have a L-periodic even solution ($\mathscr{L}$ is not hermitian). This has been determined for the case $c_1 = 1$ and the curve of such cases is denoted by B in Figure 5. (We parameterize the curve with values of q^2 rather than Ω_0.) The primary bifurcation curve (also see Figure 2) is denoted by A. The remaining parts of this figure are described in the discussion.

6. Discussion

When the Floquet matrix $\mathscr{F}$ discussed in the previous section possesses the eigenvalue negative one (actually a double root) spatial period doubling occurs (see Ref. 26). The locus of such cases is indicated in Figure 5 by the curve C. For values of Ω beyond these (34) possesses exponentially growing solutions and the basic Newton procedure for finding limit cycle solution runs into trouble. As a result we have terminated the secondary bifurcation curve B at this intersection.

There is only one reliable case of secondary bifurcation which resulted from the numerical experiments. This was by Keefe [21] and is indicated by the filled triangule in Figure 5. Since Keefe made a relatively fine sampling of cases in the vicinity of this value one can wonder why the agreement is not better. Since Keefe solved the initial value problem, (12) and (13)), it was necessary for him to wait for substantial periods of computation time to decide whether a solution resided on a one or a two torus. Thus a possibility is that the computation accumulated considerable truncation and roundoff error. Another possibility is that the modal termination scheme used by Keefe [21] (32 harmonics) introduced an error which is felt in such a calculation. Still another possibility is that both the one and two torus motions have different basins of attraction in this vicinity.

Another, more significant departure from the numerical experiments occurs for $c_0 \downarrow 0$, the cubic Schrodinger limit. The only numerical experiments in this limit were obtained by Moon et al. [17,18,23]. Their sampling of cases turned out to be sparse, and in fact they missed the secondary bifurcation found by Keefe. Moon et al. [17,18,23] claim that no limit cycle solutions exist in the neighborhood of $c_0 = 0$ and draw the curve B in Figure 5 so that it terminates in the intersection of the A-curve and $c_0 = 0$. This claim appears to be based on calculations at $c_0 = 0$. However, in this case the cubic Schrodinger equation is obtained, and this is well-known to be a Hamiltonian system. It follows from this that motions on torii of arbitrary genus are possible. Thus the solutions found by Moon et al. [17,18,23] are really a reflection of the initial data used by them and cannot be used as an indication of the situation which is obtained as c_0 tends toward zero.

Appendix

The coefficient matrix **F** is given by:

(A.1) $\quad \mathbf{F} = \begin{bmatrix} 0 & 1 \\ f & 0 \end{bmatrix}$

where

$$f = \begin{bmatrix} \alpha_{11} & \alpha_{12} \\ \alpha_{21} & \alpha_{22} \end{bmatrix}$$

$$\alpha_{11} = -\frac{1}{(1+c_0^2)} \left\{ (c_0\rho-\Omega) + 3\Phi_r^2(1-c_0\rho) + \Phi_i^2(1-c_0\rho) - 2c_0\Phi_r\Phi_i\left[1 + \frac{1}{c_1}\right] \right\}$$

$$\alpha_{12} = -\frac{1}{(1+c_0^2)} \left\{ c_0\left[\Omega + \frac{1}{c_1}\right] - c_0\Phi_r^2\left[1 + \frac{1}{c_1}\right] - 3c_0\Phi_i^2\left[1 + \frac{1}{c_1}\right] + 2\Phi_r\Phi_i(1 - c_0\rho) \right\}$$

$$\alpha_{21} = -\frac{1}{(1+c_0^2)} \left\{ -c_0\left[\Omega + \frac{1}{c_1}\right] + 3c_0\Phi_r^2\left[1 + \frac{1}{c_1}\right] + c_0\Phi_i^2\left[1 + \frac{1}{c_1}\right] + 2\Phi_r\Phi_i(1 - c_0\rho) \right\}$$

$$\alpha_{22} = -\frac{1}{(1+c_0^2)} \left\{ (c_0\rho-\Omega) - \Phi_r^2(c_0\rho-1) - 3\Phi_i^2(c_0\rho-1) + 2c_0\Phi_r\Phi_i\left[1 + \frac{1}{c_1}\right] \right\}$$

$$\Phi = \Phi_r + i\Phi_r$$

The matrix **M** is

(A.2) $\quad \mathbf{M} = \dfrac{1}{1 + c_0^2} \begin{bmatrix} 0 & 0 \\ -c_0 & -1 \\ 1 & -c_0 & 0 \end{bmatrix}$

References

1. L. D. Landau, C.R. Acad. Sci. U.R.S.S. **44**, 311, 1944.

2. J. T. Stuart, "On the non-linear mechanics of wave disturbances in stable and unstable parallel flow. Part I.", J. Fluid Mech. **9** (1960), 353-370.

3. J. Watson, "On the non-linear mechanics of wave disturbances in stable and unstable parallel flow. Part II.", J. Fluid Mech. **9** (1960a), 371-389.

4. A. Davey, "The propagation of a weak non-linear wave", J. Fluid Mech. (1972), Vol. 53, part 4, 769-781.

5. A. C. Newell, Envelope Equations, in Nonlinear Wave Motion, A. C. Newell ed., AMS, 1974.

6. J. D. Gibbon, M. J. McGuinness, "Amplitude equations at the critical points of unstable dispersive physical systems", Proc. R. Soc. Lond. **A377** (1981), 185-219.

7. A. C. Newell, J. A. Whitehead, "Finite bandwidth, finite amplitude convection", J. Fluid Mech. **38**, (1969), 279.

8. L. M. Hocking, K. Stewartson, J. T. Stuart, "A nonlinear instability burst in plane parallel flow", J. Fluid Mech. **51**, (1972), 705-735.

9. M. Pavlik, G. Rowlands, "The propagation of solitary waves in piezoelectric semiconductors", J. Phys. **C8** (1975), 1189.

10. R. Grimshaw, "Modulation of an internal gravity wave packet in a stratified shear flow", Wave Motion **3** (1981), 81-103.

11. D. J. Benney, A. C. Newell, "The propagation of nonlinear wave envelopes", J. Math. Phys. **46** (1967), 133-139.

12. G. B. Whitham, "Linear and Non-Linear Waves", J. Wiley & Sons, Inc., 1974.

13. V. E. Zakharov, "Stability of periodic waves of finite amplitude on the surface of a deep fluid", Soc. Phys. J. Appl. Mech. Techn. Phys. **4** (1968), 190-194.

14. A. Jeffrey, T. Kawahara, "Asymptotic Methods in Nonlinear Wave Theory", Pittman, 1982.

15. Y. Kuramoto, "Diffusion induced chaos in reaction systems", Progress of Theoretical Physics Supp. **64** (1978), 346.

16. P. K. Newton, L. Sirovich, "Instabilities of the Ginzburg-Landau equation: Periodic solutions", Quarterly of Applied Mathematics, Oct. (1985).

17. H. Moon, P. Huerre, L. G. Redekopp, Three-frequency Motion and Chaos in the Ginzburg-Landau Equation, Phys. Rev. Letters **49**(7): 458, Aug. 1982.

18. H. Moon, P. Huerre, L. G. Redekopp, "Transitions to chaos in the Ginzburg-Landau equation", Physica D **7** (1983), 135.

19. K. Nozaki, N. Bekki, Pattern selection and spatio-temporal transition to chaos in the Ginzburg-Landau equation, Phys. Rev. Letters **51** (1983), 2171.

20. C. S. Bretherton, E. A. Spiegel, Intermittency through modulational instability, Physics Letters, Vol. 96A, No. 3, 20, June 1983.

21. L. Keefe, "Dynamics of Perturbed Wavetrain Solutions to the Ginzburg-Landau Equation", Ph.D. Dissertation, University of Southern California, 1984.

22. R. J. Deissler, Noise-sustained structure, intermittency, and the Ginzburg-Landau equation, submitted for publication.

23. H. T. Moon, "Transition to chaos in the Ginzburg-Landau equation", Ph.D. Dissertation, University of Southern California, 1982.

24. P. K. Newton and L. Sirovich, Studies of the Ginzburg-Landau Equation, Bull. Am. Phys. Soc. **29**, No. 9, 1524, 1984.

25. P. K. Newton and L. Sirovich, Instabilities of the Ginzburg-Landau equation: Pt. II, Secondary Bifurcations, to appear, Quarterly of Applied Mathematics.

26. L. Sirovich and P. K. Newton, Periodic Solutions of the Ginzburg-Landau Equation (submitted for publication).

Figure Captions

Figure 1. Typical stability diagram. Abscissa represents wave number and ordinate a control parameter such as Reynolds or Rayleigh number. Solid curve represents neutral stability and divides stable base flow from bifurcated flow. Dashed curve is parabolic fit to stability curve passing through critical point (k_c, R_c).

Figure 2. Stability curves for the Stokes solution, (6). The parameters c_0 and c_1 are as indicated.

Figure 3. Limit cycle solution for $\epsilon = .93$, $c_0 = .25$ and $c_1 = 1$. Dotted curves refer to the perturbation solution through $O(\epsilon^2)$. Solid curves refer to the exact treatment gotten from integrating (19). (a), the real part; (b) the imaginary part; (c) power spectrum $|A_k|^2$. Error is less than 6 per cent.

Figure 4. Dispersion relation Ω vs. q^2 for $(c_1 = 1, c_0 = .1)$ and $(c_1 = 1, c_0 = .25)$.

Figure 5. c_0^2 vs. q^2, $c_1 = 1$: A, primary transition curve (see Fig. 2); B, secondary transition curve; C, spatial period doubling curve. Filled triangle is secondary transition computed by Keefe [21].

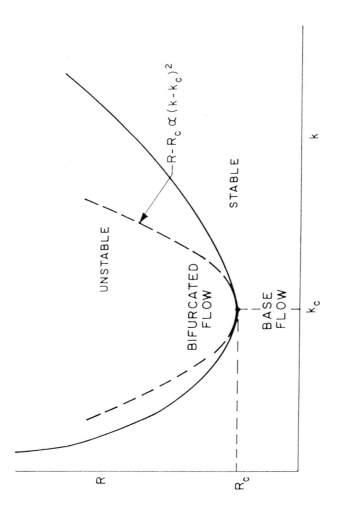

Figure 1

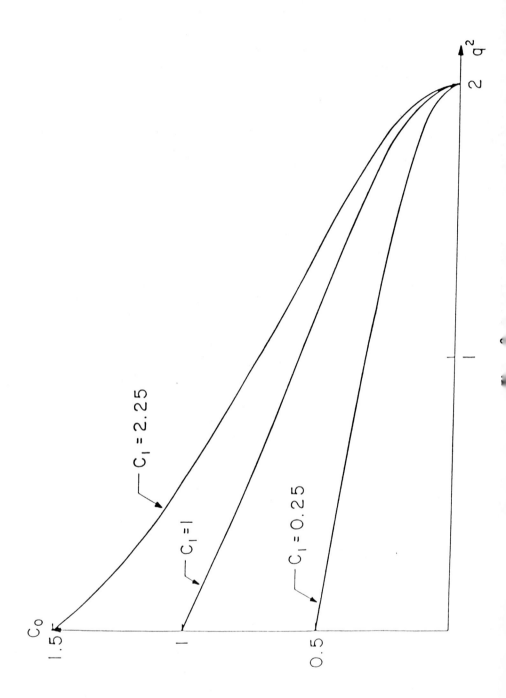

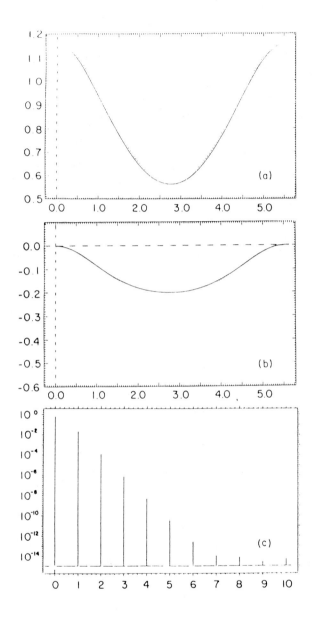

Figure 3

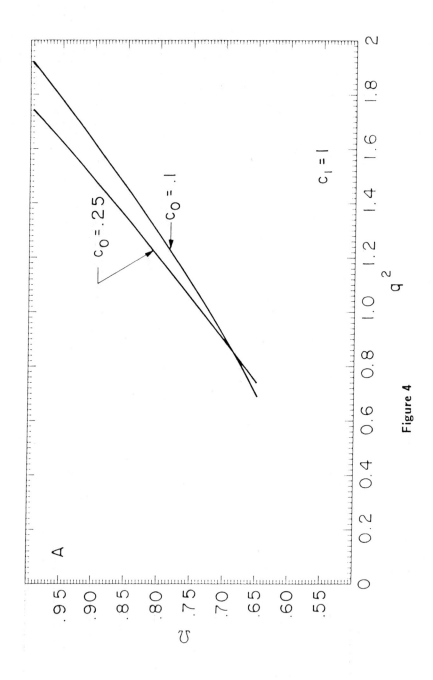

Figure 4

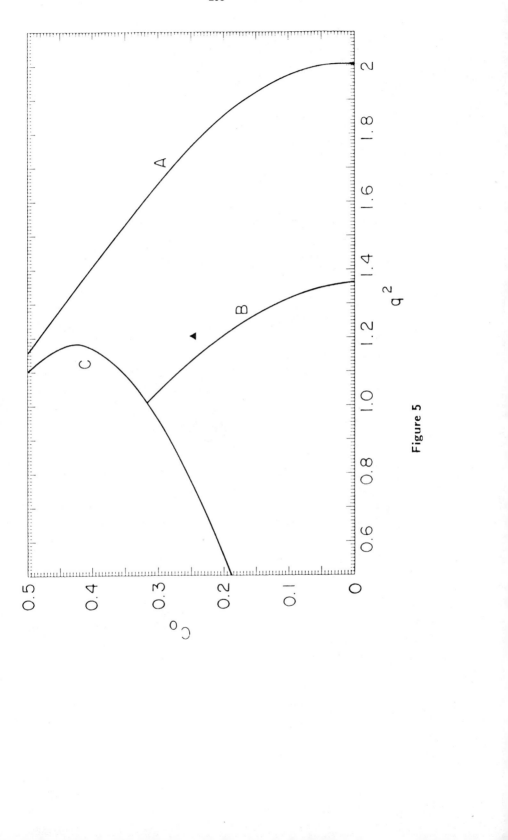

Figure 5

Stability and Resonance in Grooved-Channel Flows

NASREEN K. GHADDAR AND ANTHONY T. PATERA

Abstract

The stability and resonant response of incompressible moderate Reynolds number flow in periodically-grooved channels is investigated by direct numerical simulation using the spectral element method. For Reynolds numbers less than a critical value, the flow is found to approach a stable steady-state. The two-dimensional linear stability of this flow is then analyzed, and it is found that the least stable modes closely resemble Tollmien-Schlichting channel waves, forced by Kelvin-Helmholtz shear layer instability at the groove edge. A theory for frequency prediction based on the Orr-Sommerfeld dispersion relation is presented, and verified by variation of the geometric parameters of the problem. For Reynolds numbers greater than the critical value, the flow is shown to take the form of self-sustained oscillations similar to those observed experimentally. A preliminary investigation of the three-dimensional linear stability of spanwise homogeneous grooved-channel flows is also presented, in which a Fourier (spanwise)/spectral element algorithm is used to directly simulate the three-dimensional perturbation equations.

It is shown that oscillatory perturbation of the grooved-channel flow at the frequency of the least stable mode of the linearized system results in subcritical resonant excitation as the critical Reynolds number is approached. The importance of (nonhomogeneous) geometry in the forced response of a flow is discussed, and grooved-channel flow is compared to (straight-channel) plane Poiseuille flow, for which no resonance excitation occurs due to a zero projection of the forcing inhomogeneity on the dangerous (Tollmien-Schlichting) modes of the system. The significant effect of resonant excitation on transport, transition, and flow structure is described.

1. Introduction

Flow over grooves and in grooved channels arises in a large number of practical applications, from the design of effective cooling systems for electronic devices [1], to the analysis of buffeting due to flow over airframe cutouts [2]. Flow in grooved channels can also be interpreted as a model for the more general class of problems involving flow over rough walls. In addition to any direct practical import, flow in grooved channels serves as a simple, yet rich, example of internal separated flow, in which a thorough (numerical) study of the stability properties of a strongly non-parallel flow in complex geometry can be performed.

There is a great deal of experimental work on low-speed boundary layer flows over grooves [3,4], the primary result of which is demonstration of the existence of (hydrodynamic) self-sustained oscillations at sufficiently large Reynolds number. Analytical [5] and numerical [6-8] work to date has not isolated the physical mechanism responsible for these oscillations, the former suffering from indeterminate singularities at separation and re-attachment, the latter focussing primarily on lower Reynolds number flow phenomena. The relative effect of groove flow, external flow, and geometry on the frequency and magnitude of the self-sustained oscillations has, therefore, remained relatively poorly understood.

In addition to providing insight into the transition process in a flow, stability analyses also lay the foundations for understanding the response of the system to external disturbances. In particular, for flow over grooves, the experimental observations of overstability suggest the possibility of resonant modulatory excitation, in which the linear stability modes of the system are forced at their natural frequency. Although investigations of purely oscillatory flow in furrowed channels have been carried out [9], there has been little work on modulatory enhancement structured to exploit the stability characteristics of the steady flow. There are numerous applications (e.g., transport enhancement) in which resonant excitation could be of significant importance.

In this paper, we investigate by direct numerical simulation the flow of an incompressible fluid in a two-dimensional periodically-grooved channel. In Section 2 we describe the problem formulation, and briefly describe the numerical solution technique used. In Section 3 we present steady flow solutions, and examine their stability with respect to infinitesimal disturbances. A Tollmien-Schlichting theory for frequency determination is presented, and verified by variation of the geometric parameters of the problem. More details of these results can be found in reference [10]. In Section 4, the requirements for resonant excitation of linear modes are given, and grooved channels (for which resonance occurs) is compared to plane channels (for which resonance does not obtain). More details of these calculations, with extension to transport enhancement, are given in [11]. Lastly, in Section 5, some results are given for the three-dimensional stability of subcritical grooved channel flows.

2. Problem Formulation

The geometry to be considered is the periodically-grooved channel shown in Fig. 1, assumed infinite in extent in the streamwise (x) and spanwise (z) directions. The flow is assumed to be fully-developed in x, and to be independent of spanwise co-ordinate, z (with the exception of the three-dimensional analyses in Section 5). To put the problem in non-dimensional form, we scale all the velocities by 3/2 V, where V is the cross - channel average time - mean velocity, $V = (2h)^{-1} \int_{-h}^{h} <u(x=0,y,t)> dy$, and all lengths by the channel half-width, h. Here the bracket notation, $<\cdot>$, refers to temporal averaging. Unless otherwise stated, all variables considered henceforth are assumed to be non-dimensionalized in terms of 3/2 V and h.

The governing equations for incompressible flow in the domain D can then be written as,

(1a) $\qquad \vec{v}_t = \vec{v} \times \vec{\omega} - \nabla \Pi + R^{-1} \nabla^2 \vec{v} \qquad$ in D,

(1b) $\qquad \nabla \cdot \vec{v} = 0 \qquad$ in D,

where $\vec{v}(\vec{x},t)$ ($= u \hat{x} + v \hat{y}$) is the velocity, Π is the dynamic pressure, $\vec{\omega}$ is the vorticity, and $R = (3/2)Vh/\nu$ is the Reynolds number, where ν is the kinematic viscosity of the fluid. In addition to R, the flow is governed by the geometric parameters L, l, and a, representing periodicity length, groove length, and groove depth, respectively.

The fully-developed boundary conditions for the velocity $\vec{v}(\vec{x},t)$ are

(2a) $\qquad \vec{v}(\vec{x},t) = 0 \qquad\qquad$ on ∂D

(2b) $\qquad \vec{v}(x+mL,y,t) = \vec{v}(x,y,t)$,

corresponding to no-slip and periodicity, respectively. Here ∂D corresponds to the boundary made up of the top and bottom walls of the channel, and m is an integer periodicity index taken to be unity (see [10] for appropriate justification). For the pressure we require,

(3a) $$\Pi(\vec{x},t) = -\Pi_x(t)x + \Pi'(\vec{x},t)$$

(3b) $$\Pi'(x+mL,y,t) = \Pi'(x,y,t) \quad ,$$

where the term $\Pi_x(t)$ is the driving force for the flow, and is determined by the imposed flowrate condition,

(4) $$Q = \int_{-1}^{1} u(x=0,y,t)dy = 4/3(1+\eta \sin 2\pi \Omega_F t) \quad .$$

Note the linear pressure term in (3a) is consistent with periodicity of the velocity (2b), as only the gradient of the pressure enters into (1a). The case of $\eta = 0$ corresponds to steady forcing, whereas with $\eta \neq 0$ the flow is excited by a modulatory force of frequency Ω_F.

In addition to the full nonlinear problem described by (1-4), it is also of interest to consider the linearized problem about a steady solution to the Navier-Stokes equations, $\vec{v}_s(\vec{x})$, in which we assume solutions of the form

(5) $$\vec{v}(\vec{x},t) = \vec{v}_s(\vec{x}) + \varepsilon \vec{v}'(\vec{x},t), \quad \varepsilon \ll 1 \quad .$$

Inserting (5) into (1), and neglecting terms $O(\varepsilon^2)$ and higher, gives the following linear equation for $\vec{v}'(\vec{x},t)$,

(6a) $$\vec{v}'_t = \vec{v}_s \times \vec{\omega}' + \vec{v}' \times \vec{\omega}_s - \nabla \Pi' + R^{-1}\nabla^2 \vec{v}' \quad ,$$

(6b) $$\nabla \cdot \vec{v}' = 0 \quad .$$

The boundary conditions on the perturbation $\vec{v}'$, Π' are as in (2-3), however the flowrate condition (4) for the case of a linear stability analysis ($\eta = 0$) is now replaced with,

(7a) $$Q' = \int_{-1}^{1} u'(x=0,y,t)dy = 0 \quad ,$$

corresponding to no net perturbation flow. In considering the oscillatory forced flow ($\eta \neq 0$), it will be of interest to consider linearization in the modulatory amplitude, η, in which case we identify $\varepsilon = \eta$ in (5-6), and replace (7a) with

(7b) $$Q' = \int_{-1}^{1} u'(x=0,y,t)dy = {}^4\!/_3 \sin 2\pi \Omega_F t \; .$$

For sufficiently large times, the solution of the initial value problem (6,7a) will approach the least stable mode of the eigenvalue problem resulting from normal mode formulation of the same equation (independent of the initial conditions). In particular, the initial value problem result can be interpreted as,

(8) $$\vec{v}'(\vec{x},t) \sim \exp(\sigma t) \operatorname{Re}\{\exp(2\pi i \Omega t)\hat{\vec{v}}(\vec{x})\} \quad (t \Rightarrow \infty) \; ,$$

from which the growth rate, σ, and frequency, Ω, of the most unstable mode can be inferred. All linear (and nonlinear) results presented in this paper are based on initial value problem solution of the various forms of the Navier-Stokes equations given above using the spectral element method [10-14].

3. Steady Flow and Two-Dimensional Stability Theory

To find steady-states in the context of our initial value solvers, we integrate equations (1-4) from arbitrary initial conditions until a converged (in time) solution is obtained. The result of such a calculation for the "base" geometry (L = 6.6666, l = 2.2222, a = 1.1111) for R = 800 is shown in Figs. 2a, 2b, and 2c in the form of streamfunction, vorticity, and pressure contours, respectively. For the open cavities of interest here (i.e., grooves for which re-attachment occurs at the downstream edge and not at the cavity bottom), the flow in the channel part of the domain is very close to the plane Poiseuille flow parabolic profile. A relatively sharp shear layer is seen to have formed at the groove lip, suggesting that a Kelvin-Helmholtz type instability will, no doubt, occur at a higher Reynolds number. Our results for steady flow are similar to those reported by earlier investigators [6-8].

To investigate the stability of the flow, we integrate the linearized equations (5,6,7a) to a time-asymptotic state, perturbing around numerically-obtained steady-states such as that shown in Fig. 2. A typical result of these linear calculations is shown in Fig. 3, in which the perturbation velocity at a point in the flow domain is plotted as a function of time at R = 525 for the base geometry. The decay rate, σ, and frequency, Ω, can be read directly from this large-time result of direct simulation. Performing similar calculations at several Reynolds numbers, we can trace the trajectory of the least stable mode in the σ-Ω stability plane, as shown in Fig. 4. The modes are seen to be oscillatory, as expected, and the critical Reynolds number (defined by $\sigma = 0$) is at approximately $R = R_c = 900$.

Nonlinear calculations based on solution of (1-4) refine the critical point to $R_c = 975$ (defined by the point at which the large-time result of direct simulation first approaches an oscillatory rather than steady state), and also demonstrate that the nonlinear supercritical states correspond to a regular Hopf bifurcation [10]. ("Nonlinear" calculations are used to find the linear critical point due to the fact that our direct simulation methods can only find stable states, thus limiting our linear stability calculations to stable flows.) An example of a supercritical flow is given in Fig. 5 as a plot of (total) velocity at a typical point in the flow domain as a function of time at $R = 1200$ in the base geometry. For large times, the flow is seen to approach a limit cycle, corresponding to nonlinear saturation of the linear instability indicated in Fig. 4. In summary, our linear and nonlinear calculations confirm the oscillatory nature of groove-channel stability modes, and demonstrate the existence of self-sustained flow oscillations.

To determine the nature of the hydrodynamic instability resulting in self-sustained oscillations, we plot in Fig. 6 the time-asymptotic perturbation streamlines for $R = 525$ in the base geometry at several times during the flow cycle, $0 < t < T (= \Omega^{-1})$, obtained by direct numerical simulation of the linear equations (5,6,7a). The striking feature of this plot is the fact that, in the channel part of the domain, the grooved-channel modes closely resemble (plane Poiseuille flow) Tollmien-Schlichting travelling waves [10,15]. As there are two wavelengths per imposed periodicity length, the solution is denoted a $n = 2$ (two-wave) solution, with wavenumber $\alpha = 2\pi n/L$.

It is clear that the Tollmien-Schlichting waves seen in Fig. 6 cannot be directly responsible for the grooved-channel instability, given the disparity between the linear critical Reynolds number of approximately 1000 for our grooved geometry, and the corresponding quantity of $R_c = 5772$ for plane Poiseuille flow. Rather, the source of instability is a Kelvin-Helmholtz type instability of the groove shear layer, as can be seen in Fig. 7, a plot of the contours of the total vorticity, $\vec{\omega}_s + \varepsilon\vec{\omega}'$, at several times during the flow cycle, $0 < t < T$. (The perturbation vorticity, $\vec{\omega}'$, is the time-asymptotic result of a solution of the linear equations, (6,7a) ; ε is arbitrarily chosen to correspond to a maximum perturbation velocity of roughly ten percent.) It is seen that there is significant waviness and "roll-up" of the vortex sheet at the groove lip.

Although the occurrence of oscillatory shear layer instability in the grooved-channel flow is not particularly unexpected, it is not at all apparent how the frequency of the oscillations is chosen. This is particularly important as regards the resonant response of the system, and is, in general, an indication of how the (unstable) shear layer interacts with the surrounding flow (channel modes) and geometry. From Figs. 6 and 7 it appears reasonable to interpret the grooved-channel instability process as free shear layer destabilization of erstwhile stable Tollmien-Schlichting waves. If this be the case, it also seems plausible that the unstable, "massless" part of the flow, the shear layer, would allow the stable, "massive" part of the flow, the Orr-Sommerfeld mode, dictate the frequency of their coherent oscillations, in this way minimizing the resistive component of the system. This turns out, in fact, to be the case, as seen in Fig. 8, a plot of Ω as a function of α (= $2\pi n/L$) for seventeen different grooved-channel solutions at R = 525. The points are seen to lie exactly on the plane Poiseuille flow Orr-Sommerfeld dispersion relation for the least stable wall mode at R = 525.

Note it is not true that the shear layer plays no role in the frequency selection process; it determines the wave index n, and hence α, which does *not* always correspond to the allowable wavenumber (integer multiple of $2\pi/L$) at which the "associated" Tollmien-Schlichting wave is least stable [10]. However, once the spatial scale (α) of the disturbance is set, the dispersion relation of the channel modes determines the frequency. (Note that, in fact, our solutions need not be L-periodic, but rather mL-periodic, where m is the integer index introduced in (2b,3b). Inasmuch, α will generally be of the form $\alpha = 2\pi n/mL$; this countable set of modes clearly does not constitute a continuous spectrum, however it is also different from discrete spectra typically encountered. Numerical and experimental evidence that the choice m = 1 captures most of the essential physics is given in [10,11,16], and will not be discussed further here.)

4. Resonance Conditions

From the linear theory grooved-channel results of the previous section, it would appear obvious that perturbative forcing of the flow at the frequency of the least stable mode should lead to resonant excitation as the critical Reynolds number of the flow is approached. To indicate why this is obvious, but nontrivial, we consider formal solution of the forced linear equations, (6,7b). To this end, we reformulate the problem in streamfunction representation,

(9) $$\vec{v}'(\vec{x},t) = \nabla x \psi(x,y,t)\hat{z} \quad,$$

for which (6) can then be written as

(10) $$\mathcal{L}\psi = M\psi_t \quad,$$

with M the (negative) Laplacian operator, $M\psi$ the perturbation vorticity, and $\mathcal{L}$ the equation for convection and diffusion of vorticity (linearized Navier-Stokes) [11,20]. The boundary conditions on ψ are

(11a) $\quad\quad\quad \nabla\psi\cdot\hat{n} = 0 \quad\quad\quad$ on $\partial D \quad,$

(11b) $\quad\quad\quad \psi = 0 \quad\quad\quad$ on $\partial D_B \quad,$

(11c) $\quad\quad\quad \psi = {}^4/_3 \exp(2\pi i \Omega_F t) \quad\quad\quad$ on $\partial D_T \quad,$

where ∂D_B and ∂D_T refer to the bottom and top walls, respectively. The condition (11c) follows from the flowrate condition (7b) (we take the imaginary part of ψ as our physical solution).

If we now write ψ in the form,

(12) $$\psi(x,y,t) = \{ \phi(x,y) + \psi_s(x,y) \} e^{2\pi i \Omega_F t} \quad,$$

where $\vec{v}_s(\vec{x}) = \nabla x \psi_s(x,y)\hat{z}$, we get the following equation for ϕ,

(13) $$(\mathcal{L}-2\pi i \Omega_F M)\phi = -(\mathcal{L}-2\pi i \Omega_F M)\psi_s \quad,$$

with homogeneous boundary conditions,

(14a) $\quad\quad\quad \nabla\psi\cdot\hat{n} = 0 \quad\quad\quad$ on $\partial D \quad,$

(14b) $\quad\quad\quad \psi = 0 \quad\quad\quad$ on $\partial D \quad.$

It is clear from (13) that the condition for resonance is that the left-hand side operator become singular ($\Omega_F \Rightarrow \Omega$, $R \Rightarrow R_c$), *and* that the equation be non-solvable in this singular limit. When these two conditions are satisfied, the assumed form (12) is no longer valid, and must be augmented by a secular term in time, indicative of resonance excitation (and, ultimately, nonlinear intervention) [11].

It would therefore appear that for "generic" overstable flows, linear resonance is to be expected, and this is, in fact, true. However, many classical flows are not generic, being in some sense geometrically degenerate. This leads to a singular but solvable system in (13), and hence resonance does not obtain [11]. For instance, if we consider the case of plane Poiseuille flow, it is clear that any adjoint eigenfunction with wavenumber nonzero will annihiliate the right-hand side of (13). Thus for all the dangerous (unstable) modes of the system, equation (13) is solvable, and hence resonance will not occur (i.e., the only modes with nonzero projection are purely decaying Stokes modes). In fact, the solution to linearized forced plane Poiseuille flow is simply parallel Womersley flow; although this flow may be unstable, this is a different phenomenon (and linearization) than the linear subcritical resonance of interest here.

Upon breaking the geometric degeneracy of plane Poiseuille flow with the addition of a groove, we now make the system (13) nonsolvable [11]. We illustrate that resonance occurs in grooved channels by considering the amplitude parameter,

(15) $$A(\vec{x}) = < (v(\vec{x},t) - <v(\vec{x})>)^2 >^{1/2} ,$$

corresponding to the magnitude of the fluctuating component of the vertical velocity (note for plane Poiseuille flow $A \equiv 0$). We plot in Fig. 9 A at a point in the groove shear layer as a function of Ω_F for $R = 225$ and $R = 525$ in the base geometry (η arbitrarily set to $\eta = .2$). It is seen that the response is exactly as expected on the basis of simple linear resonance considerations. First, the frequency response is peaked at $\Omega_F = \Omega$ ($= \Omega_1$), the frequency of the least stable mode calculated from the linear theory presented in Section 3. Examination of the flow field at resonance reveals the two-wave (n = 2) Tollmien-Schlichting structure expected on the basis of our linear theory results [11]. Second, the width of the peak decreases and the magnitude increases as $R \Rightarrow R_c$ (here $R_c = 975$). Lastly, there is a secondary peak in the frequency response at $\Omega = \Omega_2$, due to excitation of the second mode of the linear system (in fact, a one-wave solution, the frequency of which is accurately predicted by the Tollmien-Schlichting dispersion relation).

Another way in which to see the singular difference between grooved and ungrooved channels is to consider the nature of the instability modes. For plane Poiseuille flow, the Tollmien-Schlichting modes are travelling waves, and thus a change of reference frame reduces the least stable mode to a purely decaying disturbance, for which we do not expect resonance. On the other hand, although the grooved-channel modes closely resemble travelling waves in the channel region of the flow, the nonhomogeneous geometry precludes "global" solutions in the form of travelling waves in the entire domain. Thus, although the Tollmein-Schlichting dispersion relation can be used to predict the resonant frequency in grooved-channel flows, the response of grooved channels is markedly (singularly) different than the response in a plane channel, for which there is no resonance, no vertical velocity generation, and no transport enhancement. This shows the very strong effect (nonhomogeneous) geometry can have on flow behavior, in particular as regards response to external forcing.

A more mathematical description of the conditions for resonance is given in [17], in which it is shown that resonance obtains for critical modes which are not common to the dual eigenproblems corresponding to the differential equation (10) with zero imposed disturbance flowrate (as in (14)) and zero imposed disturbance pressure drop, respectively. For critical modes which are common to both eigenproblems (e.g., Tollmien-Schlichting waves in a plane channel), the product of imposed oscillatory pressure drop and flowrate response (or, similarly, imposed oscillatory flowrate and pressure drop response) is zero, from which it follows that the critical mode cannot extract energy from modulatory forcing [17].

Upon solution of the full nonlinear equations (1-4) for the modulatory forcing case, it is found that nonlinearity has a quantitative, but not qualitative effect on the linear resonance results presented above [11]. In particular, nonlinearity results in shifts in peak frequency , attenuation of amplitude peaks, and the appearance of harmonics of the primary frequency; these effects, as expected, are larger at larger flow amplitudes, whether the increased flow magnitude be due to increased forcing (larger η), or operation at near-resonant conditions. Nevertheless, even at these larger amplitudes, the flow clearly retains the spatial and dynamic characteristics of the originating linear resonance mechanism.

There are numerous ramifications of the resonance mechanism described here. First, it is clear from Fig. 9 that resonant excitation will result in significant transport enhancement, an important effect that has been confirmed both numerically [11] and experimentally [16]. Second, the mechanism of resonant excitation is, no doubt, at least partially responsible for the relative instability of general rough-walled systems, as Fig. 9 demonstrates that non-homogeneous geometry can have a dramatic effect on the sensitivity of a system to external disturbances (i.e., the "receptivity" of the flow). This latter point is discussed briefly below in the context of three-dimensional instability.

5. Three-Dimensional Stability of Subcritical Grooved-Channel Flow

The major assumption in the work presented above is that of two-dimensionality, an assumption which may not be physically realizable even for a channel of "infinite" spanwise extent. Understanding the effect of three-dimensionality is important in determining the range of validity of our two-dimensional analyses, as well as in extending the investigation into the (perforce, three-dimensional) regimes of transition and turbulence. In this paper we consider the first step in this process, namely analysis of the three-dimensional linear stability of the (unmodulated, $\eta = 0$) steady subcritical grooved-channel flows presented in Section 2. We continue to assume that the groove geometry is infinite (homogeneous) in the z-direction, however we now allow for spanwise flow variation.

Combining the assumption of geometric homogeneity (separability) in the z-direction with the further assumption of reflectional symmetry ($z \Rightarrow -z$; $u \Rightarrow u$, $v \Rightarrow v$, $w \Rightarrow -w$), we write the velocity as,

$$(16) \quad \vec{v}(x,y,z,t) = \vec{v}_s(x,y) + \varepsilon \begin{cases} \hat{u}(x,y,t)\cos\beta z \\ \hat{v}(x,y,t)\cos\beta z \\ \hat{w}(x,y,t)\sin\beta z \end{cases}$$

which reduces to the two-dimensional case, (5), for a spanwise wavenumber, β, of zero. Upon inserting $\tilde{v}(x,y,t)$ into the Navier-Stokes equations, (1), and neglecting terms $O(\varepsilon^2)$ and higher, we arrive at a set of linear equations analogous to (6) for the three-dimensional modal amplitudes. As all spanwise (z-) dependence is factored out by the Fourier representation (16), the spatial dependence is essentially two-dimensional (derivatives in z reduced to multiplication by appropriate factors of β), and the numerical (spectral element) methods used are therefore very similar to the corresponding two-dimensional techniques described in the references [12-14].

On the basis of the large-time results of direct simulation of the three-dimensional perturbation equations at R = 525 in the base geometry, we plot in Fig. 10 the perturbation vertical velocity, $\hat{v}'$, at a point in the groove shear layer as a function of time for a spanwise wavenumber of $\beta = 1.0$ [18]. It is seen that this (representative) three-dimensional disturbance decays, a good indication that at this Reynolds number, the steady flow is, indeed, stable to *all* disturbances. As the flow in "open" cavity geometries is relatively parallel (despite nonparallel geometry), it is possible that Squire-type arguments [15] (non-rigorously) apply, and that the first linear instability is two-dimensional. Future work will address this issue, filling out the stability diagram with respect to three-dimensional disturbances.

It should be noted that, although Fig. 10 is indicative of three-dimensional stability for the steady flow, that need not be true for the case of flow resulting from modulatory forcing. In particular, the secondary flow patterns resulting from linear mode excitation (as well as those that naturally evolve supercritically [10]) closely resemble the secondary flows (nonlinear equilibria) found in plane Poiseuille flow. This in turn suggests that the three-dimensional secondary instability isolated in plane channels [19] may also play an important role in transition in forced and supercritical grooved-channel flows, a conjecture which will be investigated in future work.

Acknowledgements

We would like to thank Mr. George E. Karniadakis for making available his three-dimensional stability results. This work was supported by the NSF under Grant MEA-8212469, the ONR under Contract N00014-85-K-0208, NASA under Grant NAG1-574 and Contract NAS1-16977, a Rockwell Assistant Professorship (A.T.P.), and the Kuwait Foundation for the Advancement of Science (N.K.G.). The computations were performed on the NASA-Ames CRAY-XMP, and a DEC VAX 750 supplied in part by a DEC External Research Program Grant.

References

1. D. E. Arvizu and R. J. Moffatt, Experimental heat transfer from an array of heated cubical elements on an adiabatic channel wall. *Rept. No. HMT-33, Thermosciences Div. Stanford University, Stanford, California*, 1982.

2. K. Karachmeti, Sound radiated from surface cutouts in high-speed flows. *Ph.D. Thesis, California Institute of Technology, Pasadena, California*, 1956.

3. T. C. Reihman, Laminar flow over transverse rectangular cavities. *Ph.D. Thesis, California Institute of Technology, Pasadena, California*, 1967.

4. V. Sarohia, Experimental investigation of oscillations in flows over shallow cavities. *AIAA J.* 15, 984 (1977).

5. P. A. Durbin, Resonance in flows with vortex sheets and edges. *J. Fluid Mech.* 145, 275 (1984).

6. F. Pan and A. Acrivos, Steady flows in rectangular cavities. *J. Fluid Mech.* 28, 643 (1967).

7. U. B. Mehta and Z. Lavan, Flow in a two-dimensional channel with a rectangular cavity.*J. Appl. Mech.* (1969).

8. T. B. Gatski and C. E. Grosch, Embedded cavity drag in steady and unsteady flows. *AIAA Paper No. 84-0436*, (1984).

9. I. J. Sobey, On flow through furrowed channels. Part 1: Calculated flow patterns. *J. Fluid Mech.* 96, 7 (1980).

10. N. K. Ghaddar, K. Z. Korczak, B. B. Mikic, and A. T. Patera, Numerical investigation of incompressible flow in grooved channels. Part 1: Stability and self-sustained oscillations. *J. Fluid Mech.*, to appear.

11. N. K. Ghaddar, M. Magen, B. B. Mikic, and A. T. Patera, Numerical investigation of incompressible flow in grooved channels. Part 2: Oscillatory heat transfer enhancement. *J. Fluid Mech.*, to appear.

12. A. T. Patera, A spectral element method for fluid dynamics; laminar flow in a channel expansion. *J. Comput. Phys.* 54, 468 (1984).

13. K. Z. Korczak and A. T. Patera, An isoparametric spectral element method for solution of the Navier-Stokes equations in complex geometry.*J. Comput. Phys.*, to appear.

14. G. E. Karniadakis, E. T. Bullister, and A. T. Patera, A spectral element method for solution of the two- and three-dimensional time-dependent incompressible Navier-Stokes equations. In *Proc. Europe-U.S. Symposium on Finite Element Methods for Nonlinear Problems, Trondheim,* 1985 (Springer-Verlag).

15. P. G. Drazin and W. H. Reid, *Hydrodynamic Stability,* Cambridge University Press, 1981.

16. M. Greiner, N. K. Ghaddar, B. B. Mikic, and A. T. Patera, Resonant convective heat transfer in grooved channels. In *Proc. Eighth International Heat Transfer Conference, San Francisco,* 1986.

17. M. Magen and A. T. Patera, Resonance conditions for forced two-dimensional channel flows. In *Proc. SIAM Conference on Mathematics Applied to Fluid Mechanics and Stability, Dedicated to Richard C. DiPrima, RPI,* 1985.

18. G. E. Karniadakis, A numerical investigation of instability and transition in complex geometries. *Ph.D. Thesis, Massachusetts Institute of Technology, Cambridge, Massachusetts,* 1986.

19. S. A. Orszag and A. T. Patera, Secondary instability of wall-bounded shear flows. *J. Fluid Mech.* 128, 347 (1983).

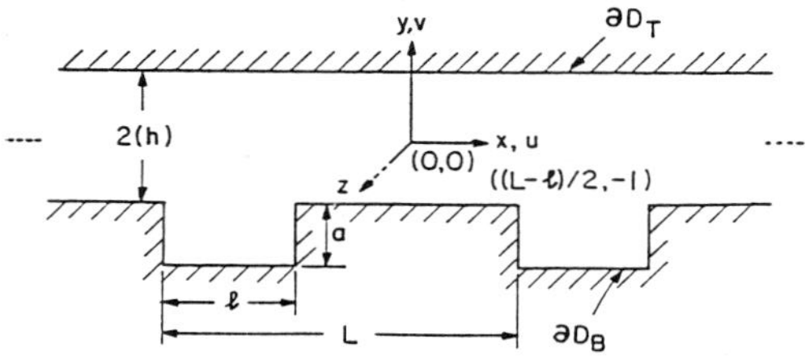

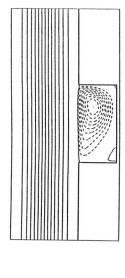

Fig. 1. The geometry of the periodically-grooved channel is described by the groove depth (a), the groove length (l), and the separation distance between the grooves (L).

Fig. 2a. Plot of the streamline contours for steady flow at R=800 in the base geometry.

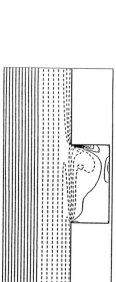

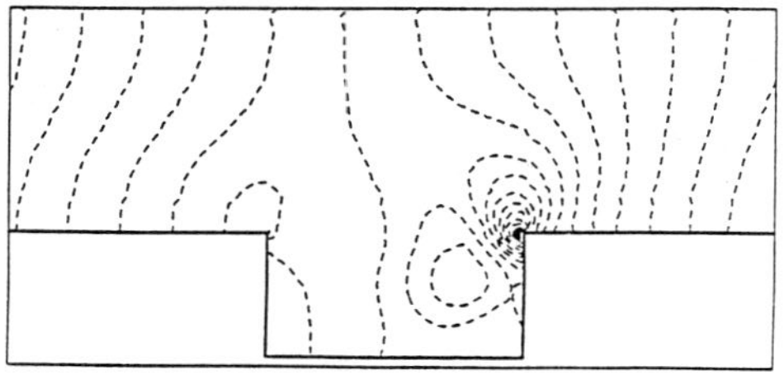

Fig. 2c. Plot of the pressure contours for steady flow at R=800 in the base geometry.

Fig. 2b. Plot of the vorticity contours for steady flow at R=800 in the base geometry.

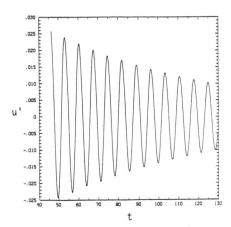

Fig. 3. A plot of the perturbation velocity, u', as a function of time at a typical point in the flow domain at R=525 (base geometry).

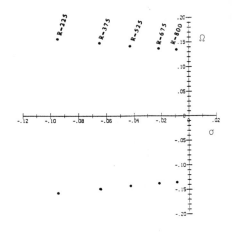

Fig. 4. A plot of the (σ,Ω) trajectory of the least stable grooved-channel mode, parametrized by the Reynolds number, R (base geometry).

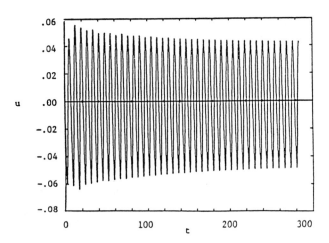

Fig. 5. A plot of the velocity, u, as a function of time at a typical point in the flow domain at R=1200 (base geometry). The stable flow is now a limit cycle.

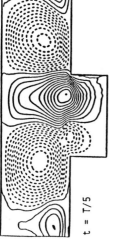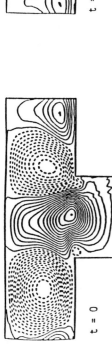

Fig. 6. A plot of the perturbation streamlines of the least stable grooved-channel mode at R=525 at two times during the flow cycle (base geometry). Note the similarity of the channel region of the flow with plane Poiseuille flow Tollmien-Schlichting waves.

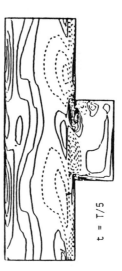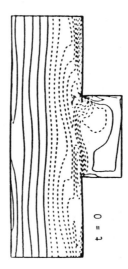

Fig. 7. A plot of the total vorticity (perturbation field scaled to correspond to a maximum velocity of ten percent) for the grooved-channel flow at R=525 at two times during the flow cycle (base geometry). The waviness of the cavity shear layer is suggestive of Kelvin-Helmholtz instability.

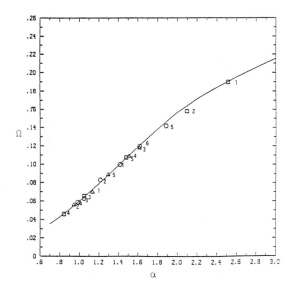

Fig. 8. A plot of the frequencies, Ω, for the various grooved channels studied versus the wavenumber of the "associated" Tollmien-Schlichting channel waves, α, at R=525. The symbols represent the grooved channel results (direct simulation), whereas the solid line is the Orr-Sommerfeld dispersion relation.

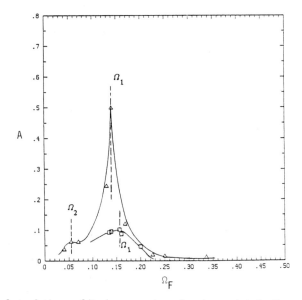

Fig. 9. A plot of the amplitude parameter, A, at a point in the groove shear layer as a function of forcing frequency, Ω_F, at $\eta=.2$ for R=525 ($\triangle$) and R=225 ($\square$). The system behaves exactly as predicted from simple resonance arguments.

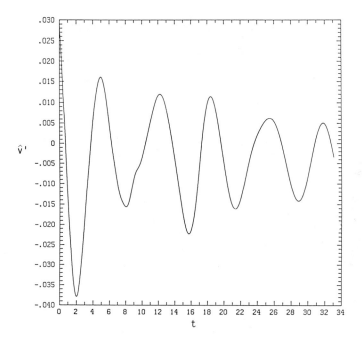

Fig. 10. A plot of the three-dimensional perturbation velocity, $\hat{v}'$, as a function of time at a typical point in the flow domain at R=525 (base geometry). At this Reynolds number it appears that the steady flow is stable with respect to all disturbances.

Finite Length Taylor Couette Flow

C.L. Streett and M.Y. Hussaini

Abstract

Axisymmetric numerical solutions of the unsteady Navier-Stokes equations for flow between concentric rotating cylinders of finite length are obtained by a spectral collocation method. These representative results pertain to two-cell/one-cell exchange process, and are compared with recent experiments.

Introduction

The Taylor experiment on Couette flow between coaxial circular cylinders has been the subject of numerous theoretical and experimental studies [1]. This flow is rich in complex phenomena; so rich, in fact, that they are still being discovered [2], and our understanding of them is far from complete. In a typical Taylor experiment, the inner cylinder rotates with a constant angular velocity while the outer cylinder, along with the top and bottom walls, are kept at rest. The relevant

---

Research was supported by the National Aeronautics and Space Administration under NASA Contract Nos. NAS1-17070 and NAS1-18107.

geometric parameters are the radius ratio, which is the ratio of the radii of the inner and outer cylinders, and the aspect ratio, which is the ratio of the length of the annulus to its width. The dynamic parameter is the Reynolds number based on the angular velocity of the inner cylinder and the annulus width. The Taylor-Couette flow is strongly dependent on all of these parameters. The theory for the infinite aspect ratio case (which neglects end wall effects) and its correspondence to the experiments in cylinders of necessarily finite but large aspect ratio are reviewed in Di Prima and Swinney [1] and Di Prima [3]. Benjamin [4] has developed a rigorous qualitative theory for the existence of nonunique steady states for confined flows and their stability and transition with particular reference to finite-length Taylor-Couette flow. His predictions have been confirmed in a series of experiments [5] in cylinders of short length with fixed end plates. They have been further confirmed by the numerical results of Cliffe and Mullin [6] who discretized the steady Navier Stokes equations by a Galerkin finite element method and solved the resulting nonlinear algebraic equations by the pseudo arclength continuation method of Keller [7].

Among other numerical studies of the finite-length Taylor-Couette flow problem, those of Alziary de Roquefort and Grillaud [8], and Neitzel [9] are worth mentioning. Both investigations were based on the time dependent vorticity-stream function formulation along with the equation for the azimuthal velocity. They used a finite difference method to solve for the steady state by a time asymptotic approach. Their numerical results show the axial structure for small Reynolds number, the smooth development of the flow with rapid increase in vortex activity for Reynolds numbers in quasi-critical range, and multiple states for sufficiently large Reynolds number. The aspect ratio being large, their results are only pertinent to the exchange phenomena beyond the two-cell and four-cell interactions examined by Benjamin.

No theoretical work on finite-length Taylor-Couette flow has incorporated the correct boundary conditions for fixed end walls. An idealized version of the end-wall boundary conditions is due to Schaeffer [10]. He introduces a parameter α in the boundary conditions, with $0 < \alpha < 1$. The parameter α interpolates linearly between the two extreme cases: $\alpha = 0$ corresponds to the infinite length problem which accommodates the Couette flow as an exact solution, while $\alpha = 1$ corresponds to the finite length problem with the correct no-slip conditions being applied on the end walls.

For $\alpha = 0$, Blennerhasset and Hall [11] have considered the linear stability problem in the small gap limit, and the key result was that the two-cell primary flow changes into four-cell at the aspect ratio of approximately 2.6. This should be compared with a value of roughly 3.7 observed by Benjamin for radius ratio of 0.615. Hall [12] has further derived the amplitude equations for this problem. An interesting feature of these amplitude equations is a quadratic term (absent in the infinite aspect ratio case) which can introduce hysteresis and soften bifurcations into smooth transitions.

For small nonzero values of α, Schaeffer [10] used a Lyapunov-Schmidt reduction procedure and the methods of singularity theory to obtain results applicable to the exchange processes between 2m and 2m+2 cells, $m > 2$. Hall [12] has studied the two cell-four cell exchange problem using amplitude equations and a perturbation method for small values of α.

The purpose of our continuing research effort is to solve the unsteady Navier-Stokes equations by a highly accurate spectral collocation method with a view to elucidate the underlying processes leading to laminar-turbulent transition in the Taylor-Couette flow. The present work is confined to the evolution of two-cell and single-cell Taylor-Couette flows with specific reference to the experiments of Benjamin and Mullin [13], Lucke, et al. [14] and Aitta, et al [15]. The main result of these studies is a second order transition from a two-cell flow, symmetric under reflection about the mid-plane, to an asymmetric single-cell flow that ensues with increasing Reynolds number beyond a certain critical value.

Governing Equations and Solution Technique

The incompressible Navier-Stokes equations for axisymmetric flow in a cylindrical geometry are, in conservation form:

$$\frac{\partial u}{\partial t} + \frac{\partial (u^2)}{\partial r} + \frac{\partial (uw)}{\partial y} + \frac{(u^2 - v^2)}{r} = -\frac{\partial P}{\partial r} + \nu [\nabla^2 u - \frac{u}{r}] \qquad (1)$$

$$\frac{\partial v}{\partial t} + \frac{\partial (uv)}{\partial r} + \frac{\partial (vw)}{\partial y} + \frac{2uv}{r} = \nu [\nabla^2 v - \frac{v}{r}] \qquad (2)$$

$$\frac{\partial w}{\partial t} + \frac{\partial (uw)}{\partial r} + \frac{\partial (w^2)}{\partial y} + \frac{uw}{r} = \frac{\partial P}{\partial y} + \nu \nabla^2 w \qquad (3)$$

$$\frac{1}{r} \frac{\partial}{\partial r}(ru) + \frac{\partial w}{\partial y} = 0 \qquad (4)$$

$$\text{where } \nabla^2 \equiv \frac{1}{r}\frac{\partial}{\partial r}(r\frac{\partial}{\partial r}) + \frac{\partial^2}{\partial y^2}$$

and u, v, w are the r, θ and y components of velocity, respectively. For the case where the outer cylinder and end plates are stationary, and the inner cylinder rotates, the boundary conditions are:

$u = v = w = 0$ on the outer cylinder $r = R_o$,

and at the top and bottom wall $y = \pm y_\ell$

$u = w = 0$ on the inner cylinder $r = R_i$

$v = \Omega R_i$.

In the present calculations, the azimuthal velocity v is split into two parts, $v = v_b + \tilde{v}$, where v_b satisfies:

$$\nabla^2 v_b = 0$$

$v_b = 0$ at $r = R_o$, and at $y = \pm y_\ell$

$= \Omega R_i$ at $r = R_i$

The quantity v_b is computed and stored at the start of a calculation, which proceeds with the computation of u, $\tilde{v}$, and w at each time step. Note that these three velocity components satisfy homogeneous boundary conditions.

A splitting method is employed to advance the solution from t^n to t^{n+1}.

Writing the Navier-Stokes equations in vector notation with $\underline{u}$ representing the velocity, $(u, \tilde{v}, w)$, we have:

$$\underline{u}_t + \underline{u} \cdot \nabla \underline{u} = -\nabla P + \nu \nabla^2 \underline{u} \quad \text{in D, the annulus} \tag{5}$$

$$\nabla \cdot \underline{u} = 0$$

$$\underline{u} = 0 \text{ on the boundary } \Gamma$$

the split scheme first advances $\underline{u}^n$ to an intermediate solution $\underline{u}^*$ by solving:

$$\underline{u}^*_t + \underline{u}^* \cdot \nabla \underline{u}^* = \nu \nabla^2 \underline{u}^* \tag{6}$$

$$\underline{u}^* = \underline{g}^* \text{ on } \Gamma.$$

The intermediate boundary condition $\underline{u}^* = \underline{g}^*$ will be discussed subsequently. Finally, the solution is advanced from $\underline{u}^*$ to $\underline{u}^{n+1}$ via:

$$\frac{\underline{u}^{n+1}}{t} = -\nabla P^{n+1}$$

$$\nabla \cdot \underline{u}^{n+1} = 0 \tag{7}$$

$$\hat{n} \cdot \underline{u}^{n+1} = 0 \text{ on } \Gamma$$

where $\hat{n}$ is unit normal to the boundary Γ. Note that the final, "pressure correction" step by itself is an inviscid calculation, and is well-posed when only boundary conditions on the normal component of velocity are enforced. At the end of the full step there exists a non-zero tangential component of velocity on the boundary. The magnitude of this slip velocity can be reduced by a proper choice of the intermediate boundary condition on $\underline{u}^*$. Marcus [16] has described the difficulties which arise from the use of $\underline{u}^* = 0$ as the intermediate boundary condition. The conditions used here are based on the work of Fortin, et al. [17]. Using backward Euler time discretization for Eq. 7 yields:

$$\underline{u}^{n+1} = \underline{u}^* - \Delta t \, \nabla P \tag{8}$$

and the slip velocity on the boundary is given by

$$\hat{\tau} \cdot \underline{u}^{n+1}|_\Gamma = \hat{\tau} \cdot (\underline{u}^*|_\Gamma - \Delta t \, \nabla P^{n+1}|_\Gamma) \tag{9}$$

If $\hat{\tau} \cdot \underline{u}^* = 0$ on the boundary, then $\hat{\tau} \cdot \underline{u}^{n+1}|_\Gamma = O(\Delta t)$ where $\hat{\tau}$ is the unit tangent to the boundary Γ. However, if ∇P^{n+1} is expanded in Taylor series about $t = t^n$:

$$\nabla P^{n+1} = \nabla P^n + \Delta t \, \nabla P^n_t + O(\Delta t^2)$$

and the second term is approximated by

$$\Delta t \, \nabla P^n_t = (\nabla P^n - \nabla P^{n-1}) + O(\Delta t^2)$$

then Eq. 8 becomes

$$\hat{\tau} \cdot \underline{u}^{n+1}|_\Gamma = \hat{\tau} \cdot \left[\underline{u}^*|_\Gamma - \Delta t \, (2\nabla P^n - \nabla P^{n-1})\right] + O(\Delta t^3).$$

Thus the slip velocity may be reduced to $O(\Delta t^3)$ through the intermediate boundary condition

$$\hat{\tau} \cdot \underline{u}^*|_\Gamma = \hat{\tau} \cdot \Delta t \, (2\nabla P^n - \nabla P^{n-1}). \tag{10}$$

Of course the boundary condition $\hat{n} \cdot \underline{u}^* = 0$ is retained.

The pressure step is actually carried out in two parts. First, the divergence of the first of Eqs. 8 yields:

$$\nabla^2 p^{n+1} = \frac{1}{\Delta t} \nabla \cdot \underline{u}^* \qquad (11)$$

where $\nabla \cdot \underline{u}^{n+1} = 0$ is enforced. Then the velocities are updated using

$$\underline{u}^{n+1} = \underline{u}^* - \Delta t \, \nabla p^{n+1} \qquad (12)$$

Note that this formulation requires a boundary condition for the pressure. This poses a problem, since there is no natural boundary condition for pressure. Deriving a condition by enforcing the normal momentum equation at the boundary is a questionable practice, as the equation need not hold on the boundary at the differential level of the equations. This inconsistency often produces explosive instability in a spectral code. Fortunately, the split scheme yields a self-consistent pressure condition,

$$\hat{n} \cdot \nabla P \big|_\Gamma = 0 \qquad (13)$$

since both $\hat{n} \cdot \underline{u}^*$ and $\hat{n} \cdot \underline{u}^{n+1}$ are zero on the boundary. The error involved in this specification is, we believe, related to the overall splitting error of the scheme.

Zang and Hussaini [18], have extensively investigated a related split scheme in which two coordinate directions were periodic and the third employed general boundary conditions. Comparison between split and unsplit codes using the same discretization yielded agreement to five decimal digits. However, they utilized a staggered grid for pressure in the non-periodic direction, obviating the need for a pressure boundary condition. Staggered grids in two dimensions either lack the ability to set both velocity components equal to zero on all boundaries, or are susceptible to an oscillatory "checkerboard" pressure mode. The unstaggered scheme used here, on the other hand, requires a pressure boundary condition. Actually, the consistent pressure equation derived by Zang and Hussaini yields exactly the same condition on pressure as used here. No instabilities were ever encountered with the present scheme.

Discretization and Solution Scheme

Since no-slip boundary conditions are enforced in both the r and y directions, Chebyshev spectral representation is appropriate in both directions. Collocation is used for a number of reasons: straightforward treatment of nonlinear terms and boundary conditions; capability to include coordinate stretchings; and ability to solve the resulting discrete equations rapidly. For further discussion of this form of discretization, see Hussaini et al. [11].

A coordinate stretching was employed in the radial direction to resolve the large gradients near the inner cylinder. The form of the stretching is:

$$r = \frac{[1 + b \exp(-a)] (R_o - R_i)}{[1 + b \exp(-a r_c)]} r_c + R_i \quad (14)$$

where r_c is the radial coordinate in the computational space. Values of a = 2 and b=5 to 50 were typical in this work. The y-direction was not stretched.

Time discretization of the first step in the split scheme involved the low-storage mixed Runge-Kutta/Crank-Nicholson scheme [20]. Writing the semi-discrete equation for the first step as

$$u_t = A(u) + D(u) \quad (15)$$

where $A(u)$ and $D(u)$ represent advection and diffusion terms, respectively, the mixed scheme advances from time step t^n to t^{n+1} using a third-order Runge-Kutta scheme for the advection terms and Crank-Nicholson for the diffusion terms:

$$u^0 = u(t^n)$$

$$H^1 = \Delta t \, A(u^0)$$

$$u^1 = u^0 + \frac{1}{3} H^1 + \frac{1}{6} \Delta t \, (D(u^0) + D(u^1))$$

$$H^2 = \Delta t\, A(u^1) - \frac{5}{9} H^1$$
$$u^2 = u^1 + \frac{15}{16} H^2 + \frac{5}{24} \Delta t\, (D(u^1) + D(u^2)) \tag{16}$$

$$H^3 = \Delta t\, A(u^2) - \frac{153}{128} H^2$$
$$u^3 = u^2 + \frac{8}{15} H^3 + \frac{1}{8} \Delta t\, (D(u^2) + D(u^3))$$
$$u(t^{n+1}) = u^3$$

The second step of the split scheme uses backward Euler time discretization, as mentioned in the previous section.

The above scheme is stable to $O(1)$ Courant numbers, which involves time steps many times larger than that desired for accuracy. Typically a time step which resulted in Courant numbers of .1 to .2 based on the smallest physical mesh spacing was used. The slip velocity resulting from the choice of time step was normally eight to ten orders of magnitude below the maximum velocity in a given direction.

Note that the above scheme reduces the problem to a sequence of uncoupled Helmholtz/Poisson equations to advance the discrete solution. Since one time step requires the solution of nine positive-definite Helmholtz equations with Dirichlet boundary conditions, and one Poisson equation with pure Neumann boundary conditions, a computationally efficient technique had to be developed if this study is to be feasible. The present scheme is fairly efficient. This scheme involves preconditioned Richardson iteration, in which an optimum relaxation parameter is chosen dynamically using either a minimum-residual (MR) or an orthogonal-residual (OR) procedure. Details of this technique are as follows.

Write the equation to be solved as

$$L(u) = f \tag{17}$$

The residual at a given iterate "m" is

$$R^m \equiv L(u^m) - f \tag{18}$$

A preconditioned iteration scheme may be looked at in the following way. At a given iteration, the goal is to compute an update such that the next residual is zero, i.e.

$$R^{m+1} \equiv L(u^m + \Delta\tilde{u}) = 0 \qquad (19)$$

or for a linear operator L (as in this case):

$$L(u^m) + L(\Delta\tilde{u}) = R^m + L(\Delta\tilde{u}) = 0 \qquad (20)$$

Approximating the operator L which is difficult to invert by a more easily-inverted operator M yields the following preconditioned scheme for the update:

$$\Delta\tilde{u} = M^{-1}[L(u^m)] = M^{-1}[R^m] \qquad (21)$$

A well-known method for computing an optimal ω^m involves minimizing the L_2 norm of the residual, R^{m+1}:

$$\min_{\omega^m} (R^{m+1}, R^{m+1}) = \min_{\omega^m} (R^m, R^m) = 2\omega^m (R^m, L(\Delta\tilde{u})) + \omega^{m^2} (L(\Delta\tilde{u}), L(\Delta\tilde{u})) \qquad (22)$$

for which a stationary point is

$$\omega^m = -\frac{(R^m, L(\Delta\tilde{u}))}{(L(\Delta\tilde{u}), L(\Delta\tilde{u}))} \qquad (23)$$

This is the usual minimum residual (MR) method.

An alternate method for chosing ω^m is to require that the successive residuals be orthogonal to each other. This yields an orthogonal residual (OR) scheme related to the method of steepest descent:

$$(R^m, R^{m+1}) = (R^m, R^m) + \omega^m(R^m, L(\Delta\tilde{u})) = 0 \qquad (24)$$

and hence

$$\omega^m = - \frac{(R^m, R^m)}{(R^m, L(\Delta\tilde{u}))} \quad (25)$$

The scheme proceeds as follows from an initial guess u^0 and residual $R^0 = L(u^0) - f$:

$$\Delta\tilde{u} = M^{-1} [R^m]$$

$$u^{m+1} = u^m + \omega^m \Delta\tilde{u} \quad (26)$$

$$R^{m+1} = R^m - \omega^m L(\Delta\tilde{u})$$

etc.

The preconditioning operator M is the second-order finite-difference operator which corresponds to the spectral operator L. Unequal-mesh spacing formulae, based on the physical-space point locations (which includes the radial stretching), are used in constructing M. A fixed V-cycle multigrid scheme driven by approximate-factorization is used to invert the preconditioning operator (Eq 21). It was found that complete convergence of this step is not required for the overall preconditioned scheme to converge: essentially no difference in convergence rate is seen between fully solving the preconditioning step and merely reducing the residual (associated with inversion of the preconditioner) by two orders of magnitude which usually requires only two or three multigrid cycles.

The overall convergence rate based on spectral operator evaluations is never slower than .35 (for Dirichlet boundary conditions); rates between .15 and .2 are typical even on the highly stretched mesh. The use of MR or OR to compute the optimum relaxation factor yields equivalent convergence rates, although the OR scheme is found to be more robust in other contexts.

However, the observed convergence rate deteriorates for this scheme when it is applied to the pure Neumann problem encountered in the Eqs (11) and (13). Since a large number of such problems are to be solved, a boundary influence-matrix technique is devised that reduces the computation time associated with this step. To compute the solution of

$$L(u) = f \text{ in } D$$
$$u_n = 0 \text{ on } \Gamma \tag{27}$$

a series of homogeneous solutions $u^{(i)}$ is obtained:

$$L(u^{(i)}) = 0 \text{ in } D$$
$$u^{(i)} = \delta(x_i) \text{ on } \Gamma \tag{28}$$

for each boundary point x_i. For each solution $u^{(i)}$, , the vector of normal gradients $[u_n^{(i)}]$ at all boundary points is computed. These vectors are collected as columns of a (square) matrix. This matrix is the influence of a unit disturbance at the boundary on the normal gradient at the boundary of the solution to the linear operator $L(u) = 0$. Denote this matrix by

$$C = \text{col } \{[u_n^{(1)}], [u_n^{(2)}], \ldots [u_n^{(N_b)}]\} \tag{29}$$

where a total of N_b points lie on the boundary. The matrix C may be inverted to yield the influence of a unit normal gradient at the boundary on the boundary value of a solution of $L(u) = 0$, with provision for setting the level of the solution required. Then the solution of the original problem (Eq. 27) proceeds by solving the related problem with homogeneous Dirichlet boundary conditions:

$$L(u^h) = f \text{ in } D$$
$$u^h = 0 \text{ on } \Gamma \tag{30}$$

The vector of normal derivatives of u^h at the boundary $[u_n^h]$ is computed. u^h must be corrected if it is to satisfy the desired Neumann boundary conditions. This correction is computed by applying the influence matrix to the boundary gradient error vector:

$$L(u^c) = 0 \text{ in } D$$

$$u^c = C^{-1} [u_n^h] \text{ on } \Gamma$$

(31)

The desired solution is $u = u^h + u^c$.

Computation and inversion of the influence matrix is done in a preprocessing step; since it is a function only of the mesh geometry, it may be stored and used whenever it is needed. To obtain a solution of the pure Neumann-Poisson problem at each time step requires that we solve just two Dirichlet-Poisson problems and the solutions satisfy the desired boundary conditions exactly.

Implementation and Performance:

The above split scheme for solving the time-dependent incompressible Navier-Stokes equations has been implemented with a view to processing on both scalar and vector computers. In scalar form on a CYBER-175, the code requires approximately 20 seconds per time step on a 17 x 17 mesh, and 56 seconds/step on a 25 x 25 mesh. A large fraction of these times is spent in computing the 11 Helmholtz/Poisson solutions required in each time step. On the VPS-32 vector processing machine at NASA Langley, the vector code requires about 14 seconds step on a mesh comprised of 65 points in the y direction and 33 points in the r direction. Almost 85 percent of this time is taken in inverting the preconditioning step of the Helmholtz/Poisson solution as the approximate factorization part of the step is not vectorizable in a manner which yields adequate vector lengths for the two-dimensional problems. This observed performance for the two-dimensional Chebyshev-Chebyshev code is in line with the performance quoted in (21) for a three-dimensional Chebyshev-Fourier-

Fourier code (both codes have similar operation counts). For a coordinate direction discretized with Chebyshev series using N points, the operation count for that direction is $O(N^2)$, whereas a Fourier-discretized direction requires only $O(N)$ operations. Thus, the present Chebyshev-Chebyshev method and the Chebyshev-Fourier-Fourier method both have total operation counts of $O(N^4)$, with the latter method having the advantage of larger possible vector lengths.

When the azimuthal direction is added to this simulation, however, the relative performance improves dramatically. Since the azimuthal direction is periodic, Fourier series is an appropriate discretization. The computations are performed in the Fourier wave-space of that direction; the discrete equations for each azimuthal mode decouple (see [21] for details)), allowing vector lengths to increase by a factor of the number of points in the azimuthal direction. This increase in vector length improves the CPU seconds/point/time step performance of the present scheme by a factor of four, and should allow the planned three-dimensional simulations to be performed.

RESULTS

The results presented here pertain to the axisymmetric two-cell/one-cell bifurcation, which occurs when the Taylor apparatus has an aspect ratio up to about 1.5. The form of the bifurcation depends sensitively on this parameter; experiments [15] show that this transition can change from supercritical to subcritical with variations in the aspect ratio of as little as eight percent.

Most of the results are obtained either by making quasi-static changes in the Reynolds number and allowing the stable, dominant mode to settle, or by slowly sweeping through a Reynolds number range, monitoring the change in a particular mode. Of course, using a time-accurate code to simulate the bifurcations of steady-state solutions is quite inefficient, owing to the extremely small growth rates near the bifurcation points. However, the eventual aim of this work is to simulate the turbulence and broadband structure exhibited by Taylor-Couette flow at moderate Reynolds numbers and the code was developed with these time dependent flows in mind. The ability to simulate accurately the sensitive, steady state bifurcations at lower Reynolds numbers is an excellent test for the numerical method.

Moreover, by using a time-dependent computation to investigate the steady state bifurcations, we can obtain information on the path which the system follows as states exchange stability. This is illustrated by the following result. A 17 x 17 grid is used in all these simulations with a few calculations on a 25 x 25 mesh for accuracy check. The time step in these simulations corresponds to a maximum Courant number of about 0.2; the time step is limited by accuracy, and not by stability. For the geometry of Benjamin and Mullin [13] with a radius ratio .615, and aspect ratio 1.05, the symmetric two-cell mode is allowed to stabilize at a relatively low Reynolds number (R = 62). The Reynolds number is then raised impulsively to 165, above the experimental bifurcation boundary of about 150 and the growth of the one-cell asymmetric mode is observed. Random machine roundoff error on the order of 10^{-14} provides the initial energy for the mode. The order parameter used here to quantify the asymmetry of the mode is due to Lucke et al. [14]:

$$\psi = \frac{\int drdy \ (u(r,y) - u(r,-y))}{\int drdy \ (|u(r,y)| + |u(r,-y)|)}$$

The integrals were performed by spectral collocation. The logarithm of this parameter is shown in Fig. 1 plotted against time in units of the diffusion time scale y_ℓ^2/ν. As can be seen, the initial instability leading to the one-cell mode appears to be linear; that is, exponential growth with time is observed with only the later stages being modulated by nonlinear effects. Also shown in Fig. 1 is a plot of the logarithm of the disturbance energy versus time. After an initial period, the disturbance energy grows at a rate which is within 2% of double the growth rate of ψ as expected.

Streamlines in a cross-sectional plane at various stages in the two-cell/one-cell exchange are shown in Fig. 2. Locations of these intermediate states on the ψ vs time curve are indicated in Fig. 1. Note that the progression between states is smooth without abrupt collapse or alteration of the flowfield structure.

The geometry of Lucke, et al. [14] has a radius ratio of .5066, and an aspect ratio of 1.05. This choice of parameters leads to a smooth supercritical bifurcation to the one-cell mode as the Reynolds number is increased beyond a critical value. A plot of the order parameter against Reynolds number for this simulation is shown in Fig. 3. This curve is traced in the direction of both increasing and decreasing Reynolds number. As we approach the initial critical value, the Reynolds number is varied very slowly at a rate of about ± .2 units based on the diffusion time scale. The growth rates are much larger on the upper branch of this plot, which permits larger changes in Reynolds number. This bifurcation diagram is validated by restarting the simulation at various points along the curve and allowing the flow to settle to eight decimal digits for a fixed Reynolds number.

Also shown in Fig. 3 are the results of Lucke et al. [14] for this case. Their results were computed using a staggered-grid finite-difference scheme on a grid of about 28 x 30 points; however, in that calculation the Reynolds number was changed at a rate of about 4.1 based on the present time scale. A large discrepancy in the critical Reynolds number as predicted by these two studies is noted.

We also investigated the geometry of Aitta, et al. [15] where the radius ratio is .5. Their experimental results relate to three aspect ratios: 1.129, 1.266, and 1.281; for which the two-cell/one-cell bifurcations are supercritical, transcritical, and subcritical, respectively. In Fig. 4, an order parameter due to Aitta et al. [15] is plotted as a function of Reynolds number.

$$\psi' = \frac{\int_{-y_\ell}^{+y_\ell} w(r,y)\, dy}{\int_{-y_\ell}^{+y_\ell} |w(r,y)|\, dy}$$

where $r = R_i + .14(R_o - R_i)$. Also shown in Fig. 4, are the experimental results of Aitta, et al. The two loci of states agree in form. They also agree as to the level at which the asymmetric branch becomes unstable and the width of the region in which both symmetric and asymmetric modes are stable. The critical Reynolds number from the simulation is within 3% of that of the experiment. The growth rates of the one-cell mode in the "hysteresis" region of the bifurcation are exceedingly small, several orders of magnitude below those observed in the first two cases, producing a flow costly to simulate accurately. Resolution requirements are also large; a 33 x 33 mesh is required for this simulation.

References

[1] DiPrima, R. C. and H. L. Swinney, "Instabilities and transitions in flow between concentric rotating cylinders," Hydrodynamic Instabilities and the Transition to Turbulence, H. L. Swinney and J. P. Gollub (eds.), Springer-Verlag, New York 1981.

[2] Pfister, G., "Deterministic chaos in rotational Taylor-Couette flow," Lecture Notes in Physics, No. 235, pp. 199-210, 1985.

[3] DiPrima, R. C., "Transition in flow between rotating concentric cylinders," Transition and Turbulence. R. E. Meyer (ed.), Academic Press, 1981.

[4] Benjamin, T. B., "Bifurcation phenomena in steady flow of a viscous fluid I. Theory." Proc. R. Soc. London A 359, 1-26, 1978.

[5] Benjamin, T. B., "Bifurcation phenomena in steady flows of a viscous fluid II. Experiments." Proc. R. Soc. London A 359, 27-43, 1978.

[6] Cliffe, K. A. and T. Mullin, "A numerical and experimental study of anomalous modes in the Taylor experiment." J. Fluid Mech., 1985, vol 153, pp. 243-258.

[7] Keller, H. B., "Numerical solutions of bifurcation and nonlinear eigenvalue problems." Applications of Bifurcation Theory, (ed. P. H. Rabinowitz), pp. 359-384, Academic Press.

[8] Alziary de Roquefort, T., G. Grillaud, "Computation of Taylor vortex flow by a transient implicit method." Comput. Fluids 6, 259-269, 1978.

[9] Neitzel, G. P., "Numerical computation of time-depenent Taylor-vortex flows in finite length geometries." J. Fluid Mech., 1984, vol. 141, pp. 51-66.

[10] Schaeffer, D. G., "Qualitative analysis of a model for boundary effects in the Taylor problem." Math Proc. Camb. Philos. Soc. 87, 307-337 (1980).

[11] Blennerhasset, P. J., P. Hall, "Centrifugal instabilities of circumferential flow in finite cylinders: linear theory." Proc. R. Soc. London A 365, 191-207 (1979).

[12] Hall, P., "Centrifugal instabilities in finite containers: a periodic model." J. Fluid Mech., 1980, vol. 99, pp. 575-596.

[13] Benjamin, T. B. and T. Mullin, "Anomalous modes in the Taylor experiment." Proc. Roy. Soc. London A 377, 221-249, 1981.

[14] Lucke, M., M. Mihelcic, K. Wingerath, and G. Pfister, "Flow in a small annulus between concentric cylinders." J. Fluid Mech., vol 140, pp. 343-353, 1984.

[15] Aitta, A., G. Ahlers, and D. S. Cannel, "Tricritical phenomena in rotating Couette-Taylor flow." phys. Rev. Lett., vol 54, pp. 673-676, 1985.

[16] Marcus, P., "Simulation of Taylor-Couette flow – numerical methods and comparison with experiment." J. Fluid Mech., vol 146, pp. 45-64, 1984.

[17] Fortin, M., R. Peyret, and R. Temam, "Resolution numerique des equations de Navier-Stokes pour un fluide incompressible." J. de Mecanique, vol 10, pp. 357-390, 1971.

[18] Zang, T. A. and M. Y. Hussaini, "On spectral multigrid methods for the time-dependent Navier-Stokes equations." Appl. Math. Comp., vol 19, pp. 359-372, 1986.

[19] Hussaini, M. Y., C. L. Streett, and T. A. Zang, "Spectral methods for partial differential equations." Proc. 1st Army Conf. on Appl. Math. and Computing, ARO Report 84-1, 1984.

[20] Williamson, J. H., "Low-storage Runge-Kutta schemes." J. Comp. Phys., vol 35, no. 1, March 1980, pp. 48-56.

[21] Zang, T. A. and M. Y. Hussaini, "Numerical experiments on subcritical transition mechanisms." AIAA Paper No. 85-0296, 1985.

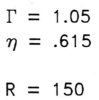

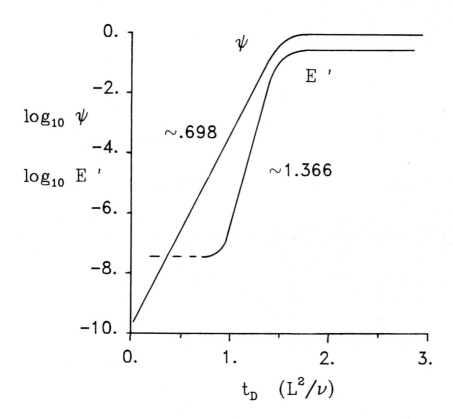

Figure 1: Growth of order parameter and disturbance energy with time; Benjamin geometry.

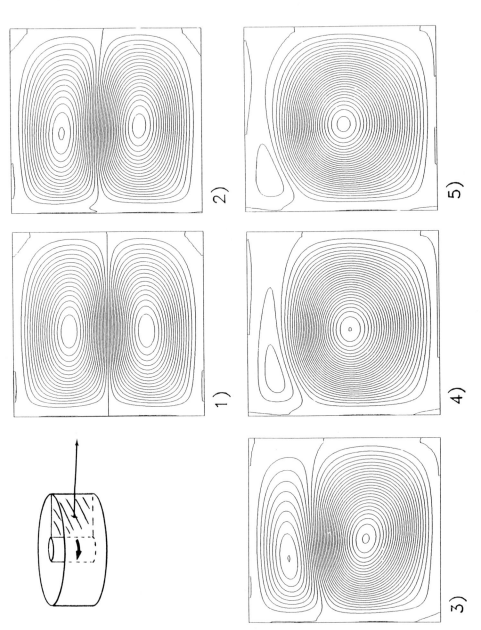

Figure 2 Evolution of streamlines in crossflow plane; Benjamin geometry.

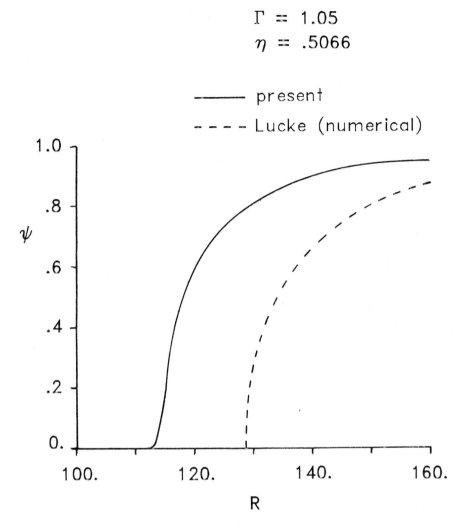

Figure 3 Order parameter vs. Reynold's number; Lucke geometry.

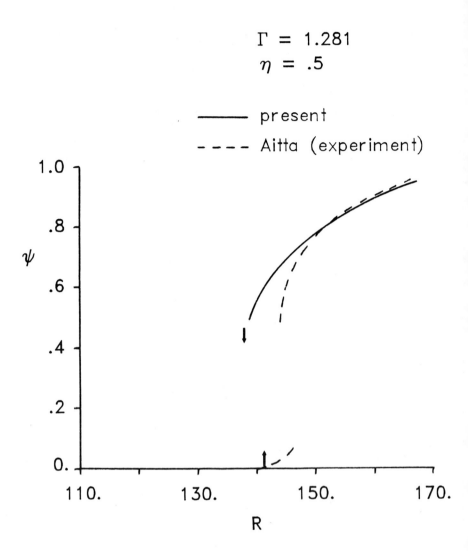

Figure 4 Order parameter vs. Reynold's number; Aitta geometry.

Vortical Structures in the Breakdown Stage of Transition

D.R. WILLIAMS

Summary

Data from an experimental investigation of boundary-layer transition has been processed to show the presence of vortical structures that have previously been undetected in the early breakdown stage. The structures, which were identified from organized patterns of vortex lines, consisted of hairpin and inverted vortices. Similarities between the breakdown process and the turbulent burst process suggest that the vorticity dynamics are similar between the two processes.

Introduction

A significant amount of progress has been made in the last twenty years toward understanding the transition of a flat-plate boundary layer from laminar to turbulent flow. Most experiments have concentrated on the development of single wavenumber disturbances that are introduced into the boundary layer with small amplitudes. When the amplitude is less than about 0.5% of the freestream velocity, the disturbance behaves like a two-dimensional Tollmien-Schlichting (T-S) wave. The linear stage of development has been carefully documented by Schubauer & Skramstad (1948) and Ross et al (1970).

However, it was recognized early on in the flow visualization experiments by Hama et al (1957), Hama (1960), Hama & Nutant (1963) and Wortmann (1977) and in the detailed hot-wire anemometer measurements by Klebanoff & Tidstrom (1959), Klebanoff et al (1962) and Kovasznay et al (1962) that the two-dimensional disturbance quickly becomes three-dimensional. Organized patterns of Λ-shaped vortices were seen to form in the flow before the fully turbulent state was reached. Recent numerical experiments by Orszag & Patera (1983) and Herbert (1983a, 1983b, 1984) have shown that the rapid growth of the three-dimensional waves is actually a secondary instability which prevails when the two-dimensional wave reaches a finite amplitude of about 1%. The growth of the three-dimensional waves distorts the field into the Λ shape. Herbert (1983a) showed that subharmonic secondary instability is also possible, which led to a staggered pattern of Λ vortices and explained the experimental results of Thomas & Saric (1981). Numerical simulations of boundary-layer transition by Wray & Hussaini (1980) have followed the three-dimensional disturbance development beyond the linear stage of the secondary instability to finite amplitudes. These investigations together with the measurements by Williams et al (1984) have verified the existence of the Λ vortex and clarified its role in the transition process.

In contrast to the preceding detailed studies, the later stages of transition are not well understood, primarily because the computations and experiments have lacked sufficient spatial resolution to follow the formation of the smaller-scale structures. The flow visualization experiments of Hama (1960) and Hama & Nutant (1963) showed that a hairpin-shaped vortex formed in the tip region of the Λ vortex, and additional hairpins developed over the leg regions. Klebanoff et al (1962) found that the hairpin vortices are associated with sharp downward "spikes" in the streamwise velocity signal, which they termed the breakdown stage. Kovasznay et al (1962) examined the early breakdown stage by using a hot-wire rake of sensors. They were able to map the contours of the du/dy shear field and showed that a "kink" formed in the high-shear layer when spikes in the velocity signal occurred. Similar division of the high-shear layer into small vortical structures was observed in the numerical experiments of Wray & Hussaini (1980).

In addition, the breakdown process in plane Poiseuille flow has shown some similarity to the boundary-layer transition. This flow has been investigated perhaps in more detail than the boundary layer, primarily by the experiments of Nishioka et al (1981) and numerical investigations by Biringen (1984), and Biringen & Maestrello (1984) and by Kleiser (1985). All of these studies gave evidence of small-scale vortical structures forming during this stage.

It is apparent that an understanding of the vorticity dynamics is required to explain the breakdown process. Up to now the quantitative experimental studies have been limited to measurements of the du/dy shear component, which is only an approximation of the spanwise vorticity, ω_z. Although the ω_z vorticity has the largest magnitude, by no means does it give a complete picture of the mechanics of the breakdown process. Therefore, it was the purpose of this investigation to improve our understanding of the vorticity dynamics during breakdown by mapping the entire three-dimensional vorticity field. This would allow the vortical structures that are present to be identified and their dynamics inferred. In particular, the relationship between the Λ vortex and hairpin vortices has been examined by studying the macroscopic vortical structures formed through organization of vortex lines. Quite surprisingly, a connection between the vortical structures found in the breakdown process and those found by Kim & Moin (1984) to be associated with the burst process in turbulent channel flow has also been established.

Experimental Procedure

The data for this investigation were obtained in the Moody Hydrodynamics Laboratory water channel at Princeton University. The boundary layer was grown on a flat glass plate which had an elliptical leading edge. In the region of measurement, the boundary-layer thickness, δ, was approximately two centimeters in height with the freestream velocity, U_o, fixed at 10 cm/sec. The displacement thickness Reynolds number, Re*, ranged from 700 to 850 over the measurement area. The displacement thickness has been given the symbol δ^*. A two-dimensional disturbance with an amplitude of $.01U_o$ was introduced by oscillating a small diameter wire at 0.263 Hz, which corresponds to a dimensionless frequency parameter $F = 2\pi f \nu / U_o^2 = 1.4 \times 10^{-4}$, where f is the frequency and ν the kinematic viscosity. The wave that developed had a fundamental wavelength of λ_f = 16.2cm. At 40 cm downstream from the oscillating wire it had developed T-S wave-like characteristics in its amplitude and phase distributions, although its maximum amplitude was approximately $0.04U_o$.

The three velocity components u, v, and w correspond to the streamwise, vertical and spanwise (x, y and z) directions, respectively. The coordinate system was oriented such that x = 0 was at the oscillating wire, z = 0 was near the spanwise center of the developing disturbance, and y = 0 was the surface of the plate. The measurement locations formed a volume of grid points which consisted of 8 locations over 4 cm in the z direction, 16 locations over 3 cm in the vertical direction and 8 streamwise locations in 5-cm increments covering a 35-cm range.

The velocity components were measured with slanted hot-film probes and constant-temperature anemometers. The instantaneous voltage signals from the anemometers were recorded together with a reference signal from the oscillating wire. The reference signal allowed phase-averaging techniques to be used, so the entire flow field could be reconstructed at any instantaneous phase angle in the cycle of the primary disturbance. Due to insufficient spatial resolution in the streamwise direction, it was necessary to use Taylor's frozen-flow hypothesis as an approximation to the streamwise coordinate. In cases where Taylor's hypothesis was used, the phase angle of the reference signal, T, has been plotted on the axis instead of the streamwise distance. When it was necessary to estimate a streamwise distance such as in the gradient dx, the phase velocity of the fundamental disturbance of 4.2 cm/sec was used. In this way the time increment, dt was converted into approximate streamwise distance for the vorticity calculation.

The three instantaneous vorticity components, ω_x, ω_y and ω_z, have also been computed from the velocity field with a second-order central differencing scheme. Additional details of the data acquisition and processing can be found in the dissertation by Williams (1982).

Results

The major features of the vorticity field during the early stage of breakdown to turbulence are discussed in this section. In order to describe the redistributed vorticity field, it is expedient to define two types of vortical structures, namely the high-shear layers and the vortices. Although the definitions are somewhat subjective, it will be shown that sharp differences in the direction of vortex lines in the structures make them easy to identify.

The high-shear layers are regions of strong du/dy, which are typically thin in the vertical dimension but have large spanwise and streamwise extensions. The high-shear layers are easiest to identify by the simple u-velocity measurements and, as such, their role in the transition process may have been overemphasized.

The second type of vortical structure is the vortex. Typically, the vortex does not have a maximum of vorticity magnitude, but it can be distinguished by a compact cross-sectional area of organized vortex lines oriented primarily in the streamwise and vertical directions. Despite the relatively low intensity of vorticity, the vortex plays a major role in the formation of the high-shear layers. The well-known Λ vortex and hairpin vortex are examples of vortices occurring in the transition process.

In this experiment the three-dimensional wave development led to the in-line Λ-vortex pattern. The details of the wave development and structure of the Λ vortex can be found in Williams et al (1984). Prior to the breakdown stage, a single high-shear layer exists away from the wall ($y = 0.6\delta$) for each fundamental wavelength of the disturbance and is associated with the Λ vortex. The streamwise components of vorticity in the leg regions intensify the high-shear layer in the spanwise center up to $\omega_z = 0.9U_o/\delta^*$ as shown in figure 1. The contours of constant ω_z vorticity have been plotted along the center plane of the disturbance, $x = 60$ cm and $z = 0.0$ cm. The single high-layer is shown by the solid contours to be centered at $y = 0.85$ cm, $T = 60^\circ$. The solid contours indicate vorticity that is stronger than the maximum mean vorticity.

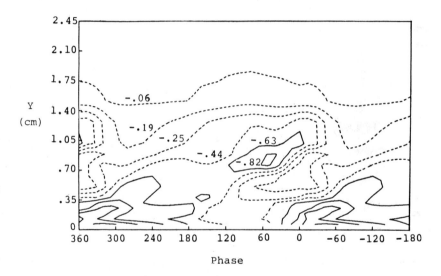

Fig. 1 Contours of spanwise vorticity prior to breakdown at x=60cm, z=0.0cm. Solid contours indicate vorticity that is stronger than the maximum mean flow vorticity. One high-shear layer centered at y=0.85cm, T=60° exists per fundamental wavelength.

Except for the wavefront warping, there are no fundamental changes in the structure of the vorticity field during the formation of the Λ vortex until breakdown is reached. One Λ vortex and corresponding high-shear layer are formed for each cycle (fundamental wavelength) of the oscillating wire. At $x = 65$ cm ($4.0\lambda_f$) downstream of the oscillating wire, the single high-shear layer breaks apart into two high-shear layers per fundamental wavelength. The breakup of the high-shear layer creates a fundamentally different distribution for the vorticity field. This reorganization process is in fact used as the definition for breakdown in this paper. Figure 2 shows the contours of ω_z at x = 65 cm, z = 0.0 cm which have broken apart into the smaller structures. The downstream high-shear layer centered at y = 1.1 cm, T = 105° associated with a hairpin vortex, and the weaker high-shear layer upstream at y = 0.7 cm, T = 270° is associated with an inverted vortex. The characteristics of these vortices will be described in a later section.

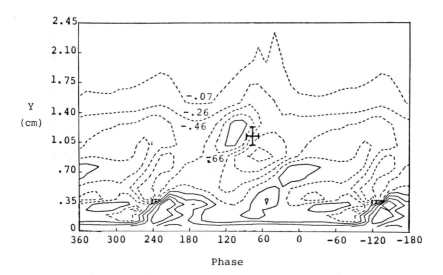

Fig. 2 Contours of spanwise vorticity at the onset of breakdown at x=65cm, z=0.0cm. Two high-shear layers exist per fundamental wavelength. The high-shear layer centered at y=1.1cm, T=105° is associated with a hairpin vortex. The approximate location of the hairpin is marked by a cross.

The u' velocity signals at x = 65 cm, z = 0.0 cm in figure 3 show downward fluctuations which occur close to the high-shear layers. It will be shown later that this region is actually the tip of a hairpin vortex. The spacing between the mean signal heights in figure 3 corresponds to the hot-film probe height in the boundary layer. Two downward spikes are seen in the u' signal at y = 1.0 cm. The first spike occurs as the hairpin vortex passes the probe. The second spike follows about 120° in phase after the first spike which indicates a spatial separation of about 5 cm.

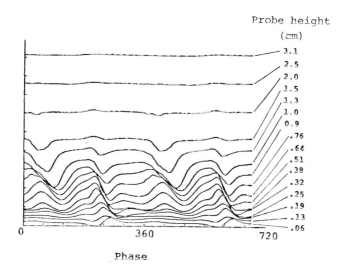

Fig. 3 Streamwise velocity signals measured at different heights x=65cm, z=0.0cm.

The phase angle of the leading edge of the high-shear layer can be used to estimate the convective speed of the hairpin vortex tip. The leading edge of the high-shear layer at x = 65 cm, y = 1.05 cm, T = 90° was determined from phase measurements in figure 4 to be convecting at $0.61U_o$. The tip has also moved away from the wall (y = 0.85 cm at x = 60 cm to y = 1.1 cm at x = 65 cm) as the hairpin vortex developed. The convective speed of the hairpin vortex is high compared to the overall disturbance speed of $0.42U_o$ measured in the freestream (see Williams et al, 1984). Thus, the tip of the hairpin vortex is rapidly overtaking the next downstream wave and moving away from the central region of the Λ vortex. Hama (1960) called this behavior the "snatching process" when the hairpin vortices were observed to move into higher regions of the boundary layer where they were rapidly convected away from the remainder of the Λ vortex.

Furthermore, since the mean flow velocity is $0.89U_o$ at the height of the hairpin vortex, a downward spike is registered in the u' signal when the hairpin passes by at its much slower convective speed of $0.61U_o$. This shows that the downward spikes are indeed associated with the hairpin vortices.

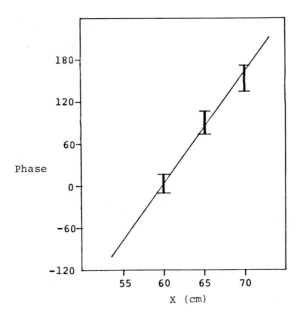

Fig. 4 Phase angle of the tip region of the developing hairpin vortex measured at several streamwise locations. The convective speed of the tip was found to be $0.61U_o$.

Vortical Structures

So far the results have focused on the spanwise centerline measurements of u' and ω_z contours. The redistribution of the vorticity field during the breakdown stage forms vortical structures that are highly three-dimensional and have small scales relative to the Λ vortex which makes discrimination difficult. The clearest way to identify these structures is by computing the three-dimensional vortex lines in the flow. The vortex lines were found by numerically integrating the equation $\vec{\omega}/|\vec{\omega}| = d\vec{x}/ds$. The experimental data provides the $\vec{\omega}(\vec{x})$ field. An initial starting point, $\vec{x}_o$, was chosen in the flow, the vortex line passing through that point was followed in two directions until a boundary was reached at each end of the line. The next line was computed by choosing a new starting point and repeating the procedure. The starting points were incremented in uniform steps while keeping the other two spatial variables constant. For example, in figure 5, the starting location of each line changed by $T_o = 5$ while

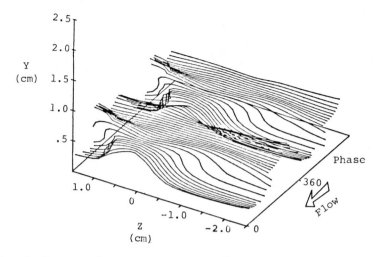

Fig. 5 Perspective view of the vortex lines showing the Λ vortex before breakdown at x=60cm, y_0=0.55cm, z_0=0cm. The streamwise coordinate has been reduced by a factor of 10 relative to the spanwise coordinate in order to fit two wavelengths in the figure. The Λ shape is seen where the lines have vertical orientation.

y_0 = 0.55 cm and z_0 = 0.0 cm remained constant. The uniformity of the starting point increments helped give a more objective view of the important organized structures in the flow.

For comparison, the vortex lines corresponding to the Λ vortex prior to breakdown at x = 60 cm have been computed for y_0 = 0.55 cm as shown in figure 5. Approximately two wavelengths of the data are shown in this perspective view. Note that the streamwise coordinate has been reduced by a factor of 10 relative to the spanwise coordinate, which causes the streamwise components of vorticity to appear weaker than the vertical and spanwise components. Because of the viscous nature of this disturbance, no single line traces the entire Λ shape. Instead, the vortex lines wind around the leg regions of the Λ vortex, which can be seen where the lines bend severely in the vertical direction. Otherwise, the vortex lines are more or less evenly distributed, and no significant amalgamation of the vorticity into discrete vortices has occurred.

Once breakdown begins the vorticity field develops small-scale structures. The first vortical structure considered is the hairpin vortex at x = 65 cm shown in figures 6a and 6b. The hairpin vortex has

developed from the spanwise center region of the tip of the Λ vortex.
The tip of the hairpin in figure 6a is centered between z = 0.0 cm and
z = -0.5 cm, T = 60° - 90° at a height of y = 1.1 cm (y/δ = 0.6). The
"core" diameter is roughly 9 mm. The position of the hairpin relative
to the high-shear layer has been indicated by a cross on figure 2. One
sees that the hairpin vortex is actually located slightly downstream (at
an earlier phase) of the high-shear layer. This explains why Kovasznay
et al (1962) found that the "spike" in the u' signal was not centered at
the maximum ω_z contour.

The hairpin is completely distinct from the high-shear layer, and
occupies only a small region in the flow field. To better see how the
hairpin vortex is situated in the streamwise direction, the starting
points for the lines in figure 6b varied T_o while y_o = 0.95 cm and
z_o = 0.0 cm were kept fixed. The accumulation of vortex lines into the
organized core of the hairpin vortex can be seen by comparing figures 5
and 6b. It can also be seen that the hairpin has developed from the tip
region of the Λ vortex.

A second vortical structure with an inverted or "V" shape was
discovered upstream of the hairpin in the vicinity of T = 240°. The
vortex lines in figures 7a and 7b show two views similar to those in 6a
and 6b. The V vortex has a small "core" diameter of about 4 mm. The
magnitude of the vorticity in the core is about $/\bar{\omega}/ = 0.2U_o/\delta^*$. The
lines that agglomerate to form the V shape are connected with high-shear
layers higher up in the boundary layer and off the spanwise center at
z = -0.55 cm and z = 1.0 cm. The dark regions in figure 7a at the top
of the V vortex (for example, y_o = 0.8 cm, z_o = 0.5 cm) consist of
strong vorticity $/\bar{\omega}/ = 0.6U_o/\delta^*$. These off-center regions are
located above the legs of the Λ vortex. It is highly probable that
the secondary hairpin vortices observed by Hama (1960) are forming in
these high-shear regions over the legs of the Λ vortex.

The hairpin and inverted vortices have shapes similar to the
vortical structures found by Kim & Moin (1984) to be associated with the
turbulent channel flow burst process, which is known to be a major
contributor to turbulence production. In addition to the shape of the
vortical structures, the u' signal near the wall in figure 3 shows a
similarity to the type of signal used by Blackwelder & Kaplan (1976) for
detection of the bursting process in a turbulent boundary layer. The
burst signal consists of a slow deceleration follow by rapid
acceleration in the streamwise velocity component near the wall. The
implication is that since the breakdown process in the transitioning
boundary layer shows similarity to the turbulent bursting process, then
the vorticity dynamics of one will apply to the other.

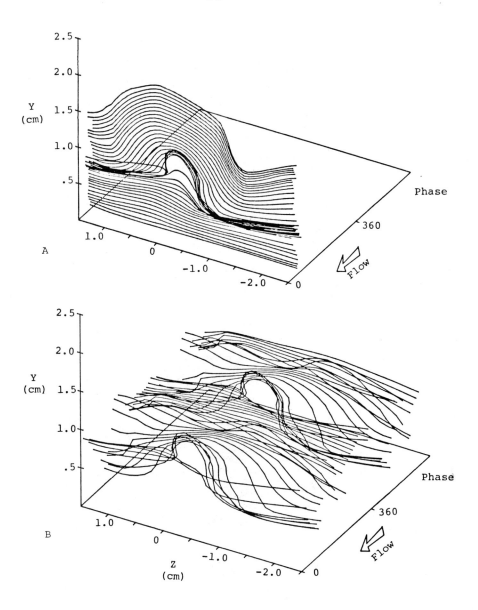

Fig. 6 Vortex lines computed at x=65cm which show the hairpin vortex. The lines in figure 6a have starting coordinates $z_o=0$cm, $T_o=67°$ while y_o varies in 0.05cm increments. This emphasizes the vertical extent of the hairpin vortex. The lines in figure 6b have starting coordinates $z_o=0$cm, $y_o=0.95$cm, while T_o varies in $13°$ increments. The streamwise orientation of the hairpin vortices is emphasized in this view.

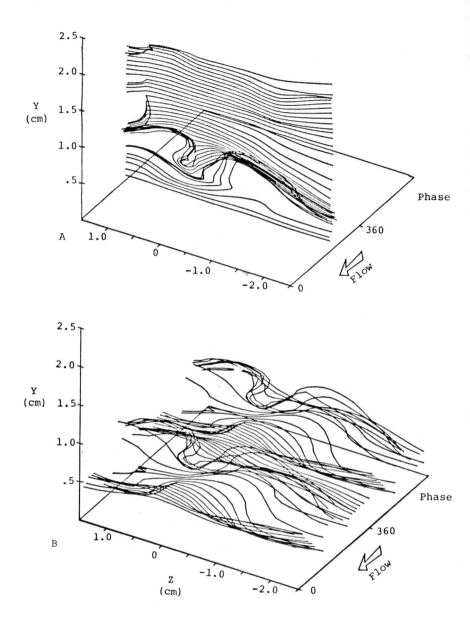

Fig. 7 Vortex lines at x=65cm which show the inverted vortex. The starting coordinates in figure 7a are z_o=0cm, T_c=253°, which shows the vertical orientation of the V vortex. Figure 7b has starting coordinates z_o=0cm, y_o=0.55cm, which emphasizes the streamwise orientation of the V vortices.

The burst process in a turbulent flow consists of two events. First, the "ejection" of low-speed fluid away from the wall causes the deceleration in the velocity signal. This is closely followed by the "sweep" of high-speed fluid towards the wall, which forms the rapid acceleration. The numerical investigations of Kim & Moin (1984) have shown the ejection to be associated with the hairpin-type vortex, and the sweep to be associated with an inverted vortex. The instantaneous velocity fields associated with the hairpin and V vortices during the breakdown process at x = 65 cm, z = 0.0 cm in figure 8 support the concepts discussed above. In this figure the velocity vectors are observed from a reference frame moving with the overall disturbance wavespeed, $0.42U_o$. The tips of the velocity vectors are marked by a small x, and the instantaneous velocity profiles have been superposed. A downward "jet-like" region of fluid stretches across the boundary layer for $180°< T <240°$, which is associated with the inverted vortex. This is believed to be responsible for the sweep-like event in the breakdown. The convection of the high-speed fluid toward the wall is responsible for the acceleration in the u' signal. Likewise, the flow field between $90°< T <150°$ shows a strong upward convection region, which agrees with the expected induced velocity field of the hairpin vortex and deceleration of the u' velocity.

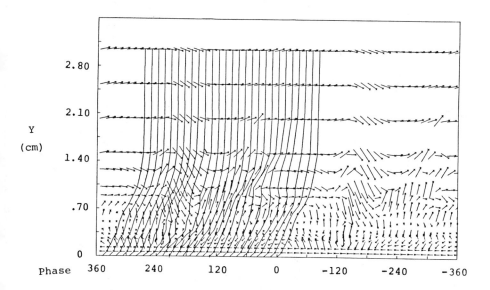

Fig. 8 Instantaneous velocity vectors at x=65cm, z=0cm seen from a moving reference frame. Instantaneous velocity profiles have been superposed over one cycle. A downward "jet" of fluid near T=180° is responsible for the sweep-like behavior.

Conclusions

Exploration of the three-dimensional vortical structures formed during the breakdown process in boundary-layer transition has been accomplished by maps of vorticity contours and vortex lines. The vortex lines became organized in narrow regions of the flow to form a hairpin-shaped vortex and an inverted vortex. The inverted vortex is located upstream of the hairpin. The tip region of the hairpin was convected at a faster rate than the overall disturbance speed, which supported the "snatching process" observed by Hama (1960). The hairpin vortex was shown to be a physically distinct vortical structure from the high-shear layer observed in the contours of ω_z.

Similarity with the turbulent burst process in channel flows was discovered. The hairpin vortex created an ejection-like motion, and the inverted vortex generated a sweep-like motion. As a result the u' velocity signal near the wall showed a rapid acceleration quite similar to the turbulent burst signal. Thus, the vorticity dynamics of the breakdown process are also similar to the turbulent bursting process.

The original data was acquired under sponsorship by the Office of Naval Research. Subsequent data processing was supported by the Alexander von Humboldt Foundation and the Air Force Office of Scientific Research. The author is grateful to E. S. Nelson for her help in preparing this paper, and to F. R. Hama for reminding him of the "snatching process".

References

1. Biringen, S.: Final stages of transition to turbulence in plane channel flow, J. Fluid Mech. 148, 413-442 (1984).

    Biringen, S. & Maestrello, L.: Development of spot-like turbulence plane channel flow, Phys. Fluids 27, 318-321 (1984).

3. Blackwelder, R. F. & Kaplan, R. E.: On the wall structure of the turbulent boundary layer, J. Fluid Mech. 76, 89-112 (1976).

4. Hama, F. R.: Boundary-layer transition induced by a vibrating

ribbon on a flat plate, Proc. 1960 Heat Transfer & Fluid Mech. Inst., Stanford University Press, 92-105 (1960).

5. Hama, F. R., Long, J. D. & Hegarty, J. C.: On transition from laminar to turbulent flow, J. Appl. Phys. 28, 388 (1957).

6. Hama, F. R., & Nutant, J.: Detailed flow-field observations in the transition process in a thick boundary layer, Proc. 1963 Heat Transfer & Fluid Mech. Inst., Stanford University Press, 77 (1963).

7. Herbert, Th.: Secondary instability of plane channel flow to subharmonic three-dimensional disturbances, Phys. Fluids 26, 871-874 (1983a).

8. Herbert, Th.: Subharmonic three-dimensional disturbances in unstable plane shear flows, AIAA Paper No. 83-1759 (1983b).

9. Herbert, Th.: Analysis of the subharmonic route to transition in boundary layers, AIAA Paper No. 84-0009 (1984).

10. Kim, J. & Moin, P.: The structure of the vorticity field in turbulent channel flow. Part 2: Study of ensemble-averaged fields, NASA Tech. Mem. 86019 (1984).

11. Klebanoff, P. S. & Tidstrom, K. D.: Evolution of amplified waves leading to transition in a boundary layer with zero pressure gradient, NASA Tech. Note D-195 (1959).

12. Klebanoff, P. S., Tidstrom, K. D. & Sargent, L. M.: The three-dimensional nature of boundary-layer instability. J. Fluid Mech. 12, 1 (1962).

13. Kleiser, L.: Three-dimensional processes in laminar-turbulent transition, in Nonlinear Dynamics in Transcritical Flows (ed. H. L. Jordan, H. Oertel, K. Robert), Springer Lecture Notes in Engineering (1985).

14. Kovasznay, L. S. G., Komoda, H. & Vasudeva, B. R.: Detailed flow field in transition, Proc. 1962 Heat Transfer & Fluid Mech. Inst. 1, Stanford University Press (1962).

15. Nishioka, M., Asai, M. & Iida, S.: Wall phenomena in the final

stage of transition to turbulence, in Transition and Turbulence (ed. R. E. Meyer), 113, Academic Press (1981).

16. Orszag, S. A. & Patera, A. T.: Secondary instability of wall-bounded shear flows, J. Fluid Mech. 128, 347-385 (1983).

17. Ross, J. A., Barnes, F. H., Burns, J. G., & Ross, M. A. S.: The flat plate boundary layer. Part 3. Comparison of theory with experiment, J. Fluid Mech. 43, 819-832 (1970).

18. Schubauer, G. B. & Skramstad, H. K.: Laminar boundary layer oscillations on a flat plate, NACA Rep. 909 (1948).

19. Thomas, S. W. & Saric, W. S.: Harmonic and subharmonic waves during boundary layer transition, Bull. Am. Phys. Soc. 26, 1252 (1981).

20. Williams, D. R.: An experimental investigation of the non-linear disturbance development in boundary-layer transition. Dissertation, Princeton University (1982).

21. Williams, D. R., Fasel, H. & Hama, F. R.: Experimental determination of the three-dimensional vorticity field in the boundary-layer transition process, J. Fluid Mech. 149, 179-203 (1984).

22. Wortmann, F. X.: The incompressible fluid motion downstream of two-dimensional Tollmien-Schlichting waves, AGARD Conf. Proc. 224, 12-1 (1977).

23. Wray, A. & Hussaini, M. Y.: Numerical experiments in boundary-layer stability, AIAA Paper No. 80-0275 (1980).